Military and Humanitarian Health Ethics

The interdisciplinary book series Military and Humanitarian Health Ethics fosters an academic dialogue between the well-established disciplines of military ethics on the one hand and medical ethics, humanitarian ethics and public health ethics on the other hand. Military and Humanitarian Health Ethics have emerged as a distinct research area in the last years, triggered among other things by the unfortunate realities of armed conflicts and other situations of humanitarian disasters - man-made or natural. The book series focuses on the increasing amount of ethical challenges while providing medical care before, during, and after armed conflicts and other emergencies. By combining practical first-hand experiences from health care providers in the field with the theoretical analysis of academic experts, such as philosophers and legal scholars, the book series provides a unique insight into an emerging field of research of high topical interest. It is the first series in its field and aims at publishing state-of-the-art research, illustrated and enriched by field reports and ground experiences from health care providers working in armed forces or humanitarian organizations.
We welcome proposals for volumes within the broad scope of this interdisciplinary and international book series, especially proposals for books that cover topics of interest for both the military and the humanitarian community, and which try to foster an exchange between the two often separate communities of military and humanitarian health care providers.

More information about this series at http://www.springer.com/series/16133

Daniel Messelken • David Winkler
Editors

Ethics of Medical Innovation, Experimentation, and Enhancement in Military and Humanitarian Contexts

Springer

Editors
Daniel Messelken
Zurich Center for Military Medical Ethics, Center for Ethics
University of Zürich
Zürich, Switzerland

David Winkler
Center of Reference for Education on IHL & Ethics
International Committee of Military Medicine
Bern, Switzerland

ISSN 2524-5465 ISSN 2524-5473 (electronic)
Military and Humanitarian Health Ethics
ISBN 978-3-030-36318-5 ISBN 978-3-030-36319-2 (eBook)
https://doi.org/10.1007/978-3-030-36319-2

This Springer imprint is published by the registered company Springer Nature Switzerland AG.
The registered company address is: Gewerbestrasse 11, 6330 Cham, Switzerland

Contents

Chapter 1
Ethics of Medical Innovation, Experimentation, and Enhancement in Military and Humanitarian Contexts. Introduction to the Volume

Daniel Messelken and David Winkler

1.1 Introduction

The topic of this volume, the *Ethics of medical innovation, experimentation, and human enhancement in military and humanitarian contexts*, is a vast subject area that gives rise to many ethical issues of very different kinds. The purpose of this introductory chapter is to give a panoramic overview and to provide points of reference together with an initial outline of the ethical questions that arise. It is not intended to contribute to the discussion of the volume's subject-matter but shall rather introduce the thematic terrain in which the other chapters of the present book are situated.

Innovations like the use of new medical methods and forms of experimental treatment (e.g. with drugs not (yet) approved for a specific disease) occur in military or humanitarian contexts for different reasons: one rationale may be to cope with scarce resources when the usual means are just not available; another strong motivation can simply be to gain military advantages. Historically and still today, military medicine has generated a number of novel and innovative treatments which have advanced medicine and which have gone on to be widely used in general medicine as well (Givens et al. 2017; Kitchen and Vaughn 2007; Ling et al. 2010; Ratto-Kim et al. 2018). Thus, conflict may even serve as a catalyst for the development of novel

D. Messelken (✉)
Zurich Center for Military Medical Ethics, Center for Ethics, University of Zürich, Zürich, Switzerland
e-mail: messelken@ethik.uzh.ch

D. Winkler
Center of Reference for Education on IHL & Ethics, International Committee of Military Medicine, Bern, Switzerland
e-mail: melac@bluewin.ch

D. Messelken, D. Winkler (eds.), *Ethics of Medical Innovation, Experimentation, and Enhancement in Military and Humanitarian Contexts*, Military and Humanitarian Health Ethics,
https://doi.org/10.1007/978-3-030-36319-2_1

approaches "with the casualty imperative driving innovations that have subsequently been adopted in civilian practice" (Hodgetts 2014: 86). Obviously, military and civilian emergency medicine also share a central goal (viz. saving lives) and can mutually learn from each other (Blackbourne et al. 2012; Haider et al. 2015; McManus et al. 2005; Reade 2013; Schrager et al. 2012). For example, "[t]he lessons of Vietnam and the development of trauma systems, the 'golden hour,' and air medical services provide additional reminders of the mutual benefits gained by military and civilian practice." (De Lorenzo 2004: 129) At the same time, the military medical context is often said to be problematic with regard to the conduct of research because of hierarchies, mixed interests, or also the difficult control by the public due to nondisclosure of certain information. These doubts are also based on a number of precedencies as "[h]istorically, military researchers have been negligent in protecting the rights of research subjects" (Frisina 2003: 538). We will look into this issue in more detail further below.

The working context of humanitarian actors may give rise to similar practical challenges and medical research in the environment of, for example, a disaster aftermath can be ethically problematic as well. Dealing with scarce resources, being confronted with rare diseases, and the unavailability of better medical care can lead to accepting innovative but unproven methods. This being said, medical research and the use of innovative technologies in both military and humanitarian contexts need not necessarily be unethical or more problematic than it would be in clinical contexts. Much depends on the organization, the study design, and the people involved.

Innovation and medical research in military and humanitarian contexts clearly may yield positive effects and a mutually beneficial spillover of new knowledge between the different environments does exist. Still, the development of new therapeutic approaches in a military or humanitarian environment and the use of medical enhancement also raise important ethical questions related to human experiments in medicine. This volume contributes to the debate of the ethical issues related to innovation and research in military and humanitarian contexts. It does so by looking at some specific challenges from both fields and by analyzing some more fundamental ethical questions. By integrating both the military and humanitarian perspectives we do not want to negate a difference between these two worlds. In practice, military and humanitarian health care personnel however often work close to each other and confront similar ethical challenges. Thus, even if their motivation or the motivation of their organizations may differ fundamentally, they can still learn from each other with regard to specific ethical issues in health care.

To start the discussion, this introductory chapter is meant to provide an overview on some key concepts and ethical rules and regulations that have been developed to guide research in medicine. The present chapter thus aims at providing some fundamental knowledge on which the remaining chapters can build.

1.2 The Distinction Between Medical Practice, Research & Innovation

First of all, we would like to clarify a few relevant concepts and terms. Namely, the topic of the volume calls for a clear understanding and distinction of (ordinary) medical practice compared to medical research and innovation.

1.2.1 Medical Practice and the Limits of Research

The 1979 "Belmont Report" introduces the distinction between ordinary practice and research. It defines practice as "interventions that are designed solely to enhance the well-being of an *individual patient or client* and that have a reasonable expectation of success." In contrast, "'research' designates an *activity designed to test an hypothesis*, permit conclusions to be drawn, and thereby to develop or contribute to generalizable knowledge". (National Commission for the Protection of Human Subjects of Biomedical and Behavioral Research 1979: Part A, Section A. Emphasis added.) Practice thus is the application of proven interventions at the benefit of an individual, whereas research consists in interventions that are equally applied to individual patients (who are part of a larger cohort) but come with the additional aim of testing the effectiveness of new medical interventions. Whereas research is necessary to generate new knowledge and advance medical practice, it is also problematic for one cannot know, by definition, the effectiveness and possible adverse effects of experimental treatments. The boundaries between ordinary practice and research are of course not always clear cut. Frisina (2003: 550f.) illustrates this point with the example of using investigational agents for preventive treatment to protect soldiers against the effects of possible biological and chemical warfare during the Persian Gulf War in 1990/91. This example shows that criteria like intent and risk can equally be factors that need to be taken into account.

1.2.2 Innovation – Practice Beyond the Ordinary But Not (Necessarily) Research

When something is considered as innovative, this generally means that it introduces a new idea, method, or device without being a completely new intervention. In medicine, innovation thus refers to the blurry area in between practice and research as it belongs neither to standard practice nor to the area of medical research; accordingly, innovative methods from an ethical viewpoint merit special attention. An example for an innovative method in medicine is the development of surgical tools, when existing knowledge and experience is combined with a new and less harmful way of surgery aiming to improve the overall outcome and quality of an intervention.[1]

Innovation in medicine can also consist in using a non-standard treatment *in individual cases*. This is not considered to fall in to the domain of research as such an individual experimental treatment does not aim at proving a general hypothesis, but rather to offer a treatment to an individual patient when proven methods have been exploited unsuccessfully. These so-called "unproven interventions in clinical practice" or "experimental treatment in individual cases" refer to a treatment which differs from the standard treatment, or are employed in the absence of a standard treatment (cf. Tom L Beauchamp 2008; Swiss Academy of Medical Sciences 2015, 2017; World Medical Association 2013 (1964)). They have of course to respect (ordinary) medical ethics and must be administered with the informed consent and to the best interest of the patient. However, when such treatments prove to be successful in individual cases, doctors may want to prove the innovative method by applying it to a greater number of people and by letting it undergo the usual research process so that it can become a new standard treatment. Such a process then needs to be designed and approved according to usual research ethics standards.

Depending on the context, the use of a new method in medicine can thus fall into the categories of research, innovation, or practice and be subject to the different ethical frameworks that apply respectively. Much depends on factors like the context, available alternatives, experience and intention of the medical team whether one would rather speak of the reasonable use of an innovative approach or an incautious recourse to unproven methods that would need to be explored in a much more strictly regulated research setting.

When in doubt whether an intervention has to be considered as general practice, innovation, or research physicians may ethically be on the safer side if they consider it as research and apply the more restricted framework and rules of research ethics. Depending on the context, innovative interventions are of course ethically still justifiable if they are, for example, the only available chance to save a life. Still, the continuous quality control of proven interventions and, more importantly, the advancement of medicine by inventing new treatments has to be based on research which ultimately has to include studies involving human research subjects. This is explicitly recognized by the World Medical Association's Declaration of Helsinki (cf. World Medical Association 2013 (1964): § 5&6). It is not the involvement of human research subjects *per se* that may render these studies ethically problematic, but rather the study design and lacking respect for ethical principles. In the next section, we will therefore look into research ethics to elaborate and present the most important rules and regulations.

[1] See the chapter of Jack Taylor in this book for more examples of innovation in (military) medicine and for a discussion of related ethical issues.

1.3 Ethical Restrictions of Research in Medicine

During medical research on human beings, ordinary medical ethics obviously remains valid and must be respected in the first place. Thus, the health of the (individual) patient has to be the first consideration of medical interventions and the principles of respect for patient autonomy and dignity remain fundamental. In addition, the conduct of research in medicine is governed by a large number of documents and legal regulations that also set ethical limits to how research on human beings may be conducted. The main aim of this ethical framework is to protect patients and other potential research participants from becoming part of medical experimentation or research without their knowledge and without their explicit and informed consent. In addition, those who willingly and knowingly participate in medical research are protected by research ethics against different kinds of abuse.

Historically, the most important documents that propose basic principles for medical research ethics are probably the Nuremberg Code from 1947 (see e.g. George J Annas and Grodin 2008), the World Medical Association's Declaration of Helsinki from 1964 amended the last time in 2013 (World Medical Association 2013 (1964)), and the Belmont Report from 1979 (National Commission for the Protection of Human Subjects of Biomedical and Behavioral Research 1979). These codes have often been drafted and later evolved as a reaction to misconduct and unacceptable precedencies (like the Nazi doctors' experiments or the Tuskegee syphilis experiment). Even if it is interesting to know where the current regulation and legislation come from (Emanuel et al. 2008: Ch. 12–14), the discussion of historical bad practice should not dominate the current debate. Historical accounts of medical research in the military can be revealing with regard to what has gone wrong in the past and can thus serve to caution with regard to current and future research (George J. Annas and Grodin 1992; Emanuel et al. 2008: Ch. 1–10; Harris 2003; Lederer 2003). Nevertheless, one should not conclude that medical research in military or humanitarian contexts is necessarily and always ethically untenable. Rather one should take the guidance offered by professional ethics guidelines seriously and conduct research (in any context) while respecting the clear and well-founded limits that medical research ethics sets.

In the remainder of this section, we will briefly summarize the most important international guidelines of professional ethics that govern the conduct of medical research and "that promote and ensure respect for all human subjects and protect their health and rights" (World Medical Association 2013 (1964): § 7). In doing so, we want to provide a summary of the broadly accepted ethical principles for research in medicine. In addition to the documents referred to above, our summary also takes into account the Council of Europe's Additional Protocol to the Convention on Human Rights and Biomedicine concerning Biomedical Research (Council of Europe 2005) and the World Health Organization's Standards and operational guidance for ethics review of health-related research with human participants (World Health Organization 2011).

1.3.1 A Precondition: Scientific Design of the Research

A precondition for the ethical conduct of medical research with human beings is that the research is no "wild" experimentation. Studies involving human research subjects must build on hypotheses that are based on solid assumptions and prior research. The World Health Organization, for example, states that "[r]esearch is ethically acceptable only if it relies on valid scientific methods. Research that is not scientifically valid exposes research participants or their communities to risks of harm without any possibility of benefit" (World Health Organization 2011: 13). It is assumed that the persons who are conducting (and supervising) the research have the necessary expert knowledge in their field to evaluate the chances for a benefit of the research and to weigh them against the risk that is inevitably imposed on research subjects (cf. World Medical Association 2013 (1964): § 12, 21–22).

This understanding of medical research implies that it cannot be conducted spontaneously but has to be planned in order to be properly designed and documented. As we have seen above, non-standard interventions can be ethically justified in certain circumstances as "non-standard treatment in individual cases"; if such treatments shall however be tested systematically on a larger cohort of patients, this must be done in a planned and well-designed research study. Therefore, military and humanitarian health emergencies (like wars and natural disasters) very often are not the apt circumstances to perform medical research.

1.3.2 Respect for Persons: The Necessity of Voluntary Informed Consent

The ultimate aim of medical research ethics and the paramount principle behind its provisions consists in guaranteeing respect for persons: "It is the duty of physicians who are involved in medical research to protect the life, health, dignity, integrity, right to self-determination, privacy, and confidentiality of personal information of research subjects." (World Medical Association 2013 (1964): § 9) Accordingly, the principle of informed consent, which demands to adequately inform a patient and to then respect her or his decision, is of greatest importance. The requirement of informed consent should be understood as a process that consist in an interactive dialogue between the research staff and (prospective) research participants. It focuses on informing and protecting research participants from misunderstandings with regard to what they can and have to expect. It must therefore include a concise presentation of the research and the key information that is most likely to assist a prospective research participant in understanding why he or she might want to participate in the study or not. Facts and expectations of the study should be presented in a way that is understandable for lay person and must provide the information that a reasonable person would want to get (cf. World Medical Association 2013 (1964): § 25–32).

The idea of informed consent does not imply that no risk may be imposed on individuals or on patients who take part in research. As Beauchamp (2008: 151) puts it, the "purpose of consent provisions is not protection from risk [...] but protection of autonomy and personal dignity, including the personal dignity of incompetent persons incapable of acting autonomously". However, the health of individual patients may not be put at an undue risk. In this context, a particular ethical issue with regard to informed consent is to ensure that patients (or communities) "understand the difference between receiving medical care and being involved in research." (Schopper et al. 2009: 3) This is particularly important in contexts and communities where this difference is not generally known and understood.

Two other aspect related to the principle of respect for persons that are worth mentioning are respect for privacy and medical confidentiality. Someone who agrees to participate in medical research still has a right to as much privacy as possible and also that all collected data is treated confidentially (e.g. anonymized) or that personal information only disclosed with the consent of the patient.[2] In military and humanitarian health care contexts, particular challenges regarding voluntary informed consent, as well as the aspects of privacy and confidentiality may occur, and we will say more about this further below.[3]

1.3.3 Beneficence: Distribution of Risk and Benefit

As with medicine in general, some risks and burdens for the patient come with any kind of medical intervention that promises a benefit. This is not *per se* an ethical issue. Nevertheless, medical research must not involve risks and burdens to the human research subjects that are *disproportionate* to the potential benefits. Thus, burdens must be outweighed by the importance of the research objective, it must be clear that "risks are reasonable in relation to probable benefits" (Tom L Beauchamp 2008: 151), and measures to minimize risks for patients must be implemented. In the words of the World Medical Association, "[p]hysicians may not be involved in a research study involving human subjects unless they are confident that the risks have been adequately assessed and can be satisfactorily managed." (World Medical Association 2013 (1964): § 18) When assessing the risk potential of research, it is furthermore important that "risk" and associated harm can occur in different dimensions that are not always obvious. The World Health Organization, for example, explicitly refers to physical, social, financial, and psychological risks (World Health Organization 2011: 13) and cautions to consider all of them.

---

[2] Exceptions may apply (as in ordinary clinical contexts) when a medical condition poses an immediate risk to the patient or his/her environment as for example with highly infectious diseases or certain psychological conditions.

[3] In more detail, the issue is also discussed in the chapters of Coleman and of Eagan/ Eagan in this volume.

On the other hand, one should also acknowledge that research must not always and only be considered as a burden for potential participants; the inclusion in a research study can also offer an individual patient the chance to benefit, for example, from the latest advances in medicine and to contribute a share to the development of new forms of medical treatment. Nevertheless, a fair and transparent distribution of (potential) burdens of medical research studies must be guaranteed and risks must be communicated to the potential harm bearer.

What has to be avoided by all means and may never be tolerated is a purely utilitarian justification of (abusive) research that would put few at a great risk for the (potential) benefit of many. "The interests and welfare of the human being participating in research shall prevail over the sole interest of society or science." (Council of Europe 2005: 2) Anything else would go against the principle of respect for persons and be an example of violating a person's dignity by using her as a means rather than an end in itself.

1.3.4 Justice: Selection of Research Participants and Distribution of Resources

Contemporary medical ethics requires respect for the principle of distributive justice, according to which both (scarce) resources as well as risks and benefits must be distributed fairly among patients (e.g. Tom L. Beauchamp and Childress 2009: Ch. 7). Similarly, ethics for research in medicine requires "fairness in the distribution of both the burdens and the benefits of research" (Tom L Beauchamp 2008: 151). According to the World Health Organization this means that as a result "no group or class of persons bears more than its fair share of the burdens [… and] no group should be deprived of its fair share of the benefits of research" (World Health Organization 2011: 13). In practice, this principle can be specified in many different ways, one of which is that researchers must not abuse dependencies but consider that "[s]ome groups and individuals are particularly vulnerable and may have an increased likelihood of being wronged" (World Medical Association 2013 (1964): § 19). This is especially important in settings where power is distributed asymmetrically and where knowledge is not available and accessible to everyone equally. Military and humanitarian health care settings undeniably have a significant *potential* for such situations, even though for very different reasons. This does not imply that abuse and exploitation necessarily happen, but that the principle of (distributive) justice and a fair selection of research participants should be given extra consideration and caution. One point to consider with this regard is that nobody should be excluded from (standard) treatment when he or she is refusing to participate in a research study.

1.3.5 *Ethics Review Boards: Apply the Abstract Principles on Project Level*

One way of trying to guarantee "coherent and consistent application of the ethical principles articulated in international guidance documents" (World Health Organization 2011: 12) during medical research on human beings is the independent and *ex ante* evaluation of the research project by an ethics review board. The purpose of such a board has to be seen in an institutionalized implementation of checks and balances in order to avoid malpractice, define optimal research practice and "protect the dignity, rights, safety and well-being of research participants." (Council of Europe 2005: 3). In addition, it may sensitize medical personnel to ethical issues that otherwise could be overlooked "in the heat of the battle". However, the "heat of the battle" or simple time constraints may also make a long review process problematic and it has to be avoided that the ethical review is overly complicated or that even multiple reviews have to be undergone (cf. Gilman and Garcia 2004). A solid ethics review should thus be conducted on the lowest possible institutional level (of course depending on scope of the research, risks etc.) to avoid critical delays. If well executed, such reviews will foster transparency and accountability for the research projects that are envisaged without putting too much of a burden or obstacles in the way of well-designed research projects. When the system functions well, it will be accepted over the time and the necessity of the review become a usual factor. The experiences within MSF illustrates these claims (Schopper et al. 2009).

After this brief review of fundamental ethical principles in medical research, the next section will turn to some of the challenges that medical research must deal with in military and humanitarian contexts.

1.4 Some Specifics of the Military and Humanitarian Contexts Compared to Clinical Research

It is quite well established that that military medical and/ or humanitarian contexts do not need and do not have special ethics. A recent guidelines document of a number of high-profile organizations[4] starts by stating that "[e]thical principles of health care do not change in times of armed conflict and other emergencies and are the same as the ethical principles of health care in times of peace" (International

[4] The original document has been signed and endorsed by the International Committee of the Red Cross (ICRC), the World Medical Association (WMA), the International Committee of Military Medicine (ICMM), the International Council of Nurses (ICN) and the International Pharmaceutical Federation (FIP). Meanwhile, other organization and bodies like the World Health Organization (WHO) but also the Chiefs of Medical Services (COMEDS) within NATO have joined as advocates of the principles proposed by this document.

Committee of the Red Cross (ICRC) et al. 2015). A difference to ordinary (civilian peacetime) medical practice does obviously exist but it is to be found in the different circumstances, contextual restrictions and the like. It is understandable then that even if fundamental ethical principles do not change, their application poses different challenges in military medical, humanitarian, or emergency contexts.

Military health care providers have to confront ethical challenges both within their own organization (treating soldiers, "offensive medicine," enhancement) and when treating external patients like enemy soldiers or civilian casualties. Humanitarian health care providers are mostly confronted with ethical challenges when treating patients within a population they have come to support. Thus, military, humanitarian, and emergency situations differ in many respects, but they also share important features (like scarce resources and an unsecure environment), and this is why discussing them in a joint volume (and book series) promises to provoke interesting discussions and mutually fruitful insights.

This being said, one major difference between military and humanitarian health care shall not be negated or ignored. While humanitarian actors are motivated solely or at least primarily by the principle of humanity and seek to improve well-being as well as to reduce suffering, the same is not completely true for military health services. The latter have at least mixed tasks and can be clearly associated with a specific party to a conflict (from an organizational perspective). Even though military health services have a legal obligation and do in fact treat wounded soldiers from all sides, they cannot always be neutral and impartial in the same way as their humanitarian counterparts. This is at least so with regard to research on (for example) human medical enhancement to develop better medical support for soldiers. Such research and the resulting (non-therapeutic) treatment options would certainly be supplied to own soldiers exclusively. Hence, one can legitimately ask the question when and to what extent military health services can sometimes be said to contribute to the war efforts of one side only. This question shall however not be further pursued in this volume as it would go far beyond our main topic. The moral and legal implications of partial healthcare or of a medical contribution to an (unjust) war effort can however be important (see for example Fabre 2009; Liivoja 2017; Messelken 2017).

With regard to the more restricted topic of the present volume, both military and humanitarian actors are interested in innovation (although for different reasons and motivations), both can be implicated in research studies, and both work in similar fields of intervention (like conflict settings, disasters, etc.). To illustrate these points, we propose to look at some issues that notoriously problematic in the context of medical research and innovation from an ethical perspective: the influence of military interest, some challenges of doing medical research in conflict settings, and working with vulnerable populations.

1.4.1 A First Concern: Military Interests

A first concern that we would like to mention with regard to medical research in a military context is that of mixed purposes and interests. According to Frisina (2003: 540) "military biomedical research does present a double-edge sword. Most often what is learned in the area of biomedical research has potential uses for both good and evil." By this statement the author alludes, for example, to the contribution of developing new kinds of weapons or the use of medicine to contribute to the soldiers' ability to fight or their resilience to the detrimental effects of war. Indirectly, by providing protective means against biological or chemical weapons, biomedical research could also contribute to making the use of these (banned) means of warfare more likely. Other examples will be discussed in more detail in the book chapters on medical human enhancement.[5]

A different but related issue consists in mixing military interests with that of patient-soldiers. In both medical research and medical care, military interests may play a role. Gross (2006: 103) argues that "[w]hile medical personnel work to provide good medical care, they are obligated to provide the care necessary to maintain soldiers as a fighting force – that is, a corporate personality. [...] Soldiers do not receive medical care to guarantee their health as individuals but to preserve the health of a larger organism, a common good quite distinct from the interests of the soldier as patient." Even though the collective interest will not prevail under normal circumstances, its possible influence on justifications of questionable practice should not be underestimated. With regard to doing medical research in the military context or with military interests, Frisina (2003: 538) cautions to acknowledge "a tension, if not competition, between protecting the rights of research subjects on the one hand and conducting research that some view essential to national security interests on the other." Undoubtedly, a lot has changed since the unacceptable Nazi experiments and other reprehensible medical research in the military. The acceptance of ethical restrictions of medical research is now much more widely accepted. Nevertheless, "the sense of necessity and urgency that motivated many of the military's human experiments with unconsenting and uninformed soldiers has not disappeared." (Bonham and Moreno 2008: 417) History should serve as a warning and caution researchers to better think twice whether a specific study meets the ethical requirements and that it actually respects the interests and rights of the research subjects concerned.

[5] See namely the chapters of Eisenstein/ Draper, of S. Eagan, and of Stevens/ Gilbert.

1.4.2 Context: Research in Conflict and Disaster Settings

In a different way, medical research of both military and humanitarian actors may (unintendedly) infringe upon the rights and interests of research participants when it is undertaken in the difficult environments of conflict and disaster. One of the reasons is that patients, because of scarce resources and no alternative options, do not have a free choice of their doctor but rather have to trust themselves into the hands of those who offer medical attention. Thus, it is important that the latter clearly distinguish between urgent emergency field response and research (Brown et al. 2008).

An interesting example in this regard are epidemics like for instance the 2014–2016 outbreak of Ebola in Western Africa (Messelken and Winkler 2018). In the catastrophic circumstances of such a major public health emergency it is almost impossible to conduct scientifically sound clinical trials. A fast reaction is needed, which means that the usual way of developing medication does not work and circumstances are "pushing researchers towards pragmatic solutions and prudent transgressions from conventional models of drug development and research ethics." (Calain 2016: 2) In the case of the Ebola outbreak, one of the questions was whether to offer unproven (new) interventions untested on human beings as there was no curative treatment available (Rid and Emanuel 2014). As a result, individuals may well be put at risk (with, on the other hand, no effective cure available anyway) to further the collective interest of testing new interventions. The question then is "to what extent do emergency circumstances justify derogations or particular regimes in the application of common ethical standards of research?" (Calain et al. 2009: 3) For such cases, benchmarks for research in developing countries and resource-scarce environments have for example been formulated by Emmanuel et al. (2004), Ford et al. (2009) or also by Leaning (2001) with a focus on refugee populations. They all caution to take ethical restrictions seriously and namely to include local actors and people experienced in the kind of local setting expected (Gilman and Garcia 2004; Schopper et al. 2009).[6]

In military contexts and namely in the theatre of war when research may include wounded military personnel, similar difficulties can be encountered. For example, "[c]an a wounded soldier provide truly voluntary, informed consent?" As options are limited and the need for help urgent, "one could hardly argue that a wounded and suffering young service member, thousands of miles from home, is not somehow vulnerable in this context." (De Lorenzo 2004: 129) In addition, working conditions in a "forward-deployed, austere, and hostile war zone" (De Lorenzo 2004: 128) are probably not adapted to the testing of new methods and treatments. This is why international humanitarian law insists that "medical or scientific experiments or any other medical procedure not indicated by the state of health of the person

---

[6] In this volume, the chapters contributed by Sandvik and also by Smith et al. discuss similar questions that arise within the context of humanitarian action.

concerned and not consistent with generally accepted medical standards are prohibited." (Henckaerts and Doswald-Beck 2005: 320).

All of this should of course not be read to completely ban research and innovation from either disaster or conflict settings. A lot can be learned in these contexts and civilian medicine eventually profits as well (McManus et al. 2005). However, ethical restrictions must apply, and pitfalls for malpractice and possible abuse be avoided even if this means that in some situations no medical research can be conducted.

1.4.3 A Second Concern: Vulnerable or Captive Populations

One of the major ethical concerns of doing research in conflict and disaster settings is the risk of disrespecting the rights and dignity of so-called vulnerable or captive populations. This term refers to people who are not able to give a truly free and informed consent because of dependencies in their relationship to the health care providers who might treat then. The vulnerable patients lack "in some critical sense [...] the ability to exercise free choice" (Bonham and Moreno 2008) for several reasons. People can for example be "desperately poor and frightened" (Leaning 2001: 1432) and thus (feel) not to have a choice, or they can be in a superior-subordinate relationship when they are military service members (Amoroso and Wenger 2003). In the first case, there is then a "tension between the need to develop evidence-based emergency health measures and the need to protect vulnerable populations from possible exploitation or harm" (Leaning 2001: 1432). In the second case, there is a danger that "military expediency may be used, albeit sub-consciously, to authorize research in soldiers that would not be permitted in the general population." (Bonham and Moreno 2008: 472) Or, as De Lorenzo (2004: 129) puts it "in the context of a research study, this special physician-researcher and service member–subject relationship is complex and not completely understood, even by those in uniform." In both cases, the ethical issue is based in the fact that "they are unequal players and may thereby have a diminished ability to exercise free choice" (Bonham and Moreno 2008).[7]

After these rather general challenges for medical research in military and humanitarian contexts, we will briefly touch on a current example in the next section: medical human enhancement for military purposes.

[7] This issue will be discussed in some detail in this volume by Coleman and by S. Eagan in their respective chapters.

1.5 Human Medical Enhancement as a Specific Example of Research and Innovation

One example for both innovation and research that will be discussed more extensively in this volume is the use of human medical enhancement methods in the military. Recently, the possibility to improve soldiers' capabilities of surviving during conflict and for defeating their enemies by medically enhancing them has emerged and seems to become more and more realistic or even order of the day. It remains unclear, however, to what extent medical human enhancement should be considered as research or as a new form of (ordinary) medical treatment.[8]

The distinction between therapy and medical enhancement can be similarly blurry than the one between practice, research, and innovation that we have discussed above. In many cases, it can probably only be decided on a case-to-case basis whether a treatment is a therapy or has to be considered an enhancement as too many factors have to be taken into account (see e.g. C. L. Annas and Annas 2008; Daniels 2000). In a very broad sense, human medical enhancement can be defined as those "biomedical interventions that are used to improve human form or functioning beyond what is necessary to restore or sustain health." (Juengst and Moseley 2016: 1) A historical (but still relevant) example that is often referred to is the use of amphetamines by air force pilots ("pilots on speed") to cope with sleep deprivation and/or fatigue on long-distance mission that last longer than one can normally endure (Bower and Phelan 2003; Caldwell and Caldwell 2005; Ko et al. 2018; Rasmussen 2011). Another example from the civilian domain are different forms of doping in sports or the use of stimulants, which can similarly be found in the military context (Friedl 2015). In contrast to treating injuries or illness, enhancement more generally "is about boosting our capabilities *beyond the species-typical level or statistically-normal range* of functioning for an individual" (Allhoff et al. 2010: 3) and, consequentially, "enhancement interventions aim to improve the state of an organism beyond its normal healthy state." (Bostrom and Roache 2008: 120) One can only suspect that many studies and results about tests and the use of enhancing subjects have been published academically so far (Ko et al. 2018). These few points may suffice here to give a first understanding of what is meant by human medical enhancement.[9] There may even more powerful or invasive tools for human enhancement be available in the future, with CRISPR technology being a much-debated example. Such technologies might irreversibly alter parts or aspects of the human being, and thus change the game drastically (Greene and Master 2018).

It is somewhat obvious that "[a]rmed forces that aim for peak physical and cognitive performance under adverse circumstances naturally take a special interest in

[8] A general overview on these questions and some related ethical challenges is given in the chapter of Puscas. More concrete examples will be discussed in the three chapters of Miletic, of Stevens/ Gilbert, and the chapter of Marijn/ Liivoja.

[9] For a discussion of several definitions of enhancement, see also the chapters of Vongehr and of Fischer in this volume.

the prospects of such 'biomedical human enhancement'" (Liivoja 2017: 421). This raises however some ethical issues that go beyond the ethical concerns of a civilian use of enhancing technologies. One criticism has been formulated by Frisina (2003: 538) according to whom "the potential for ethical conflict is considerable when medical researchers conduct studies that do not focus solely on the welfare of a human being but focus also on maintaining and sustaining a person's physical and psychological efficiency as a soldier – a human weapon system." When a soldier is medically enhanced, one could summarize, medicine is used more as a weapon development tool than as a cure for an illness. The same author (three decades ago already) cautioned doctors to distinguish between offensive and defensive medical research (Frisina 1990) warning that at least some forms of offensive medical research ethically are highly problematic. In a similarly problematic way, some doctrinal documents distinguish between a clinical and an operational use of certain drugs, namely so-called go/no-go pills used for fatigue management of deployed soldiers (cf. Liivoja 2017: 437). The (experimental) use of medication in the military can be ethically all the more problematic as "battlefield exigencies" may be used to override personal preferences and informed consent requirements. Soldiers, as discussed earlier, may be counted among the captive or vulnerable populations namely because they "live in circumstances in which the command structure may force them to participate and the needs of the whole may override the interests of the few." (Bonham and Moreno 2008: 472) These are only a few of the ethical concerns related to human medical enhancement in the military and they are discussed in more details by the chapters in the last section of the present book.

Another ethical issue with regard to using human enhancement for military purposes that shall briefly be mentioned here has been elaborated upon by Wolfendale. She urges that the (moral) responsibility of military personnel must not be undermined by the consequences of enhancement. According to her, if "performance-enhancing technologies dissociated military personnel from their actions, their awareness of the moral import of their actions and their ability to understand and learn from the moral consequences of those actions would be severely compromised." (Wolfendale 2008: 37) This illustrates well that human enhancement can involve long-ranging effects on human beings and may eventually alter the human condition.[10]

Human medical enhancement, to sum up this brief outline of the problem, has a vast potential for far-reaching consequences both for individuals and society at large some of which already glimpse up in reality. It is, therefore and for its topicality, a very interesting and illustrative example for research and innovation in the military context.

[10] Similar questions are also raised in the chapters of Fischer and of P. Gilbert.

1.6 Synopsis – Outline of the Volume

The 14 chapters of this volume treat the *Ethics of Medical Innovation, Experimentation, and Enhancement in Military and Humanitarian Contexts* from quite different angles and with a broad variety of backgrounds. They combine analyses and experiences from an international group of experts with an interdisciplinary approach to point to a range of ethical issues related to the main topic. The book gives voice to military and humanitarian practitioners but also to academic scholars with the aim of combining the direct experience from the ground with the rather distant reflection within the "ivory tower" of academia. We are convinced that an exchange between these two very different worlds brings advantages and new perspectives to the reader and all actors involved.

1.6.1 Part I – Research and Research Ethics in Military and Humanitarian Contexts

The first section of the present book looks at ethical issues that arise when medical research is conducted on human beings in military and humanitarian contexts. The six chapters thus give insights, from both practical and theoretical perspectives, on ethical issues related to research, experimentation, and innovation in non-clinical contexts.

Jack Taylor brings in his long-term experience as a Navy surgeon who treated patients during many missions around the world. He looks at what innovation and research can mean in a military environment and context that is very different from a controlled research environment. As experience shows, practical issues like the lack of resources, new types of injury, etc. emerge in military missions as factors leading to the search for new methods of treatment and that favor innovative approaches. Taylor analyzes some of the ethical challenges that inevitably come with the application of new methods and thus medical research or innovation in such non-clinical contexts and illustrates them with both historical cases and his vast own experience.

Nikki Coleman equally looks at medical research within the military but from a different angle: she departs from the fact that obedience is a defining feature of the military and discusses ethical issues related to that feature when medical research is conducted on military service members. As soldiers have to obey the orders of their military doctors and may not seek medical care outside the military health system, we cannot speak of fully autonomous patients and/or research subjects. Nevertheless, military personnel are often used in medical research. The chapter discusses ethical issues relating to the duty to obey orders and the impact it has on military personnel in relation to their health care, particularly when they are involved in medical experimentation.

Neil Eisenstein and **Heather Draper** introduce a more concrete example of medical prophylaxis in the military context (and also an example of what could be labeled "enhancement" if one accepts a wider definition of the term). They look at preventive medical interventions like vaccinations and ask how they are different from non-medical prophylaxis (e.g. body armor). The authors question the current regulations according to which soldiers can be ordered to use non-medical prophylaxis but can refuse to give consent to be vaccinated even though the protective value of the intervention for the individual and the unit is well established. In discussing the question whether some "old-school" medical prophylaxis could be compulsory, the chapter provides groundwork for potential discussions on ordering the use and application of future forms of medical enhancement. The authors offer several ways to deal with the problem and conclude in arguing for a greater sensitivity to the circumstances and requirements of military medical practice.

Another practical example is discussed by **Paul Eagan** and **Sheena Eagan** who discuss research ethics in the context of vaccine development and subsequent trials on human populations. On the one hand, they give an overview of novel vaccine technologies and recent developments in research. Building on that they then analyze the most important ethical issues surrounding the use of unproven vaccines in military personnel or during humanitarian crisis missions: namely, the question of informed consent and human experimentation in vulnerable military and civilian populations.

With the chapter of **Kristin Bergtora Sandvik** there is a change of focus to innovation in a humanitarian context. In her chapter, Sandvik looks at a specific recent example in humanitarian technology experimentation that is the use of "wearables" for tracking the health, safety and nutrition of aid recipients. The main concern that she elaborates is that if the main purpose of wearables is to collect and return large amounts of intimate personal data, the product is not the wearable itself but rather the data that it returns to aid agencies. Thus, large (and mostly vulnerable) populations who do not have much of a choice may be used, even in good faith and with a real benefit to them, as research subjects against their knowledge.

The chapter contributed by **Allister Smith, John Pringle and Matthew Hunt** also deals with ethical challenges of innovation in humanitarian action. As humanitarian action occurs during crises and to support vulnerable populations, the application of innovative technology offers a great potential but also comes with a risk to harm those requiring aid. The authors thus propose to use a "value-sensitive design" approach when innovations are developed and applied in humanitarian settings. The proposed framework is illustrated by the possible application of "refugee biometrics". This example is also of particular interest for the present volume, as military actors may equally be interested in using such technologies when operating in humanitarian missions.

1.6.2 Part II –Human Medical Enhancement in the Military: "Science-Fiction" in Reality

The second part of the volume narrows the focus from research ethics in military and humanitarian medicine to a more concrete example, namely human medical enhancement in a military context. The different means and methods that are summarized under the heading "medical enhancement" constitute particularly topical examples of using cutting edge medical research and innovative methods to improve the performance of soldiers on the battlefield. However, as new, invasive, and often highly experimental medical techniques they also engender important ethical issues and concerns, especially as they are *by definition* used on human beings without a medical indication. In humanitarian contexts, human medical enhancement does not (yet?) play a role and the focus of the chapters in part II is therefore on military applications.

To introduce the second part, **Ioana Maria Puscas** gives an overview on the ethical and operational issues related to military human enhancement. She covers issues such as how human enhancement may affect traditional military values (like courage, merit, and sacrifice) and concludes that one of the main challenges for the military will be to find ways to implement medical human enhancements (if it wants to use them) in a way that renders them legitimate and acceptable in the eye of war fighters.

Dirk Fischer starts from the assumption that human enhancement is one of the most challenging subjects for current and future research in the domain of military medical ethics. According to his, the use of enhancement techniques will have serious consequences for our general understanding of warfare. Fischer goes so far as to state that conflicts in which enhanced soldiers fight should be labelled differently as "transhuman warfare" in order to distinguish them from classical warfare. As the possible consequences of applying medical human enhancement also directly affect the role of military medical personnel, Fischer proposes the separation of the medical role from the new role of an enhancer. The goal of this distinction would be to not involve the medical personnel in medical interventions with the aim of enhancing soldiers.

Sheena Eagan looks into a more specific example of an enhancement technique, namely gene-editing technology. This technology has lately received important funds from the Defense Advanced Research Projects Agency (DARPA) with the aim of developing new forms of therapies and disease prevention, but also with the aim to make better soldiers. Eagan analyzes the most important ethical concerns of gene-editing and regarding the permissibility of genetic enhancement in the military context and looks at issues such as spillover-effects from the military to the civilian sphere, ownership of gene-modifications, the (ir-) reversibility of gene-editing etc. She concludes that there is still time to begin addressing ethical issues in this domain before they grow into ethical problems and that we should not only concentrate on effects on soldiers but need to take a broader perspective that respects soldiers beyond their role and time as warriors.

Tomislav Miletić proposes to envision a specific form of enhancement of soldiers that would consist in a smart exoskeleton able to monitor their health status, report it back to medical experts, and also to apply medical treatments autonomously. His chapter looks into the ethical issues arising with such an exoskeleton and that could relate to like for example data privacy, autonomy of the human patient and the controlling AI, the transparency of the automation processes and the question of trust into AI technology. Beyond these questions related to the soldier as a patient, Miletić also asks how such a technology would influence the role of military medical personnel and if they could inadvertently become combatants by administrating the soldier-enhancing exoskeleton.

Ian Stevens and **Frederic Gilbert** also use a concrete example of medical enhancement to illustrate chances and risks of research and the use of new technologies. They look into novel medical brain implants that are operated by AI and constitute a self-adapting technology, which controlled by AI and where no human-in-the-loop is necessary (or foreseen) to adapt the technology to individual patient's needs. The aim of their chapter is to explore how closed-loop stimulation undermines safety standards and results in skewed risk assessments for complex phenomenon such as a patient's personality or autonomy.

Marijn Kroes and **Rain Liivoja** discuss a technique from the field of neuroscience to treat PTSD by permanently modifying trauma memories in order to avoid returning symptoms. These kinds of memory modification techniques certainly have an enormous potential and are interesting for both military and humanitarian actors. At the same time, brain-modification obviously raises ethical and other concerns and the authors address three major sets of issues: safety and social justice concerns, concerns about threat to authenticity and identity, and possible legal and moral duties to retain certain memories. The authors conclude that given current scientific reality the concerns can be regarded as limited and do not outweigh the potential benefit to develop treatments for patients. With further developments and insights, this assessment may however need to be revised.

Case studies are often used to illustrate ethical issues in discussions. In his chapter, **Frederik Vongehr** proposes a fresh look on the ethical issues of human enhancement by means of examples taken from science fiction. Not only does enhancement sometimes remind us of science fiction but using this popular genre in discussions of related ethical issues indeed is a fruitful approach. Science fiction presents us with possible (and improbable) future scenarios and illustrates how new technologies may alter social conventions. In doing so, these fictitious examples may help us to reflect on ethical implications of technology without being restricted to actual and real-world examples. Vongehr's chapter may thus serve as a reminder to the necessity of thinking out of the box and to keep both an open and vigilant mind when it comes to future developments and their possible implications.

To close the volume, **Paul Gilbert** offers a genuinely philosophical perspective on differences and commonalities of what he calls "superagers" and "supersoldiers". More specifically, his contribution is about medical interventions to avoid death as a result of getting to an old age on the one hand and to avoid death as a result of performing the tasks of a soldier on the other hand. He argues that, as soldiering is a social role and old age a life-stage, the administration of potentially character changing interventions that some forms of enhancement consist in, they have to be regarded differently in the two contexts. As a result, the role of doctors and medical staff involved in their application also has to be regarded from different perspectives and may well lead to different ethical obligations.

1.7 Research and Innovation in Military and Humanitarian Contexts: Necessary But to Be Approached with Caution

The present volume contributes to an analysis of some of the ethical issues that may occur when new medical methods and forms of experimental treatment are used or tested in non-clinical contexts, namely in military and humanitarian settings. By combining analyses from both of these areas the authors of this book aim at elaborating both the commonalities and differences of military and humanitarian health work when it comes to research and innovation. Historical experience and current examples illustrate that important progress of medicine can indeed be made not only in clinical research, but also in less formal settings that deployment, conflict, and disasters bring with them. However, it is important to be aware of the risks, potential for abuse, and ethical pitfalls of these contexts but without exaggeration and with the potential benefits in mind as well. Most importantly, one has to acknowledge that no special (medical) ethics applies during armed conflict and other emergencies and that, as a result, "ethical principles do not change" (International Committee of the Red Cross (ICRC) et al. 2015). Centrally, this means that doctors and health care personnel more generally should make sure that respect for patients and their dignity is upheld in any context and under any circumstances. When involved in research or research-like medical activities, sensitivity about what it means for an individual (a patient) to consent to participating in medical research or to receiving care by means of innovative methods should thus govern the actions of health care providers. Where patients are respected in their individuality and where they are truthfully informed about possible treatments (be it research, innovation, or general practice) the risk of unethical behavior decreases.

Acknowledgments We could not have completed this book without the continued support of a number of people to whom we would like to express our gratitude.

Our first and profound thanks go to the contributors of this volume who wrote and revised their chapters with diligence and who were open enough to share their knowledge and own experiences with a broader audience.

We are thankful to Major General Dr. Andreas Stettbacher, Major General Dr. Roger van Hoof, and Prof. Peter Schaber under the patronage of whom the workshop was organized during which most of the chapters of this volume were first discussed. Financial support for the work on this volume was granted by the Centre of Competence for Military and Disaster Medicine of the Swiss Armed Forces to the Center for Military Medical Ethics at Zurich University.

Finally, we would like to thank the two anonymous referees for their feedback and constructive comments on the first manuscript, as well as Floor Oosting and Christopher Wilby from Springer for their advice and support throughout the conception and production of this volume.

References

Allhoff, Fritz, et al. 2010. Ethics of human enhancement: 25 questions & answers. *Studies in Ethics, Law, and Technology* 4 (1): 1.

Amoroso, Paul J., and Lynn L. Wenger. 2003. The human volunteer in military biomedical research. In *Military medical ethics*, ed. Thomas E. Beam and Linette R. Sparacino, vol. 2, 563–603. Washington, DC: Office of The Surgeon General, United States Army.

Annas, Catherine L., and George J. Annas. 2008. Enhancing the fighting force: Medical research on American soldiers. *The Journal of Contemporary Health Law and Policy* 25: 283.

Annas, George J., and Michael A. Grodin, eds. 1992. *The Nazi doctors and the Nuremberg code human rights in human experimentation*. Vol. XXII, 371 S. New York/Oxford: Oxford University Press.

———. 2008. The Nuremberg code. In *The Oxford textbook of clinical research ethics*, 136–140. New York: Oxford University Press.

Beauchamp, Tom L. 2008. The Belmont report. In *The Oxford textbook of clinical research ethics*, 21–28. Oxford: Oxford University Press.

Beauchamp, Tom L., and James F. Childress. 2009. *Principles of biomedical ethics*. 6th ed. Oxford: Oxford University Press.

Blackbourne, Lorne H., et al. 2012. Military medical revolution: Military trauma system. *Journal of Trauma and Acute Care Surgery* 73 (6): S388–SS94.

Bonham, Valerie H., and Jonathan D. Moreno. 2008. Research with captive populations: Prisoners, students, and soldiers. In *The Oxford textbook of clinical research ethics*, ed. Ezekiel J. Emanuel, 461–474. Oxford: Oxford University Press.

Bostrom, Nick, and Rebecca Roache. 2008. Ethical issues in human enhancement. In *New waves in applied ethics*, ed. Jesper Ryberg, Thomas Petersen, and Clark Wolf, 120–152. Basingstoke: Palgrave Macmillan.

Bower, Eric A., and James R. Phelan. 2003. Use of amphetamines in the military environment. *The Lancet* 362: s18–s19.

Brown, Vincent, et al. 2008. Research in complex humanitarian emergencies: The Médecins Sans Frontières/epicentre experience. *PLoS Medicine* 5 (4): e89.

Calain, Philippe. 2016. The Ebola clinical trials: A precedent for research ethics in disasters. *Journal of Medical Ethics* 44 (1): 3–8.

Calain, Philippe, et al. 2009. Research ethics and international epidemic response: The case of Ebola and Marburg hemorrhagic fevers. *Public Health Ethics* 2 (1): 7–29.

Caldwell, John A., and J. Lynn Caldwell. 2005. Fatigue in military aviation: An overview of US military-approved pharmacological countermeasures. *Aviation, Space, and Environmental Medicine* 76 (7): C39–C51.

Council of Europe. 2005. *Additional protocol to the convention on human rights and biomedicine, concerning biomedical research*, Council of Europe treaty series–no. 195. Strasbourg: Council of Europe.

Daniels, Norman. 2000. Normal functioning and the treatment-enhancement distinction. *Cambridge Quarterly of Healthcare Ethics* 9 (3): 309–322.

De Lorenzo, Robert A. 2004. Emergency medicine research on the front lines. *Annals of Emergency Medicine* 44 (2): 128–130.

Emanuel, Ezekiel J., et al. 2004. What makes clinical research in developing countries ethical? The benchmarks of ethical research. *Journal of Infectious Diseases* 189 (5): 930–937.

———. 2008. *The Oxford textbook of clinical research ethics*. Oxford: Oxford University Press.
Fabre, C. 2009. Guns, food, and liability to attack in war. *Ethics* 120 (1): 36–63.
Ford, N., et al. 2009. Ethics of conducting research in conflict settings. *Conflict and Health* 3: 7.
Friedl, Karl E. 2015. US Army research on pharmacological enhancement of soldier performance: Stimulants, anabolic hormones, and blood doping. *The Journal of Strength & Conditioning Research* 29: S71–S76.
Frisina, Michael E. 1990. The offensive-defensive distinction in military biological research. *Hastings Center Report* 20 (3): 19–22.
———. 2003. Medical ethics in military biomedical research. In *Military medical ethics*, ed. Thomas E. Beam and Linette R. Sparacino, vol. 2, 533–561. Washington, DC: Office of The Surgeon General, United States Army.
Gilman, Robert H., and Hector H. Garcia. 2004. Ethics review procedures for research in developing countries: A basic presumption of guilt. *Canadian Medical Association Journal* 171 (3): 248–249.
Givens, Melissa, Andrew E. Muck, and Craig Goolsby. 2017. Battlefield to bedside: Translating wartime innovations to civilian emergency medicine. *The American Journal of Emergency Medicine* 35 (11): 1746–1749.
Greene, Marsha, and Zubin Master. 2018. Ethical issues of using CRISPR technologies for research on military enhancement. *Journal of Bioethical Inquiry* 15 (3): 327–335.
Gross, Michael L. 2006. *Bioethics and armed conflict. Moral dilemmas of medicine and war*, xv, 384. Cambridge, MA: MIT Press.
Haider, Adil H., et al. 2015. Military-to-civilian translation of battlefield innovations in operative trauma care. *Surgery* 158 (6): 1686–1695.
Harris, Sheldon H. 2003. Japanese biomedical experimentation during the World-War-II era. In *Military medical ethics*, ed. Thomas E. Beam and Linette R. Sparacino, vol. 2, 463–506. Washington, DC: Office of The Surgeon General, United States Army.
Henckaerts, Jean-Marie, and Louise Doswald-Beck. 2005. *Customary international humanitarian law*. Cambridge: Cambridge University Press.
Hodgetts, Timothy J. 2014. A roadmap for innovation. *Journal of the Royal Army Medical Corps* 160 (2): 86–91. (British Medical Journal Publishing Group).
International Committee of the Red Cross (ICRC), et al. 2015. *Ethical principles of health care in times of armed conflict and other emergencies*. Geneva: ICRC.
Juengst, Eric, and Daniel Moseley. 2016. Human enhancement. In *The Stanford encyclopedia of philosophy*, ed. Edward N. Zalta. Stanford: Stanford University.
Kitchen, Lynn W., and David W. Vaughn. 2007. Role of US military research programs in the development of US-licensed vaccines for naturally occurring infectious diseases. *Vaccine* 25 (41): 7017–7030.
Ko, Henry, et al. 2018. A systematic review of performance-enhancing pharmacologicals and biotechnologies in the Army. *Journal of the Royal Army Medical Corps* 164 (3): 197–206.
Leaning, Jennifer. 2001. Ethics of research in refugee populations. *The Lancet* 357 (9266): 1432–1433.
Lederer, Susan E. 2003. The cold war and beyond: Covert and deceptive American medical experimentation. In *Military medical ethics*, ed. Thomas E. Beam and Linette R. Sparacino, vol. 2, 507–533. Washington, DC: Office of The Surgeon General, United States Army.
Liivoja, Rain. 2017. Biomedical enhancement of warfighters and the legal protection of military medical personnel in armed conflict. *Medical Law Review* 26 (3): 421–448.
Ling, Geoffrey S.F., Peter Rhee, and James M. Ecklund. 2010. Surgical innovations arising from the Iraq and Afghanistan wars. *Annual Review of Medicine* 61: 457–468.
McManus, John, et al. 2005. Informed consent and ethical issues in military medical research. *Academic Emergency Medicine* 12 (11): 1120–1126.
Messelken, Daniel. 2017. Medical care during war: A remainder and prospect of peace. In *The nature of peace and the morality of armed conflict*, ed. Florian Demont-Biaggi. Cham: Palgrave Macmillan.

Messelken, Daniel, and David T. Winkler, eds. 2018. *Ethical challenges for military health care personnel: Dealing with epidemics*, Military and defence ethics. London: Routledge.
National Commission for the Protection of Human Subjects of Biomedical and Behavioral Research. 1979. *The Belmont report: Ethical principles and guidelines for the protection of human subjects of research*. Washington, DC: Department of Health, Education, and Welfare, National Commission for the Protection of Human Subjects of Biomedical and Behavioral Research.
Rasmussen, Nicolas. 2011. Medical science and the military: The Allies' use of amphetamine during World War II. *Journal of Interdisciplinary History* 42 (2): 205–233.
Ratto-Kim, Silvia, et al. 2018. The US military commitment to vaccine development: A century of successes and challenges. *Frontiers in Immunology* 9: 1397.
Reade, Michael C. 2013. Military contributions to modern trauma care. *Current Opinion in Critical Care* 19 (6): 567–568.
Rid, A., and E.J. Emanuel. 2014. Ethical considerations of experimental interventions in the Ebola outbreak. *Lancet* 384 (9957): 1896–1899.
Schopper, D., et al. 2009. Research ethics review in humanitarian contexts: The experience of the independent ethics review board of Médecins Sans Frontières. *PLoS Medicine* 6 (7): e1000115.
Schrager, Jason J., Richard D. Branson, and Jay A. Johannigman. 2012. Lessons from the tip of the spear: Medical advancements from Iraq and Afghanistan. *Respiratory Care* 57 (8): 1305–1313.
Swiss Academy of Medical Sciences. 2015. *Research with human subjects. A manual for practitioners*. Bern: Swiss Academy of Medical Sciences.
———. 2017. *Distinguishing between standard treatment and experimental treatment in individual cases*. Bern: Swiss Academy of Medical Sciences.
Wolfendale, Jessica. 2008. Performance-enhancing technologies and moral responsibility in the military. *The American Journal of Bioethics* 8 (2): 28–38.
World Health Organization. 2011. *Standards and operational guidance for ethics review of health-related research with human participants*. Geneva: WHO.
World Medical Association. 2013 (1964). *WMA Declaration of Helsinki – Ethical principles for medical research involving human subjects*. Geneva: World Medical Association.

Part I
Research and Research Ethics in Military and Humanitarian Contexts

Chapter 2
Innovation or Experimentation? Experiences from a Military Surgeon

Jackson B. Taylor

2.1 Introduction

"He who wishes to learn surgery should go to war." – This saying often attributed to Hippocrates, contains a certain amount of truth. Even within civilian training programs, the trauma service is a both a classroom and a proving ground for young surgeons. Skills learned are put to the test under the immediacy of the patient's, often critical, injury. There is no time for pre-operative conferences or consultation with colleagues for a patient whose life expectancy is now measured in minutes, not days. Once the operative surgeon begins the procedure, he or she (for simplicity "he" or "his" will be used with no bias implied) is often confronted by unexpected injuries and unique patterns of tissue damage. The decisions on how to proceed are based in time honored principles but the specifics of how those principles are applied are left to the surgeon and modified to meet the needs of the scenario encountered. In dealing with an unexpected finding, the surgeon might need to be "innovative" in his approach to the problem. No reasonable person would confuse this use of innovation with experimentation or even the intentional generation of new knowledge, though that might occur as a byproduct. Military surgeons taking care of combat injured troops might find themselves placed in this position more frequently than their civilian counterparts who care for civilian trauma patients as the wounding pattern in combat usually has no analogous injury in civilian trauma.

Surgeons, both, civilian and military, caring for patients in the controlled environment of elective surgery might also apply innovative techniques to their patients during operative care. For example, in my chosen field of plastic surgery, the reconstruction of defects left after trauma or cancer resection often have several options for treatment. In any given patient, the "reported" options may be substan-

J. B. Taylor (✉)
NATO Military Medicine Centre of Excellence, Budapest, Hungary

D. Messelken, D. Winkler (eds.), *Ethics of Medical Innovation, Experimentation, and Enhancement in Military and Humanitarian Contexts*, Military and Humanitarian Health Ethics,
https://doi.org/10.1007/978-3-030-36319-2_2

dard and the operative surgeon must be "innovative" to find an acceptable option. The likely clinical scenarios and the options for treating each scenario as well as the risks and benefits of each treatment option are discussed with the patient preoperatively. The surgeon is usually given wide latitude to make decisions for the patient based on the patients stated ideal outcome. This latitude is applied and based within accepted surgical principles which may be applied in new or different ways to achieve an acceptable outcome. And again, no expectation of generating new knowledge has occurred, only the fulfilment of the doctor-patient contract of providing the best care for the patient.

But surgeons do innovate, and they do experiment. And in both cases, new knowledge will be generated with the expectation of advancing or improving care for subsequent patients. How do we determine what falls into the category of innovation that is required to provide safe patient care, innovation that advances surgical knowledge without being experimental, and clinical research? Is there a difference? Should there be? After attempting to define the differences we will look at the safeguards in place to protect the patient and surgeon.

2.2 Definitions

The Oxford English Dictionary defines innovation as "the introduction of new things, ideas or ways of doing something" whereas an experiment is "a scientific test to study what happens and to gain new knowledge" (Oxford English Dictionary 2019). Both are clear enough in their own right but how do they relate to the more nuanced field of surgery and where is the line between the two drawn? In their book "The Ethics of Surgical Practice: Cases, Dilemmas, and Resolutions" Jones et al. state experimentation occurs when the outcome of one therapy over another is not known (Jones et al. 2008). Equally important is the question of how a new procedure will affect the patient. Experimentation clearly occurs when the operative surgeon cannot, with great certainty, state that the outcome will not endanger the patient. In either of these situations the procedure should be treated as research and the standard measures of study review by an Institutional Review Board (IRB), continuous study review, and detailed patient informed formed consent be followed in accordance with accepted guidelines such as the Nuremberg Code and Declaration of Helsinki (Annas and Grodin 2008; World Medical Association 2013 (1964)). Conversely, innovation may result in significant improvement in patient care yet pose no risk to the patient.

Examples of experimentation in surgery are extremely difficult to find. In my experience, surgeons are notoriously skeptical of the gold standard in clinical research, the randomized clinical trial (Das 2011). We will discuss this issue in some more detail later.

2.3 The Grey Area of Innovation and Experimentation in (Military) Surgery

There are many examples of innovation that benefits but has no evidence of patient risk. One such example is Jean Larrey's implementation of ambulances, treatment at or near the point of injury, and triage. These practices clearly benefited the patient without posing any additional risk. Larrey's principles are still used today in modern civilian trauma systems and military combat field care (Trauma.org 2019).

But what about the grey area in between? Ambroise Paré, a surgeon in the sixteenth century, who, out of necessity when he ran out of boiling oil, bandaged his patient's amputation stumps with an ointment of egg yolks and turpentine. He found that the patients treated with the egg and turpentine ointment were in less pain and healed more quickly. From that observation, he advocated for discontinuing treatment of the stumps with oil. Paré also "re-introduced the use of blood vessel ligature during amputations" (wikipedia 2019). He again observed that patients with ligated vessels versus vessel cauterization with hot irons, had better hemorrhage control during surgery and less bleeding post-operatively. His patients were noted to have a higher rate of wound infections due to the ligatures. Despite this, vessel ligature remains the accepted method of treatment today.

Harvey Cushing, a neurosurgeon practicing in the early twentieth century introduced the use of cranial decompression during WW-I to relieve intra cranial pressure in patients with cranial trauma. Surgeons following Cushing were unable to achieve his results and the technique had been widely rejected by surgeons as increasing mortality. During our most recent conflicts in Iraq and Afghanistan the technique was used in selected patients with encouraging results (Bell 2010). It has now returned to use in both the military and civilian medical in very select scenarios.

Dr. Ralph Millard Jr. is considered by some to be one of our greatest reconstructive surgeons. He was nominated as one of the 10 Plastic Surgeons of the Millennium in 2000 by the American Society of Plastic Surgeons (ASPS). It would be difficult to overstate his importance to the field of reconstructive surgery. His technical innovations in the 1950s still provide the basis for how cleft lips are repaired today. Indeed, he literally wrote the book on care of patients with cleft lips (Millard 1976). I personally have a great amount of respect for Dr. Millard. However, his description of performing that first truly innovative cleft lip repair has always been an ethical dilemma in my mind. In 1955, Dr. Millard was serving in Korea with the US Army. By this point in his career he was already very competent in the cleft lip repair techniques of the day. He also realized their shortcomings and had devised a way to improve outcomes based on solid surgical principles. I will paraphrase his description of how he obtained one of his first patients. He saw a young boy with a cleft lip near his camp and he lassoed the boy, took him back to the camp, and under general anesthesia and without parental permission, performed the first rotation-advancement flap, repairing the child's cleft lip (cf. Millard 1976). The surgery went well, and it was only days before he had more cleft lip patients than he could care for show up at the camp.

In 1993 Michael Rotondo and William Schwab published a paper on what they termed "Damage Control Surgery" (Rotondo et al. 1993). They proposed taking a traumatically injured, clinically unstable patient to the operating room and performing only those procedures needed to stop life threatening bleeding and ongoing contamination of the abdomen. Once the bleeding was controlled, the patient was resuscitated– given intravenous fluids and blood and warmed– until physiologic parameters were normalized, and the patient stabilized. This differed from the prevailing treatment which was to take the patient to the operating room and repair all injuries while concurrently resuscitating the patient. Rotondo and Schwab's treatment plan resulted in lower mortality and better patient outcomes when compared to the current dogma of repairing all injuries at the first operation. They were so successful that Damage Control Surgery is now the standard of care in traumatically injured patients in the United States.

During recent and ongoing conflicts, tourniquets have been used extensively to stop bleeding from traumatically amputated extremities. The first documented use of the tourniquet was during the middle ages (Welling et al. 2012). Since that time, tourniquet use has come in and fallen out of favor. During my own surgical training in the mid-nineties for example, the use of a tourniquet instead direct pressure to stop bleeding was considered close to malpractice. In the current conflicts, it is not unreasonable to say that this inexpensive nylon strap saved as many lives as all other technologic or surgical advancements combined. There was no new research carried out which showed clinical superiority of tourniquet use compared to non-use, there was only the reasonable application of a device to a new wounding pattern in a new situation (Welling et al. 2012).

These are but a few examples of the grey area between innovation and experimentation and all have military surgery in common. But this dilemma is not limited to the military context. In civilian surgery, there are many examples such as the development of laparoscopic surgery and its subsequent application to virtually all intra-abdominal procedures. More recently we have seen the rise of "robotic surgery" as a "state of the art" technology. Neither of these techniques has undergone rigorous clinical testing prior to introduction into mainstream medical care. Even now, many of the new procedures we consider to be the "gold standard" or "standard of care" have not been proven through to be superior to the older surgical procedures. Even the acquisition of new surgical skills by an established surgeon may constitute innovation. There is a period of time before a surgeon gains enough clinical experience, that the outcome of a given surgical procedure may not be predictable with the same certainty as if were performed by a more experienced surgeon.

In each of these cases, we have shown changes made to advance the care provided to patients. Each example meets the definition of innovation in that new things, ideas or ways of doing something have been introduced into the care of the patient without meeting the definition of experimentation. There is clearly a spectrum of innovation from changing how a procedure is done due to limited resources to the introduction of new procedures based on need and the perceived understanding of the likely outcome both if the procedure is performed and if it is not. I would argue that, given the wide spectrum of what qualifies as innovation, trying to determine when innovation has occurred is the wrong question. Rather, we should look at the ethical implications of how we deliver surgical care.

2.4 Innovation in Practice

In medicine, as opposed to surgery, trainees are taught by example. They see what their instructors do in each situation. They study the effects and possible side effects of medications. They evaluate the merits of new medications. They recommend lifestyle choices and changes. The art of medicine comes in the diagnosis and the prescription of treatment. There is always a possibility of error but there is also a significant margin for error.

In surgical training that margin for error is much narrower. A trainee can watch his mentor perform surgery. He can practice on simulators and discuss the surgery. But at some point, he must perform the operation on a real, live patient. A person. There is no other way to learn. No other way to train each successive generation of surgeons. And society, whether it is explicitly acknowledged, understands this. No one is born with the skills to be a surgeon. And despite all the safety checks we put into place, there are still complications, mistakes, that occur when training. But at some level we accept that risk, for the greater good of society. The risk inherent in training and learning. And surgeons understand both the risk and the trust placed in them. Surgical training has a long history of self-policing. The cornerstone of this policing process is the morbidity and mortality conference where surgical complications are presented and critiqued– or harshly criticized– by a group of surgical peers. It forces the operative surgeon to take full responsibility for his mistakes while providing an opportunity for learning. It is in this training atmosphere that surgeons come to understand the profound responsibility they have and the trust their patients place in them. They understand that every time they operate on a patient they can improve their patient's health but they also accept the risk of causing permanent disability or death. Most importantly, every surgeon realizes that it is not they, but their patient, that lives with the risk.

As a surgeon, I, as most surgeons, accept that I am obligated to do the best that I can for my patient. This means both performing safe surgery using principles that I know will give me a highly reproducible and safe outcome. But I also feel that it means modifying my technique for my patient if I feel I can improve the outcome by using someone else's "How I Do It", modifying a suturing technique, encouraging better pre-operative nutrition or earlier a post-operative ambulation. Each of these changes should both improve outcome and advance surgical care. They do not represent significant deviations from the "usual" care and they do not represent a significant risk to the patient. In the previously discussed examples of innovative changes, there may have been a level of increased risk to the patient but none of the innovations required the testing of a hypothesis against a standard treatment or a null hypothesis to test its validity. And each example was felt to represent a change with little increased risk to the patient and often applied when obtaining specific consent would have been difficult or impossible. So how do we rectify the need for patient autonomy while still giving the surgeon the flexibility to modify surgical procedures through innovation? In many ways, the surgical consent provides an avenue. Most consent forms contain the statement "and all indicated procedures" meaning that during a surgical procedure the surgeon is given the latitude to address unexpected findings and to modify the surgical procedure to meet the patient's needs. It is during the consent process that the patient can make the choice not to proceed with surgery or discuss any limitations they would like to place on the surgeon and the surgical procedure. In my practice, I also recognize that there is a limitation to this latitude. When the intra-operative findings require a significant deviation from the operative plan, meaning the addition of risks that were not discussed prior to surgery, I immediately discuss the new plan and it's risks with the next of kin and obtain the necessary permission to proceed or not. This is a standard procedure for all the surgeons I work with and for those I trained under.

A more compelling question though is, how does a surgeon know that his modification of the surgical procedure is beneficial? Or rather, how does he know it is not harmful? Certainly, there are examples of procedures that were performed in the past which were later shown to be of no benefit–ligation of the internal mammary artery for chest pain is often cited as such an example. Boards certifying surgeons in their respective specialties have tried to address this through continuous practice evaluations. At present, these are semi-voluntary although the boards do require randomly selected patient files be submitted for evaluation. This is supposed to improve surgeon compliance with the self-evaluation but there is no clear evidence this is the case. Further, even if the individual surgeon is diligently participating in practice evaluation, there is no mechanism for the sharing of the knowledge gained from this evaluation. In attempting to follow the principles of beneficence and non-maleficence, the surgical community owes it to the patient, and society as a whole, to better evaluate patient outcomes, especially after application of innovative therapies.

2.5 Experimentation

We move now to the area of experimentation. As noted earlier, experimentation is a scientific test to gain new knowledge. This means that, as opposed to innovation, the outcome cannot be reliably predicted. The gold standard for clinical experimentation is the randomized controlled trial (RCT), the highest level of evidence for or against a specific treatment. In most RCT for medications there is a clear path from laboratory research to animal studies and from animal studies to human studies before a medication is deemed safe and effective. Surgical research has rarely followed this pathway. Surgeons have, with some validity, felt that RCTs are too difficult to perform for surgical procedures, require too much time to reach a conclusion, and surgeon variability which can be difficult to control for (Das 2011). One of the most difficult issues within surgical RCTs is whether there is a question that needs to be clarified. In other clinical trials this question is answered by testing the null hypothesis or that there is no difference between two variables. In surgery, this can determine of whether a surgery is better than no surgery. One begins to see the problem immediately; how do you test this? The concept of "sham surgery", or making an incision without performing a surgical procedure, raises certain ethical issues itself. Primary among these are the risk of deception and risk of patient harm without any possibility of compensatory benefit. In clinical practice, treatments have been prescribed with physician knowledge that the "treatment" should not have a clinical response but is prescribed to placate the patient. These treatments have resulted in a clinical response termed the placebo effect. Use of these treatments has its own ethical dilemmas. A surgical placebo carries with it the same risks as any surgical procedure namely infection, bleeding, anesthetic risks, and risk of death. The American Medical Association Council on Ethical and Judicial Affairs has evaluated arguments from ethicists Robert Levine, Ruth Macklin, and Henry Beecher and made the following recommendations: surgical placebo should be used only when no other trial design will yield the requisite data, informed consent must clearly delineate the surgical risks and what is meant by "placebo". Placebo surgery is not appropriate to test the effectiveness of a minor modification. In testing new surgical procedures placebo may be appropriate if the disease being treated is susceptible to placebo (Tenery et al. 2002).

Other arguments such as no two surgeries are the same owing to anatomical variation, surgical technique, surgeon skill, and anesthetic care are valid however careful study design may mitigate many of these variables and should be pursued (Das 2011).

Even though surgeons, possibly for valid reasons, have been slow to pursue RCT as a method of verifying that one surgical procedure is superior to another, it is a goal that we should work towards. It may not be possible in all cases, but an attempt should be made if we truly wish to say our specialty believes in the principles of beneficence and non-maleficence.

2.6 Safeguards

We have discussed innovation and the wide spectrum of what falls into that category of surgical advancement. We have also discussed experimentation and classical clinical research and its importance in surgery. How do we provide safeguards for our patients to ensure that they are afforded the protections required in the ethical treatment of human research subjects? I will limit my comments to research in the military setting.

Much has been written about research in the military specifically about service members as a "vulnerable population" in the same category as prisoners and those with intellectual disabilities. This is usually framed as a concern due to the hierarchical nature of the military. This argument conveniently avoids any discussion of the all voluntary nature of the military and the intense scrutiny from outside entities the military faces daily. Furthermore, there are few institutions that do not have some component of a hierarchy that could be used to influence the behavior of its junior members. In academia, the relationship of instructor to student or professor to assistant professor could easily be used to influence behavior. From my point of view and experience, the "vulnerable population" argument is a weak one at best.

The dual loyalty argument is based in some reality (Allhoff 2008). There is a loyalty to the mission and to the organization as well as to the patient. There is no argument that can defend any healthcare provider that has crossed the line in respects to torture or mistreatment of any human being. It is an ethical, moral, and legal failure and should be dealt with as such. Fortunately, the providers who find themselves on the wrong side of this moral choice are an extremely small fraction of the healthcare providers in the military. What I have witnessed are dedicated doctors, nurses, and medics who go to heroic lengths to provide the best care that they can to the most patients possible. This is done without consideration of alliance (friend or foe) or need to return to combat.

But we also have legal and ethical guidelines that we are required to follow in both clinical practice and research. All surgical specialties have statements in their membership requirements that surgeons, including military surgeons, will care for patients in an ethical manner. The American College of Surgeons Fellowship pledge states "I pledge to pursue the practice of surgery with honesty and to place the welfare and rights of my patients above all else… I will respect the patient's autonomy and individuality" (American College of Surgeons 2004). The American Society of Plastic Surgeons Ethical Guidelines state "The principle objective of the medical profession is to render services to humanity with full respect for human dignity. Members… should expose, without hesitation, illegal or unethical conduct of other Members of the profession" (American Society of Plastic Surgeons 2017).

The Department of Defense Instruction 3216.02, Protection of Human Subjects and Adherence to Ethical Standards in DoD-supported Research, provides clear guidance on human research in the military including direction that guidance in The Belmont Report (National Commission for the Protection of Human Subjects of Biomedical and Behavioral Research 1979) will be followed.

The Headquarters, United States Army Medical Research and Material Command (HQ USAMRMC), has developed a 120-page report titled "HQ USAMRMC Institutional Review Board Policies and Procedures" which outlines in detail the "strong commitment to the protection of human subjects participating in research" (HQ USAMRMC 2010). Within this document are detailed instructions for the conduct of ethical research with protection of human research subjects.

Finally, and possibly the most important, the Code of Federal Regulations (CFR) title 32, part 219- Protection of Human Subjects- outlines the federal regulations for research on human subjects for "all research involving human subjects conducted, supported or otherwise subject to regulation by any federal department or agency which takes appropriate administrative action to make the policy applicable to such research" (Code of Federal Regulations 2013). Most importantly, this document not only mandates the use of Institutional Review Boards but specifies the makeup of the board. The mandated composition includes scientific and non-scientific members, at least one member who is not otherwise affiliated with the institution and is not part of the immediate family of a person who is affiliated with the institution and excludes members who may have a conflict of interest.

These instructions carry the weight of orders. To disregard them would be no different than disregarding a direct order from a superior. While there have been lapses in judgement and questionable ethical practices in the military, just as there have been in civilian medical systems, the military has taken steps to prevent such occurrences in the future.

2.7 Conclusion

The topics of innovation and experimentation are extremely important in the military healthcare system. Defining innovation is difficult, if not impossible. The grey area that constitutes innovation, from surgeons learning new procedures to the purposeful modification of standardized procedures to achieve superior outcomes is a broad area. Learning surgery, whether a trainee doing his first surgery or an experienced surgeon incorporating a new surgical technique into his skill set comes with a learning curve and is generally accepted by society as a necessary risk. Innovation that involves an unknown outcome has become research and should be treated as such with IRB approval and oversight. Mitigating the risk of practicing innovative surgery that is of no benefit or not sharing the lessons learned can, and should, be mitigated by a more formalized process of tracking and sharing innovative approaches to standard procedures.

Hippocrates was correct, war is a training ground for surgeons. But that training ground comes with the responsibility to ensure patients still receive the best care we can provide them.

References

Allhoff, Fritz, ed. 2008. *Physicians at war the dual-loyalties challenge*. Dordrecht: Springer.

American College of Surgeons. 2004. *American College of Surgeons Fellowship Pledge*.

American Society of Plastic Surgeons. 2017. *Code of ethics of the American Society of Plastic Surgeons*.

Annas, George J., and Michael A. Grodin. 2008. The Nuremberg code. In *The Oxford textbook of clinical research ethics*, 136–140. New York: Oxford University Press.

Bell, Randy S., et al. 2010. Early decompressive craniectomy for severe penetrating and closed head injury during wartime. *Neurosurgical focus 28*(5). https://doi.org/10.3171/2010.2.FOCUS1022

Code of Federal Regulations. 2013. *Part 219, protection of human subjects*.

Das, Anjan Kumar. 2011. Randomised clinical trials in surgery: A look at the ethical and practical issues. *The Indian Journal of Surgery* 73 (4): 245–250.

HQ USAMRMC. 2010. *Institutional review board policies and procedures, Ver1*.

Jones, James W., Laurence B. McCullough, and Bruce W. Richman. 2008. *The ethics of surgical practice: Cases, dilemmas, and resolutions*. Oxford/New York: Oxford University Press.

Millard, D. Ralph. 1976. *Cleft craft : The evolution of its surgery*, 3 vols, 1st ed. Boston: Little, Brown.

National Commission for the Protection of Human Subjects of Biomedical and Behavioral Research. 1979. *The Belmont report: Ethical principles and guidelines for the protection of human subjects of research*. Washington, DC: Department of Health, Education, and Welfare, National Commission for the Protection of Human Subjects of Biomedical and Behavioral Research.

Oxford English Dictionary. 2019. *Oxford English Dictionary online*. Oxford University Press. www.oed.com. Accessed 15 Feb 2019.

Rotondo, Michael F., et al. 1993. Damage control': An approach for improved survival in exsanguinating penetrating abdominal injury. *The Journal of Trauma* 35 (3): 375–382. discussion 82–3.

Tenery, Robert, et al. 2002. Surgical "placebo" controls. *Annals of Surgery* 235 (2): 303–307.

Trauma.org. 2019. *History of Trauma: Dominique Jean Larrey*. trauma.org. Accessed 15 Feb 2019.

Welling, David R., et al. 2012. A brief history of the tourniquet. *Journal of Vascular Surgery* 55 (1): 286–290.

Wikipedia. 2019. *Ambroise Pare*. https://en.m.wikipedia.org/wiki/Ambroise_Pare. Accessed 15 Feb 2019.

World Medical Association. 2013 (1964). *WMA declaration of Helsinki – Ethical principles for medical research involving human subjects*. Geneva: World Medical Association.

Chapter 3
The Impact of the Duty to Obey Orders in Relation to Medical Care in the Military

Nikki Coleman

3.1 The Impact of the Duty to Obey Orders in Relation to Medical Care in the Military

The requirement to obey orders is one of the key features of the military. In the civilian population and even amongst private military contractors there is not the same requirement to obey orders. This makes the relationship between a military member and their doctor problematic in countries where soldiers must obey the orders of the doctor and is not allowed to seek medical care outside of the military health system. Whilst the requirement to obey orders of their doctor is important to ensure operational effectiveness, this requirement to obey their doctor goes against the bio-ethical principle of autonomy in health care. Therefore, medical experimentation on military personnel has the potential to make soldiers vulnerable to unethical or abusive experimentation.

3.2 Obedience

The concept of being obedient to another is quite foreign to those of us living in the twenty first century, unless you are a member of the military. This freedom of all people to have a choice of whether to obey the orders of those above us in social

N. Coleman (✉)
University of New South Wales Canberra Space, Canberra, ACT, Australia

Chaplaincy Department, Royal Australian Air Force, Canberra, Australia

Inamori International Center for Ethics and Excellence, Case Western Reserve University, Cleveland, OH, USA
e-mail: n.coleman@adfa.edu.au

D. Messelken, D. Winkler (eds.), *Ethics of Medical Innovation, Experimentation, and Enhancement in Military and Humanitarian Contexts*, Military and Humanitarian Health Ethics,
https://doi.org/10.1007/978-3-030-36319-2_3

status is a relatively new concept when looking at things through the lens of history from ancient Rome, medieval times, the reformation, to today. The exception to this is in regards to military personnel, who choose to 'sign up' for military service knowing that they will have to obey all legal orders given to them.

3.2.1 The Communitarian Shift

The way in which soldiers have been viewed by society has changed even over the last century. In 1916, as a part of the battle of the Somme, 1.2 million British Empire (including Australian), French, and German troops were killed and injured, many as a result of being ordered to go "over the top" into no man's land to attack the enemy trenches (Farrar-Hockley 1964, 212). Such tactics were possible because the lives of individual soldiers were not valued, with soldiers instead being seen as "cannon fodder" and expendable, following in the tradition of military leaders such as Napoleon (Reynolds 2013, xvii). Sacrificing soldiers in this way would not be acceptable to society in 2019, where we now have drones used in warfare in order to prevent casualties amongst our military personnel (Weiner and Sherman 2013; Lewis 2013). The ability of the state, through military commanders, to sacrifice so many lives of its citizens was only possible because the survival of the state was seen as more important than the survival of individuals. It was a requirement of citizenship that ALL able men must fight when required. This requirement was based on the argument that individual rights were not as important as the survival of the state, and that all citizens had a responsibility to protect the entity of the state, even at the risk of losing their own life (Shaw 2016, 152). This concept of unrestricted service to the point of death for your country was articulated by Hackett, in his 1962 discussion regarding the unlimited liability, as demanding "the total and almost unconditional subordination of the interests of the individual if the interests of the group should require it"(Hackett 1962, 140). Hackett's understanding of the subordination of the individual to the state is consistent with the tactics used in World War I, however these tactics would not be acceptable in 2019, where the life of each individual soldier is seen as important, and where military commanders go out of their way in order to prevent casualties amongst their soldiers, such as through the use of drones. The use of drones of course moves the risk away from "our" soldiers and increases the risk to non-combatants (in particular innocent civilians) in the location where the drones are being used. Thus the importance of both the individual soldiers and the state using the drones is maintained, but at the loss of rights for civilians in the locations where the drones are utilised.

This subtle but important shift in the attitude of society underlines a move away from what might be described as a "communitarian" understanding that the state is inherently more important than the individual, to the situation we have today, where the state is considered instrumentally important because of its ability to protect the rights of individuals (Bellah 1985, 142–143). In other words, this means that the importance of the state *vis-à-vis* the individual has radically changed over time. This responsibility of the citizen to the state was expressed by Hegel in 1821 when he argued that the state "has supreme right against the individual, whose supreme duty is to be a member of the state" (Hegel 1962, 155–156). Whilst the state previously held intrinsic value to those within the state, it seems that there has been a shift to where the state largely holds instrumental value, as a result of being able to protect the rights of individuals.

This shift towards the rights of individuals is also reflected in the military. As stated previously, tactics such as ordering soldiers to their certain death in trench warfare in World War I would be seen as unacceptable in western society in 2019, only 100 years after the battle of the Somme saw such high casualty rates. The change in understanding of the relationship between the individual soldier and the state is also reflected in the change in different laws internationally. These changes highlight the fact that governments are now held more accountable for the treatment of their soldiers in respect to basic human rights, both by society in general, and by their own legal systems. This change in accountability has been a significant factor in decisions in several court cases in numerous legal jurisdictions around the world. In the USA, the Supreme Court ruled in 2015 that soldiers may now sue private military companies, a ruling which had the effect of creating a partial overturning of the Feres Doctrine, which had previously banned soldiers from taking legal action against the state, due to the risks this would pose for national security.[1] In the UK, soldiers and their heirs can now sue for negligence by the Department of Defence, even on the battlefield.[2] In Australia, the Department of Defence is now being held accountable for serious occupational health and safety breaches, having been fined in 2015 for the death of service personnel in training accidents.[3] These changes to the law can be viewed as indicative of a shifting attitude in society towards military members, as well as a shift towards the importance of individual rights and away from a communitarian understanding of the supremacy of the rights of the state. The changes in law have had the effect of holding the government and various departments of defence accountable for how they treat soldiers, reflecting the view of society that soldiers are no longer expendable cannon fodder and that military organisations have an increased duty of care over their personnel. Society's perception of the increasing importance of the individual versus the state has also been reflected in a change in understanding regarding obedience in general.

Whereas military organisations were previously comprised of conscripts; modern militaries now largely consist of volunteer soldiers.[4] This subtle but important

[1] "Feres Doctrine," accessed September 12, 2015, http://www.didisignupforthis.com/p/feres-doctrine.html; *Harris v KBR* (2014) US Supreme Court, accessed September 12, 2015, http://www.supremecourt.gov/orders/courtorders/012015zor_bq7d.pdf

[2] *Smith & Ors v The Ministry of Defence* (2013) UKSC 41, accessed September 12, 2015, http://www.bailii.org/uk/cases/UKSC/2013/41.html; "Family Sues MoD over Red Cap's Death," *BBC News*, accessed September 12, 2015, http://www.bbc.com/news/uk-23518564

[3] *Comcare v Commonwealth of Australia* (2012) FCA 1419, accessed September 12, 2015, http://www.austlii.edu.au/au/cases/cth/FCA/2012/1419.html

[4] The exception to this is in the few democratic countries where compulsory national service is still practiced.

shift highlights the change in understanding towards the rights of the individual. Soldiers are no longer conscripted into service by the state. Currently soldiers voluntarily take on the extra responsibilities of citizenship through military service in order to protect the security and interests of their state. This act of taking on extra responsibilities of citizenship, makes these modern soldiers *supra civis*, that is, super citizens. In taking on the *supra civis* role, soldiers do not forfeit their rights, but rather take on extra responsibilities of citizenship, and with these extra responsibilities an extra personal risk to themselves (Coleman 2019). This situation regarding soldiers taking on extra responsibilities of citizenship and thus becoming *supra civis* means that soldiers can no longer been seen as expendable assets, as they were at the turn of the twentieth century.

3.2.2 Military Obedience

The current policy of the Australian Defence Force (ADF) regarding obedience is that all soldiers, sailors and airmen must obey all legal orders given to them by their superiors (Department of Defence 2008). These orders are to be obeyed whether they are in relation to movements on the battlefield, where the soldier may be posted or deployed, how a soldier may cut their hair or wear jewellery (depending on their gender), through to orders regarding health and safety standards (Department of Defence 2007a, b). Under the Defence Force Discipline Act (1982) and the ADF Discipline Law Manual, soldiers must obey the legal orders given to them by superior officers, or face the possibility of being charged with disobeying a lawful command, or failure to comply with a general order, for which the maximum punishment, if found guilty, is 2 years imprisonment. This requirement to obey orders led Ned Dobos to conclude in 2012 that soldiers have relinquished their rights and as such have waived their claim to be treated as people (Dobos 2012). Dobos argued that soldiers are "mere assets to be used in the pursuit of the state's objectives", echoing the language of Michael Walzer of soldiers as being in a position of servitude to the state (Dobos 2012; Walzer 1992, 37). The response of the Acting Chief of Defence Force, Air Chief Marshal (ACM) Mark Binskin, to Ned Dobos' opinion piece in the *Sydney Morning Herald* was that Dobos' opinion was "incorrect and indeed, offensive" (Binskin 2012). ACM Binskin argued that soldiers do not forfeit their humanity or relinquish their rights, however he did not state how this is possible under the current requirements of the Defence Force Discipline Act which requires soldiers to obey all legal orders. ACM Binskin stated that the ADF "takes its Workplace Health and Safety very seriously" (ibid.) and that this was reflected in the 2012 Safe Work Australia Awards, where a Navy member won an award for the best individual contribution to workplace health and safety. It may be true that the ADF takes the safety of its soldiers very seriously, however the ADF is exempt from large parts of Workplace Health and Safety (WHS) law in Australia. This situation is a result of an exemption to current WHS law given by the Chief of Defence Force (CDF), which states that "Australian Defence Force members do not have the right to cease work where they are concerned about risks to their health or safety, including from an immediate or imminent exposure to a hazard" (Chief of the Defence Force 2012, 3; Orme 2011, 24). This situation highlights a disparity between ADF doctrine regarding the obeying of orders and the public declarations regarding a commitment to workplace health and safety. Whilst ACM Binskin (2012) stated that "any suggestion that we regard our people as 'instruments' is wrong and objectionable", current ADF doctrine regarding obedience makes it difficult to draw any other conclusion when soldiers are unable to disobey orders, even when they are concerned about risks to their health or safety.

The doctrinal basis for obedience in the ADF is that soldiers 'sign up' for unrestricted service, as understood through the unlimited liability contract (Department of Defence 2008). The interpretation by the ADF of unrestricted service means that soldiers can be ordered to do anything, anywhere, at any time, as long as the order is legal. By this I mean that some orders are illegal because they go against clearly articulated conventions under international humanitarian law. The requirement to obey orders also has a wide-ranging scope, from dress standards, how superiors should be addressed, orders on the battlefield, through to orders regarding medical treatment (particularly in regards to rehabilitation therapy).

There are many activities that military service personnel reasonably can and should commit to when they decide to join the military. It is not unreasonable to expect that, by signing up to join the military, one will be expected to take part in regular military training, field exercises, physical training sessions and endurance sessions (designed to test or further develop one's physical and mental limits), and other routine interventions (such as regular inspection of one's uniform or living quarters). One can also reasonably be expected to be deployed overseas to potentially dangerous situations (thus putting one's life at risk), and, of course, expect to be ordered to attack and kill "the enemy."

Members of other professions are sometimes exposed to similar occupational risks, and, like military personnel, may also be asked (and even required) to take part in activities that are not usually imposed upon ordinary citizens. One occupation that closely resembles the military with respect to their "unlimited liability" in this regard is that of the police officer.

Members of police forces are routinely exposed to situations which put them at a dramatically increased risk of injury or death. Quite often (as in South Africa, or in major U.S. cities), the risk imposed upon them is equal to, or even greater than that imposed upon military personnel. At times the duties expected of police entail that officers are not only subject to increased risk of harm to themselves, but (unlike ordinary citizens) that they are also authorized to use lethal force, and to kill people. Police are likewise under the command of their superiors, and are expected, like military personnel, to follow orders. Unlike military personnel, however, they are not under any obligation to follow orders without question that might put them in extreme danger. Significantly, if a police officer doesn't like, let alone sharply disagrees with the orders that they are being given or the expectations of risk to which they are held, they can simply resign. This is not an option routinely available to serving military personnel, who may only resign prior to the expiration of their period of enlistment under very restricted circumstances, and almost never in the midst of an altercation or armed conflict.

In contrast to police, then, who have the option of refusing orders, military personnel seem to be in a position in which they unable to refuse any orders, *unless those orders are manifestly illegal.* Thus even if an order is suicidal, or manifestly unjust, if that order is nonetheless legal, then a military member is normally expected to obey it. This contrast between the two occupations then raises the question of what intrinsically makes a serving military member different not only from an 'ordinary citizen,' but even from persons otherwise reasonably thought to be serving in a similar "unlimited liability" occupation, such as a police officer. What exactly is it that renders it acceptable for a police officer to refuse an order that might subject that individual to grave risk, but not acceptable for a military member to do likewise? Surely it must seem as if the difference is the nature or the degree of liability to risk that finally characterizes both occupations, and to which the policeman or woman may hypothetically withhold consent, while military personnel apparently may not.

Military service personnel are often surprised to discover that there are a number of other activities in which they are compelled to participate. On the whole, ordinary citizens would not be compelled to participate in such activities. For example, members of the Australian Defence Force are not permitted to refuse medical treatment, and on the whole do not get to choose who provides that medical treatment. This lack of choice of health care provider is usually not a problem for routine medical treatment. For the treatment of illnesses that are more problematic or controversial (such as mental health issues), however, this means that military personnel cannot choose who provides their medical treatment nor, at times, can they exercise full control over the course of that treatment, even if they object on religious or ideological grounds.

3.2.3 Military Obedience and Medical Treatment

Discovering that they do not have this right to choose medical treatment can come as a shock to serving military personnel. One of the most controversial consequences of not being able to refuse medical treatment is that military personnel may thereby become potentially vulnerable to being used in medical experimentation. During World War II, various military institutions conducted medical experimentation on civilians, captured enemy soldiers, and their own military personnel. The US Senate Committee on Veterans' Affairs found in 1994 that US military personnel were exposed to experimental drugs and vaccines (especially during Gulf War I), and that "Army regulations exempt informed consent for volunteers in some types of military research." (U.S. Senate Committee on Veteran's Affairs 1994) The VA report found that military personnel were occasionally subjects of medical experimentation involving the use of medications and vaccines intended to counter the effects of biological or chemical weapons. Quite frequently this experimentation was done without either requiring or obtaining the consent of the personnel affected. It also appears that this research was done largely without those personnel even knowing that they were the subject of research, to the point where military personnel were often not told what medications they were being ordered to take or to have injected. This means that they were often being experimented upon not only without their consent, but without even their awareness, which goes against the bioethical and legal principles of autonomy and informed consent. Furthermore, the Senate report found that "participation in military research is rarely included in military medical records, making it impossible to support a veteran's claim for service-connected disabilities from military medical research." (ibid.)

Suppose that military service personnel who are forced to take part in medical experimentation either against their will, or without their knowledge and consent, then suffer injury or illness as a consequence. They might reasonably ask themselves the question, "did I really sign up for *this*?" Military personnel are often driven by a duty to serve and protect their country. In their mind, this means to protect their fellow citizens from armed attack by some enemy. It seems highly unlikely to suggest that these personnel also imagined being used as "lab rats." It seems difficult to link such experimentation to the defense of their country, especially if they are not given critical information about the experiments and their purpose. Simply using military personnel as subjects for medical experiments, merely because they constitute a healthy, fit, conveniently available, and forcibly compliant young cohort, certainly does not provide adequate moral justification for withholding the customary right of informed consent to such involvement.

3.2.4 *Military Personnel, Obedience and Medical Experimentation*

Many of the ethical issues which affect soldiers come about as a result of the obligation to obey all legal orders, including the orders of medical personnel (who, as has already been mentioned, are always considered to outrank those under their care). In the normal medical setting, patients are afforded the right to make decisions about their own medical care, with this right only being set aside if the patient is unable to make an informed decision, such as in the case of being unconscious, or mentally unsound. In the military setting, however, the ultimate decision regarding medical care and treatment is given to the medical practitioner, not the military member who is the patient, a situation which appears at odds with the fundamental right to autonomy in regards to medical decisions. This once again highlights the competing duties of military medical personnel, who must balance the needs of the patient against military necessity and operational effectiveness. However, this situation also reveals some significant issues for the soldier-patient as well.

In the civilian medical setting, confidentiality regarding medical records is a fundamental principle in medical care. In the military setting, however, the situation is more complicated. Military medical personnel can be ordered to breach confidentiality, particularly in cases where a member of the military might be thought to be a danger to others. Thus it is not unreasonable for military personnel to assume that all their discussions with medical personnel, including psychiatrists and psychologists, are NOT confidential. Since confessing physical or psychological problems to military medical personnel may have a long-term effect on a soldier's military career, and since military personnel in many jurisdictions are routinely forbidden from seeking medical treatment from non-military sources, this lack of confidentiality may well make military personnel more reluctant to seek medical help when it is required.

Because military personnel must obey all legal orders, even in regards to medical care and treatment, they are vulnerable to exploitation, particularly in regards to being used for medical experimentation. Since a soldier who is ordered to take part in medical experimentation is legally unable to refuse, any consent the soldier may have given might well be considered null and void. As a result of this, consent for participation in medical research in the military setting may not be obtained, and in some instances military personnel are not even advised that they are taking part in a medical experiment; the trial of experimental vaccines on military personnel is a historical example of such a situation. It seems that military personnel can be used in medical experiments, not out of a military necessity, but because they form a large cohort of fit healthy young people who are unable to refuse participation. This goes against foundational principles in bioethics, as highlighted the Declaration of Helsinki (1964–2008), which has at its core the respect for individual persons, the right to self-determination and the right to make informed decisions. It also goes against the principle of justice, in regards to the fact that subjects in experiments should be recruited for reasons related to the problem being studied, not because of the ease of recruitment or the ease of exploitation. Given that civilians are protected from such exploitation, what, if anything, can ethically justify this sort of treatment of military personnel?

In the US at least, one area that stands out as being manifestly unique regarding the terms of employment for military service personnel is that they are unable to sue their "employer" (i.e., the U.S. Department of Defense) if they are injured or killed, even if such injury can be attributed to their employer's negligence. This curtailment of a legal right to sue stems from what is known as the "Feres Doctrine."

In 1950, the Supreme Court of the United States of America issued a combined ruling that resulted from its consideration of three separate but similar cases.[5] In the actual Feres case itself, the estate of 1st Lt. Rudolph J. Feres sued the Department of Defense for negligent death resulting from an army barracks fire. A second suit, labelled the Jefferson suit, alleged grievous medical negligence: a soldier was obliged to have a towel surgically removed from his stomach that had been left there following previous abdominal surgery administered by military surgeons. Finally, the Court ruling also encompassed the so-called Griggs case, brought by the estate of Lt. Colonel Dudley R. Griggs, who had also died allegedly as a result of military medical negligence.

The Supreme Court ruling, now known as "the Feres Doctrine," was largely based on an underlying concern, that allowing individual military personnel to sue the government for injury or death, even due to negligence, might "undermine good

[5] "Feres v United States, Jefferson v United States, United States v Griggs. "*United States Supreme Court*", 340 U.S. 135 (71 S.Ct. 153, 95 L. Ed. 152), 1950.

order and discipline in the military." (Uniformed Services University of the Health Sciences 2003) This curtailment of the right to sue has then and since been taken to apply not only to events that occur during a time of combat (when it might be argued that the threshold for occupational health and safety concerns is much lower), but even during the entire time of a military service member's period of enlistment. This doctrine effectively bans all claims against the U.S. Department of Defense for injury or death, no matter what the cause or origin, for the entire period of service for military personnel.

This ruling effectively relegates members of the military (at least in the U.S.) to the status of "second class citizens" in the eyes of the law. Alexander Hattimer, reviewing the Feres Doctrine in 2012 in relation to a negligent death suit filed against the Department of Navy, objected that "the United States military can continue mistreat soldiers and sailors with almost complete immunity from civil suits." (ibid.) While one might plausible argue that, at least during times of war or supreme emergency, there are grounds for believing that military service members must sacrifice some of their rights in order to protect their country as a whole, it seems difficult to argue that, in times of peace, the obligations of military service remain just as onerous, especially when the resulting curtailing of the individual rights of military personnel seem suspiciously to have to do more with economic concerns than with "maintaining good order and discipline" in the military. It seems that it is economically much more advantageous to have military members obliged to perform any task (even if that task is not central to the role of the military, such as using military personnel for medical experimentation) and be immune from any consequence (such as being sued for negligent practices or death), and that this is the guiding factor in the curtailment of the rights of military personnel, rather than any valid concerns regarding military necessity.

The situation in the U.S.A. stands in contrast to that in Britain, where, in 2013, the Supreme Court ruled that families of soldiers killed in Iraq could sue the Ministry of Defence (MoD) for damages, under the European Convention on Human Rights. Article Two of that convention imposes a duty on authorities (in this case the Army) to exercise due care to protect the individual's right to life.[6] This landmark British case pertained to military personnel who were killed while driving poorly-armoured Land Rovers, as well as others who were killed or injured in a "friendly fire" incident. The effect of this case has been that families and military members are now allowed to sue for compensation, and to allow families and military members to lobby for better equipment and training as a part of that compensation. The MoD argued in their defense that there was "combat immunity where troops in action were concerned, and that it was not fair, just or reasonable to impose a duty of care on the MoD when soldiers were on the battlefield." (Wyatt 2013) The Supreme Court of Great Britain, however, did not find that this was a compelling argument for absolving the MoD of all consequences stemming from negligence. Family

[6] Supreme Court of the United Kingdom, "Smith, Ellis, Allbut (and others) v the Ministry of Defence." (London: Supreme Court of the United Kingdom 2013): UKSC 41: 72.

members, and their representing counsel, had argued that it was not appropriate to "treat soldiers as sub-human with no rights." (ibid.) On the one hand, this ruling specifically addresses consequences stemming from negligence, thus limiting its scope. However, this ruling does relate to events which occur even during combat, and not just in training or during ordinary operations in the UK. Hence, it holds broad significance for protecting the welfare and rights of individual personnel in the U.K. military services.

The effect of this ruling likely will be that the Ministry of Defence will deem itself to have a more wide-ranging duty of care to military members, not just during peace time, but also during combat operations. This has far-reaching implications for the underlying traditional moral concept of the "unlimited liability contract" or unrestricted service. This ruling specifically calls into question the reigning traditions that previously permitted the military hierarchy to order military members to take part in any military action, even those deemed tantamount to suicide missions, for example, as well as orders to participate in events such as experimental medical trials. This ruling casts into serious doubt the traditional idea that military members surrender most or even all of their individual rights merely as a consequence of their voluntarily entering military service.

Because soldiers in most western countries must obey orders in relation to not only their battlefield medical treatment, but also their routine medical care, it makes them excellent candidates for certain types of medical experimentation. As a cohort they are young, fit, healthy and compliant, which makes them ideal subjects for medical research that might not otherwise be undertaken within the wider population. The general principles of valid consent are that consent must be fully informed, not coerced, and not given by mentally incompetent persons, such as children, brain damaged adults, people who are intoxicated etc. (Kleinig 2010; Eyal 2012) If and when a soldier is ordered, rather than volunteering to take part in a medical study, the second requirement, that they not be coerced, is immediately breached, and so it can be argued that valid consent has not been given. In addition to this many military members subject to medical experimentation are not even informed that they are taking part in a study. One example of this is the on-going law suit against the US Department of Defence (DOD), Central Intelligence Agency (CIA), and the US Army in relation to troops "exposed to testing of chemical and biological weapons at Edgewood Arsenal and other top secret sites" such as "Fort Detrick as well as several universities and hospitals across America" (Morrison & Foerster LLP 2009, 2010). This case started in 2009 on behalf of veterans, sponsored by two veterans' rights organisations, the Vietnam Veterans of America and Swords to Plowshares.[7] As of 2016, the Edgewood case is still before the courts. The case against the defendants is that they deliberately exposed US military personnel "to chemical and biological weapons and other toxins without informed consent" from the 1950s until 1976, and thus the veterans have asked for the redress of "several decades of the US

[7] "Vietnam Veterans of America: Veterans Advocacy," accessed 07 June 2016, www.swords-to-plowshares.org; "Swords to Ploughshares: What we do," accessed 07 June 2016, www.swords-to-plowshares.org

Government's use of them as human test subjects in chemical and biological agent testing experiments, followed by decades of neglect" (Morrison & Foerster LLP 2016). This neglect included the use of troops to test nerve gas, psycho-chemicals, and hundreds of other toxic chemical or biological substances (ibid.).

As can be seen from the Edgewood case, soldiers were used in medical experimentation without their consent and, at times, without their knowledge. Compounding this institutional neglect was the fact that the DOD, CIA and US Army refused to provide health care to the veterans who were disabled as a result of their involvement in the experiments, refused to reveal to the veterans what substances they were exposed to so that they could seek treatment elsewhere and apply for compensation, refused to notify all the test subjects that they were experimented upon, and refused to release the soldiers from their oath of secrecy so that they could discuss their exposure with their own medical practitioners and lawyers. The US Government claimed in their defence that they were exempt from claims by the soldiers under the *Feres* Doctrine, which is an exception to the waiver of sovereign immunity created by the US Supreme Court in 1953. However, the US District Court dismissed the US Government's claim to the *Feres* doctrine exception in 2010.[8] Even if the court had not handed down this ruling, the fact that the actions of the US government were legal with regard to the soldiers in the Edgewood case most certainly does not show that those actions were in any way ethical.

Medical experimentation on military personnel is not limited to the US military, with experimentation conducted on most military personnel in the western world over the past 60 years, usually without their ostensible consent. This situation was made possible because soldiers had to obey orders to take part in experimentation. This has led to the current situation in Australia where experimentation on military personnel is now highly regulated by the ADF through the Departments of Defence and Veterans' Affairs Human Research Ethics Committee (DDVA HREC), which approves all research on military personnel in Australia (Department of Defence 2005). This committee is widely seen as the most difficult ethics committee in Australia to obtain ethics approval from. Currently approval from DDVA HREC is required in order to conduct research on any military personnel in Australia. One thing which DDVA HREC now insists upon is informed consent by all participants, who must be volunteers and not ordered to take part in research (Department of Defence 2007a, b).

Because of the impact that medical experimentation has on military personnel, such as those involved in the Edgewood case, it is difficult to sustain the justification to obey all orders, especially those in relation to routine medical care and experimentation on military personnel in particular.

[8] *Feres v. United States, Jefferson v. United States, United States v. Griggs* (1950) US Supreme Court no. 340 US 135 (71 S.Ct. 153, 95 L. Ed. 152); *Vietnam Veterans of America v. Central Intelligence Agency* (2010), US District Court, Order Granting in Part and Denying in Part Defendants' Motions to Dismiss and Denying Defendant's Alternative Motion for Summary Judgement, *No. C 09–0037 CW. (N.D. Cal. Jan 19, 2010).*

3.3 Two Potential Solutions

I am proposing two potential solutions for the problem regarding obedience in regards to medical experimentation on military personnel. The first potential solution is the relaxing of the requirement to obey orders in relation to medical care by soldiers, through the splitting of orders into "general" orders and "life orders". The second potential solution is to include soldiers as a part of the subset of "vulnerable populations" as defined by human research ethics committees internationally.

3.3.1 General Orders vs Life Order

Orders in most military organisations are, in fact, already split into two. For example in Australia, Part III, Division 3, Section 27 of the Defence Force Discipline Act (1982) states that a defence member must obey a lawful command of a superior officer, meaning that orders which are not lawful do not have to be obeyed. This requirement to disobey illegal orders is enshrined in international law, both in the Law of Armed Conflict, and through the precedent set at the Nuremberg trials following World War II, where the defence of following orders was rejected by the court.[9] Further complicating the situation in Australia, orders are split even further into the following: (i) operational orders (OPORD) which are directives, usually formal in style, issued by a commander to subordinate commanders for the purpose of effecting the coordinated execution of an operation; (ii) administrative orders (ADMINORD), which are administrative and logistical requirements that may be included in an administrative order – administration orders support operational orders; administrative instructions, which are used to coordinate action for a particular activity; and standing operating procedures/routine orders/circulars (Department of Defence 1982, Part 3, Sections 5.10, 5.16, 5.17, 5.19, 6).

Patrick Mileham (2008) has suggested that ethics in the military should be split into two: institutional (or garrison) ethics, and operational ethics. He argues that garrison ethics relates to activities that are domestic and routine, whilst operational ethics "consist of dynamic precepts and experiences and should be clearly understood as what happens or should happen during operations" (ibid, 49). I would argue that this delineation between operational and garrison ethics provides a useful split not only in regards to ethics, but also in regards to the duty to obey orders. Under this option, orders given in operational environments would be legally binding on soldiers, whilst orders given in a garrison environment would be more akin to directions given to police by their superiors, which would be open to acceptance or rejection on reasonable grounds.

[9] *"Trials of War Criminals before the Nuremberg Military Tribunals under Control Council Law No. 10"*, 1949.

Instead of splitting orders into operational and garrison orders as proposed by Patrick Mileham, I am proposing that orders be split into two classes, the first being all orders as per the military "job" (namely operational/garrison/standard orders), and the second being those orders pertaining to regular life. Orders which would come under "life" orders would include those that pertain to the soldier's life which is largely conducted away from their role in the military. Life orders would thus include the following: firstly, orders regarding routine medical care; secondly, orders regarding who a soldier can be in a relationship with; thirdly, orders relating to gender identity and sexual preference; fourthly, orders regarding talking to the media, as long as it does not pertain to national security or the defence of Australia; and, finally, orders regarding soldiers' ability to collectively bargain with regards to pay and conditions.

Because of the limited scope of orders included in the "life" category, this delineation between the two categories, along with its clear identification of legally enforceable orders, would be relatively easy to implement – although the impact on unit cohesion and discipline would need to be examined in advice of such a military wide change in practice. I anticipate that if military organisations did adopt this option, it would become standard practice that those orders that would have previously fallen into the "life" category would simply cease to be given. There may be difficulties in regards to medical care in operational environments, particularly regarding some vaccinations, however as these would affect operational capacity for the ADF, they would become orders which would have to be legally followed. How this paradigm would affect the operational capacity of military organisations is difficult to determine.

The paradigm according to which soldiers obey all legal standard orders, both in the operational and garrison setting, whilst being given some rights back in regards to their routine medical care, relationships, freedom of speech, appears to strike a balance between giving military organisations control over their soldiers, and reducing the vulnerability of those soldiers to exploitation. The quarantining of "life orders" with the effect that they do not have to be obeyed, would bring the military more in line with the expectations of society in regards to the importance of the rights of the individual and how they are balanced against the rights of the state.

3.3.2 Classification of Soldiers as Vulnerable

If military organisations find the concept of soldiers not having to obey certain orders too radical, another option would be to classify soldiers as vulnerable populations under domestic and international law in regards to participation in medical research and experimentation.

In Australia all experimentation involving humans must have prior approval of a human research ethics committee (HREC), usually at the institution sponsoring the research. In the USA this prior ethics approval is obtained from an institutional review board (IRB). Sometimes approval is sourced from an independent ethics committee (IEC), ethical review board (ERB), or research ethics board (REB). All of these ethics approval groups have as a primary goal the reduction of risk to participants in medical research. This is done through applying the principles of bioethics, particularly that participation in medical experimentation should be voluntary (and in particular be uncoerced), and that the risk of participation shall be mitigated against as much as possible. Particular vulnerable groups such as children, pregnant women, prisoners and people with diminished capacity to make decisions are given close oversight to ensure that they are not being exploited. In Australia this extra protection for vulnerable persons is enforced under the National Health and Medical Research Council (NHMRC) Act (1992) and in the US it is enforced under the "Common Rule", which is a federal policy regarding human subject protection.

Currently soldiers are not directly named as vulnerable populations, although in Australia they sometimes qualify as a person in a dependent or unequal relationship which is a classification of vulnerable person under research guidelines (National Health and Medical Research Council 2018). One potential solution to the problem of obedience in regards to medical experimentation is for soldiers to be named as a vulnerable population in regards to medical experimentation, as a result of them having to obey orders. The effect of naming soldiers as vulnerable persons is that they would immediately come under the oversight of ethical review boards, who could ensure that soldiers are not being utilised for research purposes merely because they are compliant subjects. Oversight by ethics review boards would ensure that soldiers are not subject to involuntary research, or research which is of high risk to participants without benefit to them. Oversight by ethics review boards would also ensure that the perception of soldiers as 'lab rats' to be used for medical research is eliminated.

3.4 Conclusion

There has been a dramatic shift in society views the importance of the individual verses the importance of the state. A hundred years ago it was considered tragic but acceptable to send millions of soldiers to their certain death in the battle of the Somme, in order to protect the rights of the state. In contrast, today we use drones in combat in order to protect the lives of soldiers, in order to protect their rights as individuals. Likewise, as it was seen as acceptable in previous decades to order soldiers such as the Edgewood Vets to participate in medical research. It is no longer seen as acceptable in society to order or coerce soldiers to partake in medical experimentation, just as it is no longer acceptable to send millions of soldiers to their certain death.

Two potential solutions for the real or perceived problem regarding the requirement of soldiers to obey orders in regards to medical experimentation have been considered – to split orders into "general" and "life" orders, or to place soldiers on the list of vulnerable populations in regards to medical research and experimentation.

Military organisations need to ensure that they are putting policies in place to ensure that soldiers are not being ordered or coerced into medical research or experimentation. The need to ensure that soldiers are protected in this way is as a result of the duty of care a military organisation has to its members, and to ensure that the organisation itself is living up to the standards expected by society. If a military organisation does not ensure the safety of its members in relation to medical research and experimentation it runs the risk of having policy changes and laws enforced upon it through civil legal proceedings by soldiers, veterans and their families, or by politicians creating potentially even more onerous legislation. Thus, it is in the best interests of military organisations to be proactive in its handling of the problem of the requirement of soldiers to obey orders in relation to medical research and experimentation.

References

"Feres v United States, Jefferson v United States, United States v Griggs. 1950." *United States Supreme Court* ,340 U.S. 135 (71 S.Ct. 153, 95 L. Ed. 152).

"Feres Doctrine" http://www.didisignupforthis.com/p/feres-doctrine.html. Accessed 30 Sept 2019.

"Trials of War Criminals before the Nuremberg Military Tribunals under Control Council Law No. 10", 1949.

"Vietnam Veterans of America: Veterans Advocacy" www.swords-to-plowshares.org. Washington, DC: Vietnam Veterans of America.

BBC News. 2015. *Family Sues MoD over Red Cap's death.* http://www.bbc.com/news/uk-23518564. Accessed 30 Sept 2019.

Bellah, Robert N. 1985. *Habits of the heart: Individualism and commitment in American life.* Berkeley: University of California Press.

Binskin, Air Marshal Mark, Acting Chief of the Defence Force. 2012. Statement from acting chief of the defence force – Response to the letter to the Editor by Dr Ned Dobos. *Canberra Times*, 06 June.

Chief of Defence Force. 2012. *Work health and safety act 2012 (Application to defence activities and defence members), declaration 2012.* Canberra: Department of Defence.

Coleman, Nikki. 2019. *Why soldiers obey orders.* London: Routledge, *in press.*

Comcare v Commonwealth of Australia. 2012. FCA 1419.

Defence Force Discipline Act (Cth). 1982. Canberra: Commonwealth of Australia.

Department of Defence. 1998. *ADFP 102 defence writing standards. 1998.* Canberra: Department of Defence.

———. 2005. *Defence instruction (general). Admin 24–3. Conduct of human research in defence. 2005.* Canberra: Department of Defence.

———. 2007a. *Discipline law manual. Volume 3. Summary authority and discipline officer proceedings 2007.* Canberra: Department of Defence.

———. 2007b. *Health manual. Volume 23. Human research in defence – Instructions for researchers.* Canberra: Department of Defence.

———. 2008. *Acknowledgment of the requirements of service in the Australian Defence Force (ADF). AD 304–1.* Canberra: Commonwealth of Australia.

Dobos, Ned. 2012. Are our soldiers assets or workers? *Sydney Morning Herald*, June 4.

Eyal, Nir. 2012. Informed consent. In *The Stanford encyclopedia of philosophy.* http://plato.stanford.edu/archives/fall2012/entries/informed-consent. Accessed 30 Sept 2019.

Farrar-Hockley. 1964. *The Somme.* London: B.T. Batsford.

Hackett, John Winthrop. 1962. *The profession of arms: The 1962 Lees Knowles lectures given at Trinity College, Cambridge*. New York: Macmillan. 1983.
Harris v KBR. 2014. *US Supreme Court*. http://www.supremecourt.gov/orders/courtorders/012015zor_bq7d.pdf. Accessed 30 Sept 2019.
Hegel, Georg Wilhelm Friedrich. 1962. *Philosophy of right. Translated with notes*. Trans. T.M. Knox. Oxford: Clarendon Press.
Kleinig, John. 2010. The nature of consent. In *The ethics of consent: Theory and practice*, ed. Franklin Wertheimer and Alan Miller. New York: Oxford University Press.
Lewis, Michael W. 2013. *Drones: Actually the most human form of warfare ever*. Washington, DC: The Atlantic.
Mileham, Patrick. 2008. Teaching military ethics in the British armed forces. In *Ethics education in the military*, ed. Paul Robinson, Nigel de Lee, and Don Carrick. Aldershot: Ashgate Publishing.
Morrison & Foerster LLP. 2009. *Morrison & Foerster files suit against CIA, and US Army on behalf of troops exposed to testing of chemical and biological weapons at Edgewood Arsenal and other top secret sites*. http://edgewoodtestvets.org/press-releases/pdfs/20090107-Morrison-Files-Suit.pdf. Accessed 30 Sept 2019.
———. 2010. *Morrison & Foerster secures victory for troops exposed to chemical and biological weapons testing in case against the US Government*. http://edgewoodtestvets.org/press-releases/pdfs/20100120-Morrison-Secures-Victory.pdf. Accessed 30 Sept 2019.
———. 2016. *Edgewood test vets. What this case is about*. http://edgewoodtestvets.org. Accessed 30 Sept 2019.
Nation Health and Medical Research Council. 2018. *National Statement on ethical conduct in human research*. Canberra: National Health and Medical Research Council, Commonwealth Department of Health.
Orme, Major General C. W. 2011. Beyond compliance: Professionalism, trust and capability in the Australian profession of arms. In *Report of the ADF personal conduct review*. Canberra: Department of Defence.
Reynolds, David. 2013. *The long shadow: The legacies of the great war in the twentieth century*. London: Simon & Schuster.
Shaw, William H. 2016. *Utilitarianism and the ethics of war*. New York: Routledge.
Smith & Ors v The Ministry of Defence. 2013. *UKSC 41*. http://www.bailii.org/uk/cases/UKSC/2013/41.html. Accessed 30 Sept 2019.
Supreme Court of the United Kingdom. 2013. *Smith, Ellis, Allbut (and others) v the Ministry of Defence*. London: Supreme Court of the United Kingdom.
Uniformed Services University of the Health Sciences. 2003. *The Feres doctrine*. Bethesda: Uniformed Services University of the Health Sciences.
US District Court. 2010. *Vietnam Veterans of America v. Central Intelligence Agency*, Order Granting in Part and Denying in Part Defendants' Motions to Dismiss and Denying Defendant's Alternative Motion for Summary Judgement, *No. C 09–0037 CW. (N.D. Cal. Jan 19, 2010)*. Washington, DC: US District Court.
U.S. Senate Committee on Veteran's Affairs. 1994. *Is military research hazardous to veterans' health?* Washington, DC: United States Senate.
US Supreme Court. 1950. *Feres v. United States, Jefferson v. United States, United States v. Griggs*. no. 340 US 135 (71 S.Ct. 153, 95 L. Ed. 152). Washington, DC: US Supreme Court.
Walzer, Michael. 1992. *Just and unjust wars: A moral argument with historical illustrations*. 2nd ed. New York: Basic Books.
Weiner, Robert, and Sherman, Tom. 2013. Drones spare troops, have powerful impact. *San Diego Union-Tribune*, 09 October, 2014.
Wyatt, C. 2013. *Iraq damages cases : Supreme Court rules families can sue*. London: BBC. http://www.bbc.com/news/uk-22967853. Accessed 30 Sept 2019.

Chapter 4
Medical Prophylaxis in the Military: A Case for Limited Compulsion

Neil Eisenstein and Heather Draper

4.1 Introduction

Respect for individual patient autonomy is a core principle of civilian medical ethics in the western liberal democracies (Beauchamp and Childress 2001; WMA 2006). Although the notion of autonomy that defines the principle is debated at a theoretical level in bioethics, patients' right to self-determination informs legal frameworks and drives a shared decision-making model where doctors support patients to make their own decisions about their healthcare. The resulting process of consent may be an informal, even non-verbal assent to a relatively simple test or examination, or more formal, accompanied by written information and recorded in the patient's notes after a long discussion of the advantages and disadvantages of an intervention (Chan et al. 2017). Respecting patient autonomy has resulted in some benefits for patients, including improved understanding of risks and benefits of treatments and protection from abuses (Manson and O'Neill 2007). It has, however, shifted some of the burden of responsibility onto patients, including at a

This work is the personal opinion of the authors and does not represent British military doctrine or policy

N. Eisenstein
Royal Army Medical Corps of the British Army, Birmingham, UK
e-mail: eisenstein@doctors.org.uk

H. Draper (✉)
Health Sciences, Warwick Medical School, University of Warwick, Coventry, UK
e-mail: h.draper@warwick.ac.uk

D. Messelken, D. Winkler (eds.), *Ethics of Medical Innovation, Experimentation, and Enhancement in Military and Humanitarian Contexts*, Military and Humanitarian Health Ethics,
https://doi.org/10.1007/978-3-030-36319-2_4

time when they may feel least able to shoulder it, even though they have decision-making capacity.[1]

Military clinical practice has mirrored the changes seen in civilian practice and many Western military healthcare regulatory systems do not acknowledge any difference between civilian and military medical ethical standards (NATO 2018). However, there are differences between these two environments, including altered landscapes of rights and responsibilities and the high-tempo and austere settings in which military medicine during deployment is practiced, alongside the extreme nature of the medical issues being dealt with. By simply accepting and applying civilian medical ethical principles *en masse* to the military setting, there is a risk that patient care and survival is compromised, especially when it comes to medical prophylaxis.

In this chapter, we will examine some of the key challenges that respect for patient autonomy presents in the deployed medical military setting. We will explore whether rigid adherence to this principle may be harmful and whether compulsion to receive medical prophylaxis may be justified in the military environment in a limited range of circumstances. Medical prophylaxis in the military setting has caused significant controversy, but the different ethical treatment of medical prophylaxis and physical protections suggests inconsistencies that should be explored and reconciled where possible.

4.2 Current Ethical Guidance

The British Medical Association updated its ethical guidance to military doctors in February 2018 (BMA 2018). This guidance makes it very clear that:

- Military doctors have to operate to the same ethical standards as civilian doctors:

  > Doctors working in the armed forces owe the same moral obligations to their patients, whether comrades, enemy combatants or civilians, and are subject to the same ethical standards as civilian doctors

- Shared decision-making applies to military patients just as civilian patients:

  > Members of the armed forces have exactly the same freedom of choice as to the medical treatment they receive as all other patients. Doctors should never impose treatment where a patient with capacity refuses or consents only under duress.

- Responsibility for dealing with the consequences of shared decision-making must be divided between medical and non-medical part of the military:

---

[1] We accept that as far as the law is concerned in England and Wales, having capacity is sufficient for a patient's decision to be authoritative, even though capacity as defined in law may fall short of autonomy as it is defined in philosophy.

> Where a patient will not comply with a military requirement to receive a particular treatment, for example a vaccination, doctors should refer the matter back to the military chain of command, with the patient's consent. It is a chain of command decision, rather than a medical decision, to determine the employability of an individual who declines a particular vaccination.

The guidance acknowledges some special circumstances that apply to military medicine, including dual loyalty, challenges of confidentiality, and working outside one's normal scope of practice. However, despite recognising the difficulties that these pose, it insists that there is no difference between the ethical principles and standards that govern military and civilian medical practice.

This appears to ensure that as far as clinical practice is concerned, the patients of military doctors – who include other military personnel (no matter which side of any conflict) and civilians who are affected by conflict – are afforded the same ethical protections as patients in the civilian health service. There is, however, no discussion of the inconsistencies and potential harms resulting from this approach for military personnel in combat scenarios.

In civilian medicine it is accepted that the way in which medical care is delivered should be modified to suit the situation and that extreme situations, mass casualties scenarios for example, demand a re-ordering of medical priorities. The use of triage is one example: some patients are left untreated to allow medical assets to be used on those with the best chance of survival if treated (Robertson-Steel 2006; Kaufman et al. 2013). In the military, the delivery of casualty care is explicitly dependent on the circumstances. UK medical doctrine describes a graded approach to how casualties are treated depending on the tactical circumstances (Battlefield Casualty Drills – Aide Memoire 2007). While a unit is under effective enemy fire, the doctrinal procedure is to win the fire fight and provide only very limited medical care to the casualty until this is achieved. Thus normal medical priorities (i.e. advanced life support for the casualty) are subordinate to unit survival. This reflects that fact that unit survival provides the best chance of casualty survival and also that, under extreme circumstances, normal civilian approaches are not appropriate. As the casualty is evacuated back to less austere environments with greater resources and lower risk, a more recognisably civilian level of healthcare priorities emerge (e.g. airway control, supplemental oxygen, intravenous fluids, damage control surgery etc.).

Just as there are circumstances where different medical priorities are justified, it could also be acknowledged that some circumstances require the ethical priorities to be re-ordered. This is not the same as accepting lesser or weaker professional standards on deployed operations, just a different order of priorities that reflect the needs of the unit and the individuals that comprise it. This is not only limited to the delivery of tactical combat casualty care but is also potentially applicable to other forms of medical care such as prophylaxis as outlined below.

4.3 Individual Autonomy in the Military Setting

The operation of individual autonomy in the general military context needs to be understood before further applying it to the specific context of military medicine. We will take a deliberately Western-centric (and, specifically, British) view here, as this is our experience. In many Western militaries, there is no conscription into the armed forces, which are made up of volunteers. We exclude from this discussion those countries where there is either true conscription or national service without a non-military option. Applicants choose to enlist knowing that they will enter an environment in which their choices will be constrained. This includes loss of choice over where one is employed geographically, when one is moved, whether one can bring family and dependents, what type of employment or training one is engaged in, and one's right to self expression in terms of dress or appearance. These types of constraint are presumably offset by the benefits, included those related to pay, skill acquisition, stability of employment, career prospects etc. The underlying need to restrict some individual freedoms to ensure the effective operation of a military service is accepted even though the exact balance of benefits, reward, compensation and societal recognition offered to offset these burdens is more contested. These constraints on choice are subject to temporal, procedural and geographic limitations, and are not permanent. For example, when a service person is off duty, they are not expected to wear uniform, and all can choose to leave the service subject contractual constraints.

4.4 Medical Autonomy in the Military Setting

In the British army, how healthcare is received is something of an anomaly to how the rest of the organisation operates: the same principles of respect for patient decision-making apply to military personnel as other patients (BMA 2018; NATO 2018). Given how crucial healthcare is to force generation and force protection, this seems inconsistent with how individual choice is constrained in other military contexts. Particularly striking is the difference in how physical means of protection and risk mitigation are managed and how medical prophylaxis is offered. Physical measures for combatants include personal body armour and hearing protection *inter alia*, which are designed to minimise the harms of ballistic injury and noise-induced hearing loss respectively. However, these protective measures also expose the user to some risks. Personal body armour is bulky, heavy, and impairs the body's ability to regulate temperature through heat loss (Larsen et al. 2011, 2012). Thus users may be less agile, have lower endurance, and are at higher risk of heat injury than those who do not wear it. Hearing protection carries a risk of reducing the wearer's situational awareness on the battlefield (Abel 2008). After careful analysis, the chain of command has decided that the potential benefits of these protective measures outweigh the potential risks. Under certain circumstances (e.g. combat operations) their use by service personnel is mandatory. No matter how heavy or hot the body armour feels or how disorienting the hearing protection is, there is no scope for personal discretion. In stark contrast, medical protection such as vaccines or other infectious disease prophylaxis can be refused by service personnel because they retain the same freedoms to make decisions about their healthcare as civilian patients. Even if the risks and benefits of the medical protection are as robustly defensible as those of the physical protections, and even if there is overwhelming consensus and evidence that its use would reduce the risk of morbidity and mortality, the individual still retains the right to refuse it (BMA 2018).

The different treatment of physical and medical protection needs careful justification. Superficially, they seem very similar: both exist in an environment where some individual autonomy has been voluntarily relinquished, both provide protection from harm, both carry risk of other harms. The risks and harms from infection and disease should not be under-estimated, as they can be greater than the risks from combat (Smith and Hooper 2016). One answer might be that physical protection is worn externally whereas medical protection is usually taken internally such as via ingestion of a tablet or injection of a vaccine. This makes the medical protection more invasive and difficult or impossible to reverse. One cannot extract a vaccine once given but one can take off body armour e.g. during off-duty periods. However, other consequences of military service may be equally long lasting and persist on and off duty, such as the beneficial physiological effects of physical conditioning or the psychological effects of experiencing combat. Yet, individual choice does not govern participation in physical fitness training or whether and when to engage enemy forces in combat. Moreover, it is not impossible to imagine non-medical protective equipment that also potentially violates bodily integrity but which personnel might also be ordered to use. Suppose for instance some hypothetical kind of gas mask or breathing equipment that required the use of a mouthpiece that is taken into the mouth. The fact that this mouth piece must be used for the equipment to work effectively would not be considered sufficient to permit personnel to have discretion not to use the protection.

The key difference between the physical and medical protection, it seems, is not what they are but who provides them. Body armour and hearing protection are provided by, and their use is mandated by, non-medical commanders. In contrast, medical prophylaxis is prescribed by doctors and delivered by doctors and other healthcare professionals, albeit ones who themselves are part of the military. It is the professional standards of the clinical staff, which are defined independently of the military chain of command, that afford individual military personnel the freedom to refuse medical prophylaxis. These standards are designed to retain trust and confidence in the profession *as a whole* by privileging the interests of patients in all their interactions with doctors. This is a wholly different objective to those of the military establishment viz. their personnel.

4.5 Individual Autonomy and Collective Consequences

Outside of their healthcare interactions the interests of service personnel lose this privileged status, as may occur in the case of civilian patients. A refusal of medical treatment, including prophylaxis, is not entirely without adverse non-medical consequences. In the case of civilians, it may invalidate their travel insurance, for example. For military personnel, the consequences of refusing medical prophylaxis could include exclusion from particular operations or exercises. Deliberate action that renders a service person non-deployable may jeopardize future promotion and advancement. However, there is no evidence from recent years that recourse to formal punishment has been used. Perhaps this is because it would be perceived as constituting coercion from the perspective of medical professional standards, making it difficult to justify that acceptance constituted a valid consent.

For the unit, there are several consequences of an individual's choice to refuse medical prophylaxis. These consequences should be viewed against the circumstances in which military units may operate. We are assuming, in what follows, a unit operating in an austere and hostile environment such as combat operations, not a unit on home soil in barracks. Here individual and collective survival are interdependent in directly inverse proportion to the size of the unit: the smaller the unit, the more critical each member is to its success. Even in larger units, the more combatants there are who are fit and able to fight, the more likely that the unit will function optimally. An optimally effective unit is likely to be able withstand threats more capably than a reduced and weakened unit, thus improving the survival prospects of its individual combatants.

If one combatant becomes unwell or more severely injured as a result of refusing prophylaxis, this directly impacts on the survival prospects of the entire unit. The unit would not only have lost that individual's fighting power but also that of those charged with his or her care. In this context, the actions of individuals are so inextricably intertwined with the prospects of the wider group that they arguably give individuals clear ethical *and* self-interested reasons to give greater consideration to the interests of others – in this case the whole unit – than might usually be required of civilian patients. It may, therefore, be inappropriate to apply norms based on idealised individuals who are independent, free-standing entities[2] (Wardrope 2014). We have argued elsewhere that norms that apply to family units may be more analogous (Eisenstein et al. 2017). This argument has only limited applicability. It may support the use of novel prophylaxis against the effects of trauma. It would not justify compelling *pre*-deployment acceptance of e.g. a

[2] Of course, there is a body of literature that questions the whether this conceptualisation of autonomy is ever appropriate and proposes a model of relational autonomy that give more weight to social context and emotional, embodied decision-making. This may be particularly true in the case of health promotion (see for instance https://academic.oup.com/phe/article-abstract/8/1/50/1590159?redirectedFrom=fulltext)

malaria prescription. It may suggest, however, that requiring adherence with malaria tablets (other things being equal) whilst on active deployment may be justified when consent was given to the prescription pre-deployment. The justification requires a close temporal relationship of the physiological insult and its grave consequences to the unit.

This interconnectedness of members of a military unit and the special relationship that results from it gives rise to specific duties. Military comrades know and understand that they have special responsibilities to one another and that their individual actions can affect unit performance and survival. This 'associative duty' that members of a unit have to each another and to the unit is equally relevant to decisions that combatants make about their medical and non-medical protection (Kirke 2010; Gross and Carrick 2016). In contrast to a civilian in peacetime, a combatant has a (moral) duty to their comrades to do everything legally possible to promote unit survival and minimise unit harm. This is a further illustration of how the civilian and military ethical environments are not equivalent and suggests that simply transcribing civilian ethical principles unchanged across to the military setting may result in harms to military personnel.

Another consideration in this setting is the extreme resource scarcity that often defines military medical provision. Despite huge resources being spent in the attempt to provide the best possible medical provision to forces, supply chain limitations inevitably lead to resource challenges, whether in terms of drugs, blood for transfusion, number of available operating tables or ITU beds, or any number of other critical assets. By failing to protect themselves against avoidable harms, not only does a combatant threaten the survivability of the unit, but he or she is likely to consume more of the scarce medical assets than they would have done had they taken all possible steps to protect themselves. These assets would subsequently be unavailable for the treatment of unavoidably injured comrades.

In civilian medical ethics, consideration has been given to individual responsibility in a range of contexts such as smoking cessation/prevention, weight reduction and vaccination, but there are well-established objections to penalising individuals who do not comply, not least of which is the medical profession's duty to treat the sick (Cappelen and Norheim 2005; Resnik 2007). There is also resistance to measures that would *compel* individuals to behave responsibly, save perhaps in times of emergency, when other civil liberties might also be curtailed for the preservation of social order and coherence. Compulsory public health measures are, however, at least considered discussable and we will return to these below. However, none of these civilian contexts quite mirror the relationship between the troop members.

The effects on the welfare of the military unit as a whole of individuals not taking measures to avoid avoidable harms applies equally to physical and medical protection. This similarity puts pressure on the privileging of individual choice in the case of medical protection when it is wholly absent in the case of non-medical forms of physical protection. The reasons that justify mandatory compliance in the latter case seem to justify mandatory compliance in relatively similar cases of the former. These circumstances would need to have all of the following components:

1. The medical threat is defined and an estimated risk of harm is quantifiable
2. The medical prophylaxis is effective against the defined threat and its risks are known
3. The risk of harm from taking the prophylaxis is clearly defined and less than the risk of harm from the threat
4. The wider unit is operating under resource constrained conditions (defined as a situation where the loss of the fighting effectiveness of the component individuals or their excessive use of limited medical resources would lead to increase risk of harm to other members of the unit)

The basis of the argument for permitting military personal choice with regard to medical prophylaxis is that military doctors must adhere to the same professional standards as civilian doctors. There are, however, civilian healthcare contexts where individual freedoms are curtailed for the benefit of the wider group. Examples here mainly flow from how infectious diseases are managed in public health, including public health emergencies. For instance, enforced quarantine or social isolation of patients with infectious diseases (Public Health (Control of Disease) Act 1984), and childhood vaccination as a precondition of gaining access to publicly funded education systems (Arie 2017). Non-communicable diseases may also be targeted by requirements to add folic acid to bread and other cereal products and the fluoridation of tap water. However, these risks and the subsequent curtailments of autonomy are not quite equivalent to the military risks and restrictions described above. In most cases, they are differentiated by the degree to which they can be avoided. For example, one could choose to drink bottled water rather than fluoridated water or one could home school children rather than have them vaccinated. These choices do have costs and consequences. Not everyone can afford to buy bottled water or is able to home school their children. Quarantine, in the case of contagious disease, is much closer to the kind of loosening of respect for autonomy in the military setting that we have been discussing above. It is a compulsory condition directed by a higher authority on an individual for the benefit of the wider group in a situation where the potential consequences are significant for all.

In all of these cases, the benefits and risks to the individual are balanced against those of the wider group and a decision has been taken to limit individual autonomy to a greater or lesser degree in order to promote group benefits. A similar argument can be made in relation to medical prophylaxis in the military. This would preserve some of the symmetry between professional norms for civilian and military doctors, but would draw on the norms in operation in civilian public health practice rather than depending solely on the norms applied in civilian clinical practice.

One final point to make about civilian examples where respect for an individual's autonomy is balanced against other factors is that those performing the actions that limit autonomy are not usually doctors or other regulated healthcare professionals. Water suppliers do not ask doctors to add fluoride to their supply, bread and flour manufacturers do not require doctors to add folic acid to their products, and doctors don't physically remove and detain quarantine patients. Doctors will certainly have been involved in collecting the medical evidence to support such actions, planning appropriate responses, and deciding upon the necessary criteria that should guide their application but they will not themselves perform the actions. As we discuss below, this model may have relevance to situations where medical autonomy may be limited in the military context.

4.6 Exploring Consent and Compulsion

So far, we have made a case that it may not be unreasonable for military personnel, in carefully defined circumstances, to be required/ordered to adhere to recommended prophylaxis. This brings up further challenges as to how the prophylaxis should be administered and by whom.

As things stand, a doctor registered by the UK GMC has to respect an explicit refusal of consent or face professional sanctions (GMC 2014). Given this situation, it is informative to discuss a range of hypothetical circumstances that might make the doctor's position more or less clear-cut:

(a) Vaccinating a protesting, competent adult by force (e.g. whilst they are being restrained by others)
(b) Vaccinating an adult who holds out their arm, whilst saying that they have been ordered to comply
(c) Vaccinating a whole group of people whom one knows have been 'advised' to report for vaccination by a senior officer
(d) Vaccinating an individual or group who have all pre-signed a consent form, without checking how robust the consent process was
(e) Vaccinating an individual or group of people who *would* ordinarily all have consented to be vaccinated but who have also been ordered to comply
(f) Vaccinating an individual or group of people who arguably *should* all consent to be vaccinated but who have also been ordered to comply

The first case (a) seems to be the most clear cut. Under normal circumstances,[3] such action would be difficult to justify since other measures are available to protect the unit e.g. excluding the reluctant personnel from deployment. Cases (b)–(d) are in a greyer area. In (b) & (c) there is some measure of implied consent that, outside of the military, might not attract criticism: an arm is offered for vaccination and people have reported at a specific place and time. Women routinely report for breast screening, having received an invitation and information through the post and the process of consent on the day is quite cursory (Gøtzsche et al. 2009). This is, however, an invitation rather than an order, but little effort may be made on the day to ensure that women understand the difference between testing and screening. There are also circumstances where individuals sign consent forms in advance and again the pro-

[3] I.e. leaving aside some philosophical thought experiment deliberately designed to place consequential pressure on the decision

cess of checking understanding may be limited. Rarely are the *reasons* for signing rehearsed when the patient presents; indeed, consent forms are often pre-circulated precisely to speed up processes on the day. Parents are routinely asked to return consent forms for childhood vaccinations with their child ahead of mass vaccination in a school (Levers 2018). It may be argued that some parents sign (or indeed fail to sign) without fully understanding the risks and benefits (Downs et al. 2008). Conceivably, however, some parents may return the form signed because they fear that if they do not there will be negative repercussions – for example refugee families or those applying for asylum. In the UK, careful consideration of the relative merits and demerits of vaccination programmes is given before they are rolled out (Hall 2010). Those performing the vaccinations presumably have no reason to believe that they are unwittingly causing harm and may, on the contrary, be very pro vaccination. A military doctor who accepted consent forms signed in advance and not in his/her presence would not be completely out of step with vaccination practice in the civilian world if s/he merely double-checked for medical contra-indications. The difference is the extent to which military doctors are supposed to be sensitive to the potentially coercive environment in which they practice. Thus, if a complaint was made against the doctor in circumstances (b)–(d), whether it would be upheld would depend on the extent to which is thought reasonable to a doctor to spend time with each of the patients discussing not just what they understand but also their *motivations* for apparently consenting. How these motives then relate to autonomy needs to be determined. In case (d), for instance, a patient may not have been compelled to report for vaccination, but may decide that on balance it is better to 'humour' the senior officer given this may be advantageous down the line. An individual who felt very strongly about not being vaccinated, might make a different decision.

In the case of (e), individuals have been ordered to do that which they would in any case have chosen to do. This is not coercion. Coercion requires a person to make a decision they would not otherwise make *because* of some threat, as opposed to as a result of being persuaded by the force of reason (Powers 2007). But it may be difficult for a clinician to demonstrate decisively that the consent process is robust, even if it is. The order complicates the *evidence* but not the *ethics* of the case.

The circumstances described in (f) reflect what we have argued is defensible position under some circumstances. It differs from (e) to the extent that people do not always want to do – and may, therefore, not do – what they know they should do. It is not the role of doctors to *compel* patients to comply with their ethical obligations, though there are instances where non-invasive action can be taken, without a patient's agreement, in the public interest (disclosure of confidential information, for example). Doctors can work within structures that have independently enforced restrictions on autonomy. For example, military doctors may evaluate and treat detainees and civilian doctors can treat patients in quarantine, though not without their consent. Thus, as things stand, under circumstances (f) any doctor, civilian or otherwise, would have to gain individual consent or decline to administer the vaccine, even if ordered to do so.

The position we have now reached is one where, in principle, it may not be wrong to order personnel to comply with both medical and non-medical prophylaxis but where it would appear to be wrong (or at least contravening existing professional standards) for doctors to administer them. We will now turn to the issue of whether one solution is for non-medical (or other healthcare professionals) staff to replace doctors where it is not possible for doctors to comply.

Here are some further hypothetical scenarios that help us to explore the role of the doctor in a situation where a patient's autonomy is being curtailed:

(g) A doctor provides patients with prophylaxis tablets fully aware that the patient will be ordered to take them, and may face sanctions for failure to comply.
(h) A doctor assesses the safety of providing a vaccination to each of a group of patients, knowing that it will be administered by an appropriately trained person in circumstances where both the vaccinator and vaccinatee will be ordered to comply, and face sanctions for a failure to comply.
(i) A doctors works with the chain of command to determine optimal prophylaxis for a specific mission, knowing that orders will be given for it to be administered.

In (g), the patient self-administers. Many patients choose to follow medical advice about how best to protect their health, and we can assume such patients are in the position of those described and discussed previously in relation to case (e). Within this group may be patients who either actively want to take the drug or at least do not object to taking it, but for whom the existence of the order (and therefore the potential of being admonished if discovered not to have complied) is a spur to compliance with the recommended dosing. We might say that these patients are like drivers who generally agree with speed restrictions, mostly drive within the limits but nonetheless check their speed when they spot a speeding detection camera. There will be another group of patients, who like the drivers, accept that they have agreed to follow certain rules mostly unconditionally as a condition of obtaining certain privileges or benefits (like the driving license). Provided that rules are reasonable (and in this case we refer to the list of conditions viz. prophylaxis above), clear at the time of the agreement, and applied and enforced even-handedly, and provided that they do not completely undermine autonomy they may be compatible with respect for autonomy. It is the voluntary act of agreeing to the rules that is crucial here.

In terms of compelling individuals to ingest drugs, literally forcing these down their protesting throats is a clear violation of autonomy, since it removes all possibility of choice and freedom of action. Those who are ordered to comply with a medication regime have the freedom of action to refuse, but the failure to comply with an order may not be entirely without adverse consequences (as discussed above). Individuals are free to decide for themselves what the least worst option for them is: taking the pills as directed, or accepting the consequences for failure to comply. We have already discussed that the existence of such consequences *per se* does not undermine autonomy since they may merely incline people to do what they were already disposed to do. Assuming that there are good reasons for the orders, the promise of just and proportionate punishment is not regarded by everyone as necessarily undermining autonomy. According to Kant, "he who wills the crime wills the punishment" (Kant 2017). On this account, it would be *failing* to deliver the punishment that frustrates the will of the perpetrator. Kant's formulation of retribution assumes that citizens have a part in formulating the rules that they may then be punished for breaking. The situation of the military does not fit perfectly within this Kantian model but it does not completely fall outside it either. In democracies, the military is an instrument of the state and the state is governed by the will of the people, expressed through elections. This does not justify a completely blank cheque in how the military operates, nor is this the case. It is however, at least arguable that provided that personnel are not conscripted, and provided that the rules and sanction are clear, just and proportionate, and understood at the time of joining, this Kantian formulation provides another reason for supposing that the existence of sanctions does not necessarily denote a failure to respect autonomy. That being the case, it is not obvious that doctors *are* party to violating autonomy if they party to a system that mandates medical prophylaxis under the conditions we have outlined.

The role of doctors in scenario (g) has not, to our knowledge, created problems with professional registration for British military doctors. This may be because it has not attracted attention. Or perhaps self-administration and the distance between the provision of the drug and its being taken provides some moral insulation, even though the order to comply with the drug regime is well known, and therefore foreseeable. This being so, there is equivalence between cases (g) and (i), especially where the general rule is that the prophylaxis is self-administered.

Turning finally to scenario (h) The use of personnel who do not have dual loyalties to both military and medical profession appears to resolve the problem of its being reasonable to insist on adherence to prophylaxis whilst not including doctors in circumstances that would contravene their professional standards. However, it may be argued that doctors' involvement in situation (h) would be unacceptable because they can foresee the consequences and are therefore materially complicit in forced medication. We also note that, contrary to the arguments provided in the previous paragraph, this may also suggest that *current* arrangements (where doctors provide prescriptions for drugs that personnel will be ordered to take) are also suspect. The difference between the two may turn on the matters of degree.

A further question relating to (h) is whether similar constraints ought in principle to apply to non-professionals trained to the standard where they can be charged with administration of medicines, including vaccinations. There already exists, in most Western militaries, a cadre of highly trained combat medical technicians who deliver advanced medical care on the battlefield including extreme interventions such as chest drain insertion, airway protection, and resuscitative drug administration. In most cases, these personnel do not hold external professional registration and are not *necessarily* bound by civilian codes of medical ethics. They are certainly sufficiently trained to administer any medical prophylaxis the chain of command decides has met the relevant criteria. Such personnel would not be bound by professional regulations because they had not undertaken to be so bound nor they do they 'profess' that they are so bound. Their actions do not bring a profession into disrepute and thereby threaten the relationship of trust and confidence between the public and that profession that is necessary for healthcare to be practiced effectively. In this respect, there is little to distinguish the orders to e.g. vaccinate a unit from other orders that they would expect and be expected to obey.

Alternatively, we could take a step further back and explore the relationship between skill and obligation that may run alongside issues of regulation and public trust. For example, it may be argued that anyone who has the skill to save a life under circumstance X therefore has the obligation to exercise that skill in circumstance X, whether or not they also have a professional obligation to respond. For example, an off-duty paramedic, doctor, or life guard would be obliged to commence resuscitation if close by to someone who collapses on a beach. The question might then be one of the extent to which providing people with very high levels of skill to preserve and protect health gives them similar obligations to protect the interests of their 'patients' as those who are professionally regulated, and whether this obligation might also extend to protecting welfare interests, including the interest in self-determination.

4.7 Conclusion: The Case for a Greater Sensitivity to the Circumstances in Which Military Doctors May Be Practicing

We have argued that there are certain specific circumstances in which the right to respect for autonomy may be justifiably curtailed by mandating that combatants take medical prophylaxis but that it is currently not possible for doctors to administer such prophylaxis due to their regulatory and other professional bodies insisting on complete equivalence between civilian and military professional standards.

One solution is for doctors to direct the administration of this mandated prophylaxis and for other, suitably skilled, practitioners without a professional registration to deliver it. It is not clear that this solution would protect doctors from professional sanctions, though we have argued that it should.

An alternative approach would be to re-open the issue of whether, under certain specified conditions (as outlined above), it would be ethically permissible to mandate that military personnel take medical prophylaxis, and to permit doctors to be party to delivering these. Clearly this would require new, tightly specified ethical guidance to be drawn up by the chain of command and the professional regulators that reflect the circumstances that pertain when military personal deploy in a resource limited, interdependent, high risk environment. Such a solution should also take into account the division of responsibilities recommended by the BMA (2018) between military doctors and the chain of command.

References

Abel, Sharon M. 2008. Barriers to hearing conservation programs in combat arms occupations. *Aviation, Space, and Environmental Medicine* 79 (6): 591–598.

Arie, Sophie. 2017. Compulsory vaccination and growing measles threat. *BMJ: British Medical Journal (Online)* 358: j3429.

Battlefield Casualty Drills – Aide Memoire. 2007.

Beauchamp, Tom L., and James F. Childress. 2001. *Principles of biomedical ethics*. Oxford: Oxford University Press.

BMA. 2018. *Armed forces ethics tool kit*. https://www.bma.org.uk/advice/employment/ethics/armed-forces-ethics-toolkit. Accessed Sept 2018.

Cappelen, Alexander W., and Ole Frithjof Norheim. 2005. Responsibility in health care: A liberal egalitarian approach. *Journal of Medical Ethics* 31 (8): 476–480.

Chan, Sarah W., Ed Tulloch, E. Sarah Cooper, Andrew Smith, Wojtek Wojcik, and Jane E. Norman. 2017. Montgomery and informed consent: Where are we now? *BMJ* 357: j2224.

Downs, Julie S., Wändi Bruine de Bruin, and Baruch Fischhoff. 2008. Parents' vaccination comprehension and decisions. *Vaccine* 26 (12): 1595–1607.

Eisenstein, Neil, David Naumann, Daniel Burns, Sarah Stapley, and Heather Draper. 2017. Left of Bang interventions in Trauma: Ethical implications for military medical prophylaxis. *Journal of Medical Ethics: Medethics* 2017 (104299).

GMC. 2014. *Good medical practice: General medical council*. London: General Medical Council.

Gøtzsche, Peter C., Ole J. Hartling, Margrethe Nielsen, John Brodersen, and Karsten Juhl Jørgensen. 2009. Breast screening: The facts—Or maybe not. *Cancer* 1: 2.

Gross, Michael L., and Don Carrick. 2016. *Military medical ethics for the 21st century*. London: Routledge.

Hall, Andrew J. 2010. The United Kingdom joint committee on vaccination and immunisation. *Vaccine* 28: A54–A57.

Kant, Immanuel. 2017. *Kant: The metaphysics of morals*. Trans. M. Gregor. Revised Aufl. Cambridge texts in the history of philosophy. Cambridge: Cambridge University Press.

Kaufman, Bradley, David Ben-Eli, Glenn Asaeda, Dario Gonzalez, Doug Isaacs, John Freese, and David Prezant. 2013. Comparison of disaster triage methods. *Annals of Emergency Medicine* 62 (6): 644–645.

Kirke, Charles. 2010. Military cohesion, culture and social psychology. *Defence & Security Analysis* 26 (2): 143–159.

Larsen, Brianna, Kevin Netto, and Brad Aisbett. 2011. The effect of body armor on performance, thermal stress, and exertion: A critical review. *Military Medicine* 176 (11): 1265–1273.

Larsen, Brianna, Kevin Netto, Daniel Skovli, Kim Vincs, Sarah Vu, and Brad Aisbett. 2012. Body armor, performance, and physiology during repeated high-intensity work tasks. *Military Medicine* 177 (11): 1308–1315.

Levers, Jane. 2018. *Children's division immunisation procedure for school nursing teams: Southern Health NHS Foundation Trust.*

Manson, Neil C., and Onora O'Neill. 2007. *Rethinking informed consent in bioethics*. Cambridge: Cambridge University Press.

NATO. 2018. Allied joint doctrine for medical support AJP-4.10(B): NATO Standardization Office.

Powers, Penny. 2007. Persuasion and coercion: A critical review of philosophical and empirical approaches. *HEC Forum* 19: 125–143. Springer.

Public Health (Control of Disease) Act. 1984. United Kingdon.

Resnik, David B. 2007. Responsibility for health: Personal, social, and environmental. *Journal of Medical Ethics* 33 (8): 444–445.

Robertson-Steel, Iain. 2006. Evolution of triage systems. *Emergency Medicine Journal* 23 (2): 154–155.

Smith, Arthur M., and C. Hooper. 2016. The mosquito can be more dangerous than the mortar round-the obligations of command. *Journal of Military and Veterans Health* 24 (3): 60.

Wardrope, Alistair. 2014. Relational autonomy and the ethics of health promotion. *Public Health Ethics* 8 (1): 50–62.

WMA. 2006. *World medical association international code of medical ethics.*

Chapter 5
From the Lab Bench to the Battlefield: Novel Vaccine Technologies and Informed Consent

Paul Eagan and Sheena M. Eagan

Vaccines represent among the first biological (or medical) enhancement technologies developed by humanity (Needham 2000). This enhancement uses controlled exposure to a bacterium or virus which then causes the human body to develop a preventive immune response to the infectious disease. Over the past few centuries, the development of this preventive technology has led to the eradication of diseases that once afflicted and killed large segments of the global population. We can now effectively control or eliminate once common childhood diseases, lessen the morbidity and mortality of many epidemics, and can even claim the elimination of smallpox and polio.

Of course, the story of vaccines is not one of unblemished success. Despite the significant progress made in vaccine development and disease control over the last century, reemerging and novel infectious diseases continue to challenge public health. As well, vaccines have long prompted public debate and ethical scrutiny. Contentious issues range from focusing on the risk-benefit analysis of purposefully infecting a healthy individual with a known disease given its known risks, the potential for severe adverse reactions to the vaccine, through to the public health authorities' ability to mandate compulsory vaccination programs for populations or subgroups that are felt to be at risk.

P. Eagan (✉)
The Royal Canadian Medical Service, Victoria, BC, Canada

Canadian Forces Health Services, Ottawa, ON, Canada

The Faculty of Medicine, Dalhousie University, Halifax, NS, Canada

S. M. Eagan
Department of Bioethics and Interdisciplinary Studies, Brody School of Medicine, East Carolina University, Greenville, NC, USA
e-mail: EAGANSH17@ecu.edu

D. Messelken, D. Winkler (eds.), *Ethics of Medical Innovation, Experimentation, and Enhancement in Military and Humanitarian Contexts*, Military and Humanitarian Health Ethics,
https://doi.org/10.1007/978-3-030-36319-2_5

The recent advances in genetic science and their promising application in the fields of infectious disease control, vaccinology, and immunotherapeutics hold much promise. Vaccines developed and produced using genetically-based technologies aim to provide fast, safe, and effective immunizations for novel or re-emerging viruses—shedding the previous time-burden associated with vaccine development. Nevertheless, these new technologies bring new ethical challenges to the forefront.

Past ethical debates have centered on the regulation, development, and use of traditional vaccines. Modern vaccinology has been propelled by both known and evolving diseases which continue to challenge public health. This paper will push the discussion further to provide an ethical analysis of these new vaccine technologies and practices. The first section of this paper will give a brief overview of the history of vaccines and their development, highlighting some of the recurring ethical issues which exist. The second section will draw attention to new advancements in vaccine production, exploring both the practical promises and ethical pitfalls of these new technologies. The third part of this paper will use ethical frameworks from the fields of both public health ethics and biomedical ethics to highlight concerns within this new field, recognizing its position at the intersection of public health and clinical medicine. The analysis will first employ the principalist framework of Beauchamp and Childress (2013) followed by the work of prominent public health ethicists including, Childress (2002), Kass (2001), and the Nuffield Council (2007). Importantly, this paper will move beyond the traditional discussion of research ethics in vaccine development to address ethical issues related to rapidly produced RNA-based vaccines and unique to the military context, explicitly exploring the context of armed conflict and the limited autonomy of military service-members.

5.1 History

Over a thousand years ago, the Chinese documented the practice of variolation. This practice involved the collection of dry scabs and exudate from the healing sores of smallpox victims to provide controlled and less lethal exposures to uninfected individuals. The variolation procedure was most commonly carried out by inserting the smallpox scabs or pustule fluid into superficial scratches made in the skin of *smallpox-naïve* individuals. The individual would usually go on to develop an attenuated (or weakened) smallpox infection when compared to the naturally-acquired disease. When successful, inoculated individuals would recover from the virus after 2 to 4 weeks with a decreased incidence of disfigurement and mortality. As well, the exposure and subsequent muted response to inoculated smallpox would then confer to the individual immunity from future disease (Needham 2000).

Variolation was first used in China and the Middle East before being later introduced into Europe and North America in the eighteenth century (Boylston 2012). In the late 1700's, Edward Jenner popularized the use of cowpox, a similar but less deadly relative of smallpox, which resulted in even less morbidity and mortality while still conferring upon the individual life-long protection against human smallpox infection (Riedel 2005). Much like the variolation of ancient China, Jenner inserted the dried scabs of infected individuals into the uninfected. However, Jenner did not use the materials of smallpox victims but rather that of milkmaids suffering from cowpox or vaccinia. The term "vaccination" entered the human lexicon as the worldwide smallpox vaccination programs in the twentieth century successfully eradicated the disease and relegated it to the history books (World Health Organization 1980).

The use of weaker or attenuated viruses as the first human vaccines was followed by the development of other vaccine production methods over the next 200 years (College of Physicians of Philadelphia 2018). Some common vaccines continue to use live, attenuated viruses, (e.g., rubella, mumps, varicella), while others use killed or inactivated bacteria or viruses to induce an immunological response (e.g., Hepatitis A, polio). In the case of bacterial diseases where the bacteria generate a toxin (e.g., diphtheria, tetanus), inactivated toxins or toxoids are used to elicit a protective response. Another effective strategy is the use of subunits or fragments of the pathogen to evoke an immunologic response, (e.g., hepatitis B, influenza, and pertussis). These vaccines have been overwhelmingly effective and have made significant inroads in disease control and prevention. We are now able to prevent or control epidemics, as well as avert common childhood diseases through widespread vaccine use in infants and children. Extensive and coordinated vaccine programs have enabled us to claim the eradication of smallpox and elimination of polio on a global scale. However, there are many diseases which defy the concerted efforts of vaccinologists and public health scientists to find an efficacious preventive vaccine (Woolhouse and Gaunt 2007). As well each year at least two novel viral pathogens are discovered which constitute an evolving and potentially threatening public health concern (Woolhouse et al. 2008). These novel viruses can cause disease on the global scale of HIV/AIDS, or more transient but equally devastating epidemics from organisms such as H1N1, Ebola or Zika. Finally, there is the spectre of genetically modified or weaponized biological agents. This is more than just science fiction, as a recent report (Noyce et al. 2018) documents in detail the relative ease with which the horsepox virus was genetically reconstructed in the laboratory. A significant concern arose in the scientific community over the public dissemination of this information and its potential use in synthesizing human smallpox or using these technologies to develop genetically-designed infectious agents that can be weaponized or used to wreak havoc on humanity (DiEuliis et al. 2017; Koblentz 2017).

Vaccines are widely accepted as an essential disease prevention modality, yet their production and use –or more specifically their safety and efficacy—remain points of concern. In fact, the ethical discussion surrounding vaccine development has prompted law, policy, and practice aimed at establishing rigorous safety and efficacy standards to ensure public health and safety (US Food and Drug Administration 2018). Vaccine development starts with the exploratory and preclinical stages which focus on isolating and characterizing the pathogen, identifying appropriate vaccine candidates, followed by animal testing. If the vaccine candidate is promising, then the three-phase, clinical development process occurs. Phase I involves testing the trial vaccine in small groups of people with the focus being on identifying adverse side effects and general safety concerns. In Phase II, the trial vaccine is tested in a larger group of test subjects who more closely resemble the intended target population. It is in Phase III that much larger numbers of people are given the vaccine, and it is tested for safety and efficacy. If the vaccine passes this phase, then the vaccine can be formally approved and licensed for use. Even after the vaccine is in commercial production, ongoing Phase IV monitoring studies will usually continue once the vaccine is in broader use to monitor its effectiveness and safety.

These standards coupled with time-intensive scientific technologies have made vaccine development a time-consuming process. From the discovery and laboratory characterization of the infectious agent, developing promising vaccine candidates, to clinical trials and regulatory processes, and then producing sufficient supplies so that they can be used at the front lines of the disease battlefield—this process takes time. Vaccine development takes years—if not decades—in the journey from the lab bench to a safe and effective product that can be used to prevent or mitigate disease. This long time-frame raises critical concerns related to our ability to react to the novel and re-emerging viruses—if we need that much time to develop a vaccine, how can one protect the public's health in a time of crisis? Additionally, how can one improve this technology, or reduce the timeline, safely and effectively?

This paper is particularly interested in the context of the military, and how vaccine use within this population warrants additional ethical discussion. Historically, military medicine has differed significantly from civilian medicine. Generally, these differences were manifest in the diversity and nature of diseases that are encountered, as well as its potential scale and scope. Military physicians, unlike their civilian counterparts, are and have been responsible for the health of large numbers of people, drawn together from diverse locations, and then placed in crowded environments such as camps and troop transports. Under the spectre of armed conflict, soldiers bring with them the epidemic and endemic diseases of their home environments thereby providing new unexposed hosts for infection. Infectious disease has had dire consequences in the military, taking more lives than battle throughout the history of military medicine (Gillett 1981). In recognition of this fact, military medicine has pioneered and promoted hygiene/public health. However, public health interventions in the military context present additional ethical concerns, which are discussed in section three.

5.2 The Future of Vaccine Production

The problems of long timelines and the need for a timely response to novel viral agents have led to new technologies and production techniques. One of the more promising is a messenger RNA-based vaccine that uses artificial genetic code sequences to stimulate the production of viral proteins in the individual's body which in turn trigger the creation of a protective antibody response (Moderna Therapeutics 2018). This process duplicates the mechanism by which some more traditionally produced antigen-based vaccines work while eliminating the requirement and inherent delays associated with the large-scale production of the viral antigen. There are eight vaccines of this type currently in development including ones for Zika virus and Chikungunya (Moderna Therapeutics 2018). In a similar vein, another vaccine developer (CureVac 2018) uses a comparable messenger RNA technology to insert new genetic information into the vaccine recipient to instruct the human body to produce antibodies or enhance the immunogenicity of other vaccines and proteins. Promising candidates using this approach include vaccines for rabies, influenza, and malaria (CureVac 2018).

5.2.1 The Military Context

A rapid and effective production method for vaccines is especially relevant to military medicine, as it promotes overall health and readiness within the armed forces. In recognition of the global nature of military work, this population is often uniquely situated to face exposure to novel/emerging disease. As militaries increasingly engage in disaster relief, global health diplomacy, and global health engagements, it is likely that military service-members will continue to be deployed to outbreak areas just as they were in the 2014/2015 Ebola virus disease outbreak in West Africa (Messelken and Winkler 2018; DND/Canadian Armed Forces 2015). Accordingly, military medicine needs to be prepared for both weaponized disease pathogens as well as emerging disease to maintain force readiness.

The Defense Advanced Research Projects Agency (DARPA) has recognized the importance of supporting new vaccine development technologies through a program initiative, called the Pandemic Prevention Platform with the objective to develop a system that can stop any viral disease outbreak before it can reach a pandemic level (Defense Advanced Research Projects Agency 2017a). The program's goal is to dramatically shorten the time from characterization of the viral pathogen to the delivery of an effective vaccine treatment to 60 days with the additional requirement that the new vaccine induces adequate protection within 3 days of administration (Defense Advanced Research Projects Agency 2017b). The proposal is that antibodies from recovering infected individuals would be isolated and reverse-engineered so that genetic constructs/blueprints can be developed, duplicated and delivered in sufficient quantities to allow for these antibodies to be produced by the disease-naïve individual's cellular machinery. It would make use of both RNA and DNA-based platforms. Alternatively, the platform could be used to generate specific antibodies to the novel pathogen which are then injected into the individual and thus provide passive immunity and protection against the disease to the individual. DARPA has established a four-year plan to achieve this goal with the expectation that each of the development teams will be able to take an unknown pathogen and 2 months later have a vaccine which has completed a Phase I clinical safety trial (Defense Advanced Research Projects Agency 2018). This approach has potential benefits for both civilians and military personnel. However, before this new type of vaccine can be made available, it must be researched and tested in human populations. In anticipation of these vaccine trials, our analysis focuses on the use of unproven vaccines developed using the new technologies discussed above.

5.3 Ethical Considerations and Novel Vaccine Technology

These new technologies hold tremendous promise if the many challenges are overcome and the production platforms upon which they are based are proven safe. However, every time unlicensed or experimental vaccines are used, particularly in the setting of armed conflict or humanitarian crises, there are potential pitfalls and ethical issues that need to be considered and addressed. While the literature on this topic is expansive, it primarily focuses on traditional vaccine development. Little attention has been paid to the use of newly developed vaccines, and even less attention to the use of this technology within a military context or in an austere setting. The final section of this paper will provide ethical analysis concerning the use of these new technologies in vaccinology, focusing on their application within the military context.

Initially, the ethical analysis will be framed by the principles of bioethics as espoused by Beauchamp and Childress (Beauchamp and Childress 2013). These principles are widely accepted within the field of biomedical ethics. According to these authors, there are four generally accepted principles of bioethics. These are as follows:

1. The principle of respect for autonomy;
2. The principle of nonmaleficence;
3. The principle of beneficence;
4. The principle of justice.

5.3.1 Respect for Autonomy

The concept of informed and voluntary consent is the essence of the principle of autonomy. To make a truly informed decision, sufficient information must be provided to the individual, and it should be presented in such a way that the recipient can understand the risks and benefits of the intervention, or in this case a potential preventive or therapeutic vaccine. It is critical that people (as moral agents) be allowed the opportunity to ask questions, get honest answers, and the time to carefully consider their choices. The final decision must be genuinely voluntary and given without duress or external pressure from others that may have some influence on them. Critically, people must be permitted the ability to choose freely the option that they prefer and want.

The principle of autonomy and its application to members of the military or individuals that are in a humanitarian crisis requires particular attention as these populations are not always able to make autonomous decisions. The Council for International Organizations of Medical Sciences (CIOMS), in collaboration with the World Health Organization (WHO), provides valuable guidance in this area through their publication, International Ethical Guidelines for Health-related Research Involving Humans (Council for International Organizations of Medical Sciences 2016). CIMOS identifies individuals who are in hierarchical relationships/organizations, such as members of the military, as a vulnerable population (Council for International Organizations of Medical Sciences 2016). In this case, the possible vulnerability is due to the subordinate relationships that exist in the military and the potential for that hierarchical relationship and associated peer group influence to adversely affect the autonomy of the moral agent. The act of a military member volunteering to be a recipient of an experimental vaccine may be influenced by the implied positive (e.g., preferred treatment, career advancement, other perks/benefits) or possible negative (e.g., ostracized from the group, superior officer disapproval, poor career progression) consequences of volunteering and thus influences the genuine autonomy of the eventual decision. The reality of this coercive military culture is well established throughout the history of human subject research on military personnel.

In the past, perceived or real military threats, particularly in times of war and uncertainty, have led to unlicensed vaccine usage in military personnel and resulted in unintended consequences. While the ethical analysis of each of these historical cases is beyond the scope of this paper, these cases warrant mention as they speak to the moral realities of being a service-member and establish historical precedence that prompts ethical concerns. Examples of ethically problematic military medical research expand beyond vaccine use but will be limited to this topic for our discussion. Examples of this are numerous, and the list that follows is not exhaustive:

- The mandated use of an unlicensed Yellow Fever vaccine in military personnel during the Second World War resulted in the most significant, single-source outbreak of Hepatitis B ever reported. The infections occurred because the vaccine was derived from human serum contaminated with the hepatitis virus (Furmanski 1999).
- A recent independent government report investigating an experimental anthrax vaccine trial administered to over 700 Israeli soldiers cites poorly informed consent processes, incomplete disclosure of vaccine risks and benefits, as well as undue pressure being applied by commanders and medical staff on the soldiers to volunteer for the vaccine trial (Siegel-Itzkovich 2009).
- Similar concerns over informed consent irregularities, abuse of authority/coercion and possible long-term health effects were expressed with the mandatory anthrax vaccination program of US military personnel in the same period (Katz 2001) and similar investigational drugs (Annas 1998; Cummings 2002).
- Allegations of abuse of authority and coercion concerning a Hepatitis E vaccine trial involving the Nepalese military (Andrews 2006).

Although not the primary focus of this paper, civilian populations in a natural disaster or an armed conflict are equally vulnerable to the issue of autonomy. With limited resources and knowledge, the potential threat of a disease which may be prevented by the offered intervention, and fears of possible retribution or denial of access to other scarce resources (food, health care, safe haven) if the vaccine is refused, the ability to be autonomous in the informed consent process is suspect. There may be other predisposing factors which heighten this vulnerability, e.g., the longstanding presence of a resource-constrained environment which predisposes the individual to compromise/concession to authority, or a hierarchical social and political structure in their country which duplicates in some way the authoritarian structure that is integral to the military. This dilemma is the case with well-tested vaccines, but what if the vaccine is merely promising and is not thoroughly tested?

When there is a research component to the use of experimental vaccines and treatments, care and attention to the principle of autonomy are paramount. The importance of autonomy has been established through various national and international codes and reports, to include the Nuremberg Code (The Nuremberg Code (1947) 1996), The Declaration of Helsinki (World Medical Association 2013), the Belmont Report (in the United States) (National Commission for the Protection of Human Subjects of Biomedical and Behavioral Research 1978), among others. These codes outline essential considerations in research ethics that prioritize respect for human dignity and thereby require uncoerced, informed consent as well as the ability to withdraw from the study. Notably, these codes and other policies highlight the need for additional protections when vulnerable populations are involved. The use of unproven (or experimental) interventions in vulnerable groups or individuals as part of a research protocol can have an increased probability of potential physical, social or mental harm despite the promised benefit; a benefit which (in the case of novel vaccine technologies) is not even known. At the time of disasters and disease outbreaks, the use of unproven but potentially lifesaving experimental vaccines in vulnerable populations who may already be suffering from significant psychological and physical trauma is problematic. How can these vulnerable populations adequately protect their rights and interests? How can the healthcare provider assure that the principle of autonomy is practiced and preserved?

The recommended approach (Council for International Organizations of Medical Sciences 2016) is to acknowledge the situational duress and the differences between humanitarian aid, where there are clear, proven benefits and similarly clear negative consequences; and research, where potential benefits and risks are less clear. If this is appropriately presented and included in the discussion, then despite the elements of stress and duress, a voluntary decision and informed consent are possible. However, this moral picture is further complicated by the involvement of military personnel.

While civilians may encounter coercion or exploitation as they decide whether or not to consent to unproven vaccines, they are unlikely to be forced to receive a vaccination. Mandatory vaccination remains contentious even when discussing long-established and well-proven vaccines. When talking about untested vaccines, civilians in conflict-zones or natural disasters are unlikely to be forced to participate as that mandate would be ethically problematic. Mandated participation in research is not accepted according to general bioethical principles regardless of military or civilian context. However, there seems to be an exemption for novel vaccines. If we use history as our guide, it is far more likely that the military will employ mandatory vaccine campaigns in at least some sub-populations of service-members. The exigencies of armed conflict have already been used to administer novel vaccines to this population (as evidenced by the list of vaccination campaigns provided in the previous section). It is the use of new technologies in vaccine development that alter our understanding of risk and benefit within this population, shifting the generally held belief that this practice is ethically acceptable. Specifically, public reaction to past research abuses must be considered, highlighting the need for transparency and honesty as new immune enhancement technologies are used. Notably, *forced administration* of vaccines to military service-members presents another set of more significant ethical issues to be discussed in the next sections. Rather, this research examines the *use* of this technology. Discussions of voluntary versus forced administration are important, but we would like to emphasize that *even if* service-members are given a choice, it is unlikely that valid informed consent will be granted due to the factors discussed earlier (coercion, institutional values, cultural group dynamics, among others).

5.3.2 The Principle of Nonmaleficence

The aphorism "first do no harm" is one of the most common axioms that health care providers learn in their professional training, and in many ways captures the essence of the principle of nonmaleficence. Health care providers are taught to not intentionally cause harm or injury to their patients, whether through intent or neglect. Maintaining an appropriate standard of care and demonstrating medical competence is not only a foundational component of medical professional morality but also forms the basis for tort law as it applies to medical malpractice and a health care provider's duty of care. This ethical principle extends beyond the world of clinical encounters to include medical research such as the use of unproven vaccines. According to established scientific practice, these experimental vaccines will have to have completed Phase I of the vaccine approval process before broader testing can be done. Phase I trials involve vaccines being tested in only a small group of otherwise healthy individuals to examine early preliminary evidence of the vaccine's safety and possible toxicity. Since these vaccines will have just been subjected to testing in non-human models and Phase I toxicity trials, it is difficult to know whether or not they can harm the patients receiving them as the ability of the vaccine to protect against illness in humans has not been determined. The new RNA/DNA-based production platforms may be proven safe to use, but there are no guarantees that the new vaccine does not have the potential to harm. It is because of this uncertainty in calculating risks and benefits, that emerging medical technologies are closely watched and controlled.

Again, the involvement of the military complicates ethical analysis as security and force health protection considerations may skew any weighing of risks and benefits. The military is unlikely to use unproven vaccines unless it is necessary to the mission. For instance, service-members may be required to accept pre-deployment vaccinations as protection against weaponized pathogens we suspect may be used by the enemy or because of a new disease outbreak in the area to which they are deploying. In these cases, the risk of exposure is likely higher than in non-military, or non-deployed personnel. Additionally, the military must employ a utilitarian calculus to weigh the possible burden that infected or exposed service-members could place on their unit as well as the overall mission since it is generally accepted that during times of conflict and crisis resources are limited. The military deploys those trained to accomplish a task and best able to do so, making it more likely that this population will be required to be vaccinated. Or, if not mandatory, it is expected that only protected troops will be deployed implicitly making the vaccination a requirement for participation in the mission. The risk-benefit calculation is also likely to be skewed by the importance of the mission, and the relevance/necessity of disease protection to mission success.

5.3.3 The Principle of Beneficence

The third concept or principle is that of beneficence. The obligation and expectation are that the health care provider is to help their patients and their actions should be focused on improving the situation of the patient as well as removing or preventing harms. Health care providers have the skills, training, and knowledge necessary to promote the welfare of their patients and are expected to assist the patient and improve their circumstances in life. This obligation includes providing the individual patient with proven interventions such as vaccines to prevent or minimize the harmful effects of vaccine-preventable diseases. This principle can also be expanded to a broader societal goal of improving population health and preventing disease through research and development of new vaccines.

The application of the principle of beneficence can be problematic with vaccines and their use in emergency situations or in cases such as disasters where there are limited available resources. In the case of unapproved or experimental vaccines, the benefit of the experimental intervention may be unknown since evidence of benefit may be based on nothing more than in vitro/cell-based observations, nonhuman clinical trials, or limited human data. This would be the case with the proposed rapid lab bench to patient delivery concepts that these new vaccine technologies are trying to provide. As already stated, the relative safety and potential toxicity of the vaccine would have been demonstrated based on preliminary Phase I studies, while the potential benefit is unknown. Many questions concerning the vaccine remain unanswered. For instance, is the vaccine effective in an actual epidemic situation? Is the possible cure worse than the disease? What is the real risk of contracting the infection and what is the likely outcome if illness occurs? Are there other strategies, preventive interventions, treatments, or proven vaccines that represent better alternatives which may mitigate or eliminate the risk of infection? These are all issues which need to be considered by the healthcare provider when determining how best to act in the interests of the patient.

Beneficence is also complicated by participation in military missions that may place an individual at a heightened risk of exposure. To act beneficently in this case, the healthcare provider must consider additional possible harms. If there is a high risk of disease exposure, it seems ethically right to vaccinate those who will face this risk, even if the vaccine is unproven. In line with the utilitarian military perspective mentioned earlier, considerations of beneficence should also consider the service-member population on a unit level or larger. There could be group-level consequences to deploying an unvaccinated individual to a high-risk environment. Specifically, the infection of one service-member would not only have implications for force readiness but could necessitate increased demand on limited operational, logistical, and medical resources. Taking on this population-level understanding of utility is more akin to public health ethics, and forces the ethical evaluation to weigh the possibility of harm caused by an unproven vaccine against the possible negative impact that a diseased military force could have on a mission and its operational objectives. While we are not arguing that the individual patient is unimportant to ethical analysis in military medicine/medical research, we must acknowledge that the institution of the military is necessarily motivated by more of a population level understanding of utility in its understanding of beneficence and nonmaleficence.

5.3.4 The Principle of Justice

The final principle is that of justice or fairness in assuring that all individuals that can potentially benefit from the intervention have the same access. According to this principle, it is imperative that the burdens and benefits of this research are equally distributed so that vulnerable populations (such as the military) are not exploited. The latter point is less relevant to our discussion, since these vaccines may be most appropriately given to military populations in light of mission requirements and increased risk. However, it is important to note that military service-members should not be forced to participate in unproven vaccine trials or to receive untested vaccines if their risk of exposure is low and mission specifics fail to alter the risk-benefit ratio.

Returning to the principle of justice as it relates to equal access, this should not be a problem when there is a plentiful, affordable and accessible vaccine or one that can be quickly produced in sufficient quantities during an emergency situation.

However, during a potential or evolving epidemic crisis this will not always be the case, and given the experimental nature of these platforms and their products, the likely demand for the vaccine may well exceed its supply. In this situation, there are a variety of frameworks from the field of public health ethics that examine how to distribute scarce resources in a just and ethical way. Concepts of distributive justice and epidemiology often come together with the primary, determining factor being the risk of contracting the infection. The presence of underlying chronic health conditions, extremes of age, pregnancy and potential risk of exposure due to normal daily activities may help inform the stratification model. Ideally, the determination of risk factors and prioritization within a given (local, national, or global) population are conducted with the input of a variety of specialists, including infectious disease specialists, epidemiologists, community/government leaders, bioethicists, local health care providers, and nongovernmental organizations to name a few.

Fairness has been identified as a fundamental value in military ethics. Closely linked to justice, the principle of fairness requires equal access or fair distribution of resources throughout the military population. In practice, the principle of fairness in military ethics serves as the foundation of concepts such as advancement based on merit and respect for rank (Mehlman and Corley 2014). Within the health care context, this principle requires equal access to adequate care. Thus, newly available vaccines must be equitably distributed among the military population in line with the principles of biomedical ethics and military ethics. The principle of distributive justice is especially relevant in cases of resource rationing such as limited vaccine quantities in a military population. How is it determined who should get the vaccine and who should not? Who is the authority that makes those decisions? The commanders are the relevant authority in a military context, but additional medical or epidemiologically balanced and objective background information should be provided to the commander as warranted to properly inform the decision-making process. Input from senior medical authorities to guide command decisions, as well as ongoing oversight from civilian regulatory bodies, will help avoid some of the past problems and missteps of military commanders in these situations. This prioritization process should be based on the likelihood of exposure and a robust risk assessment. The mission-critical nature of the individual's role is another factor to consider, particularly in the case of a small mission footprint where everyone would be seen as critical to mission success.

The use of an experimental Ebola vaccine by the Canadian military during the 2014/2015 Ebola epidemic in West Africa provides a useful case study of how to approach the issue of experimental vaccine use in military personnel during an evolving humanitarian crisis (Eagan P. C. 2018a). This humanitarian operation involved military health care professionals providing direct patient care to Ebola victims. As part of the planning and execution of the mission, senior military leaders worked proactively with civilian public health authorities, government regulators, and medical specialists in developing the research protocols and informed consent procedures surrounding the use of the vaccine. Guidance from an independent institutional review board (Veritas IRB 2014) was sought and helped to frame and inform the ethical and moral issues around the experimental vaccine's use.

5.3.5 Relevant Frameworks from Public Health Ethics & Research Ethics

The principles outlined and discussed above represent essential ethical considerations when talking about any proposed medical intervention. However, as stated earlier, this discussion can be understood to overlap different spheres of health/healthcare—research, public health, and clinical medicine. As noted throughout our discussion, Beauchamp and Childress (Beauchamp and Childress 2013) offer principles that are generally invoked in discussions of clinical medical ethics, and largely repeated in the Belmont Report (National Commission for the Protection of Human Subjects of Biomedical and Behavioral Research 1978) that focuses on research ethics. The analysis above purposefully extrapolated these ethical considerations beyond the clinical encounter to considerations in research and public health. However, an ethical study of new vaccine technologies must also involve a discussion of public health ethics—a field interested in more utilitarian or aggregate concerns when compared to the individualistic focus of clinical medicine. The use of this novel vaccine technology will likely involve vaccination campaigns, a type of public health intervention with a corresponding body of academic literature that has explored the pertinent ethical issues. Vaccines, like many other public health interventions, are often laden with ethical dilemmas. This reality is perhaps most apparent in discussions of epidemic disease since the effective control of infectious disease often necessitates liberty-limiting public health interventions. Specifically, public health policies and interventions may necessarily infringe upon the rights of individuals, in order to protect the health of the population. In this way, the principle of autonomy is less relevant to discussions of public health ethics, as opposed to clinical medical ethics, or even research ethics (where subjects must provide full and informed consent).

In previous sections, we explored the ethical dilemmas related to consent, autonomy and the use of unproven vaccines. According to the principle of respect for persons, consent of adults with decision-making capacity is vital to ethically sound medical interventions. In light of this, the possible inability of military service-members and other vulnerable populations to adequately consent may be ethically problematic. The calculation is changed when we shift to a public health perspective in which the individual is arguably less important than the society as a whole. Within public health ethics, there are situations such as epidemics of novel disease that may warrant forced treatment, quarantine, and other limitations rightly being placed on individual liberty. However, even within the field of public health, an authority's ability to infringe on individual rights is not unrestricted.

Prominent public health ethicists have proposed frameworks for ethical public health practice (Childress et al. 2002; Kass 2001; Nuffield Council on Bioethics 2007). According to the Nuffield Council, ethically justifiable public health interventions fall along an intervention ladder, or 'liberty-limiting continua' (Nuffield Council on Bioethics 2007). This continuum differentiates interventions based on the level and types of infringements placed on individuals. Vaccines are not inherently liberty-limiting unless their administration is mandated. According to the Nuffield continua, this novel vaccine technology would fall within the lower half of the ladder since their development seeks to both enable and guide the choices of individuals, while not restricting or limiting choice (Nuffield Council on Bioethics 2007). The voluntarily-provided decision to receive a vaccine is the most ethically desirable end-point. We have already offered that mandatory vaccination campaigns are unlikely amongst civilian populations when considering novel, or unproven vaccines. This possibility is much more likely within the military population, given their increasingly globalized role and presence in disaster response.

In light of this, as well as the other considerations raised throughout our analysis, the use of vaccines within a military context may change the proposed intervention's placement along the Nuffield continuum. Within the military institution, many vaccines are mandated as part of an individual's service requirement or are necessary for participation in a specific mission. Given the goals of rapid RNA-based vaccines, these would likely be made available during an epidemic to which the military was responding as part of its ever-expanding role in humanitarian assistance and disaster response (HADR). Following the historical use of vaccines in the military, these vaccines would likely be mandated, thereby placing them at the top of the Nuffield continua where individual choice is either restricted or eliminated (Nuffield Council on Bioethics 2007). When public health interventions place limitations on, or mandate actions of individuals for the benefit of the group, the benefits to individuals and society must be weighed against this erosion of personal freedom. Since many public health interventions do limit personal autonomy and liberty, mandatory vaccination is not prima facie unethical, even if the vaccine is new and technology novel. However, if compulsory vaccines may be permissible, it is necessary to consider when and in what forms it is an ethically permissible public health intervention.

Limiting our discussion to the military population, we can again look to the contextual realities of this group. As mentioned earlier, this population is considered to be vulnerable due to their restricted autonomy. The military context may also make it easier to justify mandatory vaccination due to the fundamental institutional and cultural value of obedience. However, it has been argued that the culture of compulsion within the military does not itself justify mandatory public health interventions (Eagan S. M. 2018b). Drawing from the field of public health ethics, mandating novel vaccine acceptance may be permissible within the military only if it is warranted for force health protection or the accomplishment of the mission. Additionally, the mission must itself be ethically good or at least neutral, in order to justify this infringement on liberty. If the purpose itself is ethically wrong, then an argument for infringing on personal freedom to support this mission would be problematic.

5.4 Conclusion

In closing, the use of unproven vaccines developed on an accelerated timeline by way of this new technology requires ongoing ethical analysis as we gain greater understanding concerning the risks and benefits. The critical potential of these modern technologies will address one of the major problems of the vaccine production process and holds the promise of moving effective vaccines more quickly from the laboratory to the populations at risk, decreasing disease incidence and improving population health. However, along with this speed and given the experimental nature of these vaccines will come the continued responsibility to recognize and protect those that are most vulnerable and assure that the use of these remarkable medications is done ethically and equitably. Due to the relevance of this new technology to the military, it is vital that future research and discussion acknowledge the unique ethical dimensions of involving those in military service. Based on this research it is paramount that in moving forward we learn from the mistakes of the past, and hold firm on the fundamental ethical principles and practices of vaccine research, as well as public health.

References

Andrews, J. 2006. Research in the ranks-vulnerable subjects, coercible collaboration and the Hepatitis E vaccine trial in Nepal. *Perspectives in Biology and Medicine* 49 (1): 35–51.

Annas, G. 1998. Protecting soldiers from friendly fire: The consent requirement for using investigational drugs and vaccines in combat. *American Journal of Law and Medicine* 24 (2&3): 245–254.

Beauchamp, T., and J. Childress. 2013. *Principles of biomedical ethics*. 7th ed. New York: Oxford University Press.

Boylston, A. 2012. The origins of inoculation. *Journal of the Royal Society of Medicine* 105 (7): 309–313. https://doi.org/10.1258/jrsm.2012.12k044.

Childress, J.E., R.R. Faden, R.D. Gaare, L.O. Gostin, J. Kahn, R.J. Bonnie, et al. 2002. Public health ethics: Mapping the terrain. *The Journal of Law, Medicine & Ethics* 30 (2): 170–178.

College of Physicians of Philadelphia. 2018. Retrieved October 25, 2018, from The History of Vaccines. https://www.historyofvaccines.org/content/articles/different-types-vaccines

Council for International Organizations of Medical Sciences. 2016. *International ethical guidelines for health-related research involving humans*. 4th ed. Geneva: CIMOS.

Cummings, M. 2002. Informed consent and investigational new drug abuse in the U.S. military. *Accountability in Research* 9: 93–103.

CureVac. 2018. Retrieved October 25, 2018, from http://www.curevac.com/

Defense Advanced Research Projects Agency. 2017a. *Pandemic prevention platform (P3)*. Retrieved October 25, 2018, from https://www.darpa.mil/program/pandemic-prevention-platform

———. 2017b. *Removing the viral threat: Two months to stop Pandemic X from taking hold.* Retrieved October 25, 2018, from https://www.darpa.mil/news-events/2017-02-06a

———. 2018. *DARPA names researchers working to Halt outbreaks in 60 days or less*. Retrieved October 25, 2018, from https://www.darpa.mil/news-events/2018-02-22

DiEuliis, D., K. Berger, and G. Gronvall. 2017. Biosecurity implications for the synthesis of horse-pox, an orthopoxvirus. *Health Security* 15 (6): 629–637.

DND/Canadian Armed Forces. 2015. *OP SIRONA*. Retrieved October 25, 2018, from http://www.forces.gc.ca/en/operations-abroad/op-sirona.page

Eagan, P.C. 2018a. The Canadian Armed Forces and their role in the Canadian response to the Ebola epidemic: Ethical and moral issues that guided policy decisions. In *Ethical challenges for military health care personnel: Dealing with epidemics*, ed. D. Messelken and D. Winkler, 60–70. London: Routledge.

Eagan, S.M. 2018b. Ebola response & mandatory quarantine in the U.S. military: An ethical analysis of the DoD 'controlled monitoring' policy. In *Ethical challenges for military health care personnel: Dealing with epidemics*, ed. D. Messelken and D. Winkler, 132–143. London: Routledge.

Furmanski, M. 1999. Unlicensed vaccines and bioweapon defense in World War II. *Journal of the American Medical Association* 282 (9): 822.

Gillett, M.C. 1981. *The army medical department, 1775–1818*. Washington, DC: Center for Military History.

Kass, N.E. 2001. An ethics framework for public health. *American Journal of Public Health* 91 (11): 1776–1782.

Katz, R. 2001. Friendly fire: The mandatory military Anthrax vaccination program. *Duke Law Journal* 50 (6, April): 1835–1865.

Koblentz, G.D. 2017. The De Novo synthesis of horsepox virus: Implications for biosecurity and recommendations for preventing the reemergence of smallpox. *Health Security* 15 (6): 620–628.

Mehlman, M., and S. Corley. 2014. A framework of military bioethics. *Journal of Military Ethics* 13 (4): 331–349. https://doi.org/10.1080/15027570.2014.992214.

Messelken, D., and D. Winkler. 2018. In *Ethical challenges for military health care personnel: Dealing with epidemics*, ed. D. Messelken and D. Winkler. London: Routledge.

Moderna Therapeutics. 2018, October 25. Retrieved April 28, 2018, from https://www.modernatx.com/therapeutic-modalities

National Commission for the Protection of Human Subjects of Biomedical and Behavioral Research. 1978. *The Belmont report: Ethical principles and guidelines for the protection of human subjects of research*. Bethesda: The Commission.

Needham, J. 2000. In *Science and civilisation in China*, ed. N. Sivin, vol. 6. Cambridge: Cambridge University Press.

Noyce, R.S., S. Lederman, and D.H. Evans. 2018. Construction of an infectious horsepox virus vaccine from chemically synthesized DNA fragments. (V. Thiel, Ed.) *PLoS ONE,* 13(1): e0188453. Retrieved October 25, 2018, from https://journals.plos.org/plosone/article?id=10.1371/journal.pone.0188453

Nuffield Council on Bioethics. 2007. *Public health: Ethical issues*. Cambridge: Cambridge Publishers.

Riedel, S. 2005. Edward Jenner and the history of smallpox and vaccination. *Baylor University Medical Center Proceedings* 18: 21–25.

Siegel-Itzkovich, J. 2009. IDF's anthrax vaccine trial "violated Helsinki convention". *British Medical Journal* 338: b1325.

The Nuremberg Code (1947). 1996. *British Medical Journal,* 313: 1448.

US Food and Drug Administration. 2018. *Vaccine product approval process*. Retrieved October 5, 2018, from US Food and Drug Administration: https://www.fda.gov/BiologicsBloodVaccines/DevelopmentApprovalProcess/BiologicsLicenseApplicationsBLAProcess/ucm133096.htm

Veritas, I.R.B. 2014. *Use of a recombinant vesicular stomatitis virus expressing the envelope glycoprotein of Ebola virus Zaire (rVSV ZEBOV-GP) vaccine for post-exposure prophylaxis for Ebola virus exposure*. Montreal: Veritas IRB.

Woolhouse, M., and E. Gaunt. 2007. Ecological origins of novel human pathogens. *Critical Reviews in Microbiology* 33: 1–12.

Woolhouse, M., R. Howey, E. Gaunt, L. Reilly, M. Chase-Topping, and N. Savill. 2008. Temporal trends in the discovery of human viruses. *Proceedings of the Royal Society B* 275 (1647): 2111–2115.

World Health Organization. 1980. *The global eradication of smallpox: Final report of the global commission for the certification of smallpox eradication*, Global commission for certification of smallpox eradication. Geneva: World Health Organization.

World Medical Association. 2013. Declaration of Helsinki: Ethical principles for medical research involving human subjects. *Journal of the American Medical Association* 310 (20): 2191–2194.

Chapter 6
Humanitarian Wearables: Digital Bodies, Experimentation and Ethics

Kristin Bergtora Sandvik

6.1 Introduction

This chapter reflects on the ethical challenges raised by humanitarian technology experimentation, as illustrated through the topical example of 'humanitarian wearables'. Humanitarian wearables are conceptualized as smart devices that can be placed on or inside the bodies of aid beneficiaries for many purposes, including tracking and protecting health, safety and nutrition. This can happen by delivering or monitoring reproductive health; producing security and accountability through more efficient registration or by monitoring or delivering nutrition. The chapter contributes to the topic of the volume—the ethics of experimentation in the military context—by offering perspectives on the humanitarian experimentation with the objects, subjects and delivery of aid. To that end, the chapter also answers the call for exploring military-humanitarian innovation collaborations in further depth (Kaplan and Easton-Calabria 2015).

'Humanitarian wearables' are not yet a 'thing'.[1] The impetus for writing this chapter comes from the authors previous collaborative work on humanitarian experimentation, on which this chapter draws significantly (Sandvik et al. 2017), from the authors ongoing work on humanitarian wearables (Sandvik 2020), and a personal experience on a flight to Dubai, where the author was travelling to speak on cyber-

[1] Tracking devices is frequently used by for example the World Food program in managing its truck fleet, and UNICEFs 'Wearable Innovation Challenge' focused on innovations relating to tracking devices for children.

K. B. Sandvik (✉)
Faculty of Law, University of Oslo, Oslo, Norway

Peace Research Institute Oslo (PRIO), Oslo, Norway
e-mail: bergtora@prio.no

D. Messelken, D. Winkler (eds.), *Ethics of Medical Innovation, Experimentation, and Enhancement in Military and Humanitarian Contexts*, Military and Humanitarian Health Ethics,
https://doi.org/10.1007/978-3-030-36319-2_6

security in the humanitarian space. Seated next to the author on the plane was an investor, who talked at length about the current top-secret testing of a predictive health-patch he was investing in: in his view, this patch would be a highly useful addition to health programs in refugee camps—in response to the authors questions about the ethical challenges arising if response capacity was not enhanced, his answer was that knowledge (i.e. generated data) about disease outbreaks was still worth a lot—regardless of the possibility to preempt or address the ensuing suffering.

This type of technology development and the prospect of its mass-distribution in the humanitarian space raises significant questions for the field of humanitarian ethics. Building on the work of Hugo Slim, this chapter takes an applied approach to humanitarian ethics as an emphasis on 'Standards of rights and wrongs' and as governing personal behavior or the way we conduct an activity in the context of emergencies with chaotic, difficult logistics. This includes lack of connectivity; weak or unwilling government; security issues for aid workers and a high degree of vulnerability in affected communities (Slim 2015).[2]

The chapter understands wearables as a form of *intimate digital humanitarian goods* developed at the interface of the affordances of emergency response contexts and the accelerating digitization of beneficiary bodies. The chapter is concerned with the exceptionally intrusive nature of wearables when used in the aid context, and the fact that this problematic combination is the prerequisite for the data flow. To that end, the chapter aims to put a specific type of ethics concern on the table for further discussion as well as to outline pointers for a future research agenda. The methodology is eclectic and draws on the authors experiential insights as a humanitarian ethics advisor, commentator and analyst, and available grey literature and scholarly contributions. The analytical point the chapter seeks to make is that wearables – and the discourses surrounding them at this point in time—represent a form of experimentation with the nature and direction of aid. The chapter contends that with a product whose main purpose is to collect and return large amounts of intimate personal data, the product is not the wearable but rather the data being returned to humanitarian actors and their private sector partners. This reversal—potentially turning aid into a highly transactional data collection process—prompts careful consideration of the ethical issues at stake. As a contribution to the growing body of scholarship on humanitarian technology ethics (Sandvik et al. 2014; Hunt et al. 2016), the chapter fleshes out three possible approaches to wearables: the top-down humanitarian imperatives and principles framework; a proposed revisited version of the bottom-up rights-based approaches—now involving an rights-based approach to data—and the emergent data justice approach, where stakeholders relationship to the key resource—data—is at the center of the analysis. This mapping also speaks to Smith, Pringle and Hunt in this volume, who elaborate on Value Sensitive Design (VSD) as an approach to technology innovation that explicitly incorporates values and thus an ethics lens, into the design process.

[2] For a take on the 'testimony' tradition in humanitarian action see Givoni 2011.

The chapter proceeds in five parts. The first part describes the emergence of 'digital bodies' as a concern for humanitarian ethics in the context of the technologization of humanitarian space and the turn to innovation. The second part accounts for risks and the notion of humanitarian experimentation. The third part carves out an idea of wearables as intimate humanitarian objects. The selected examples of the possible uses for wearables in aid are not fact-based: rather, they are culled from explicit efforts to promote current and emergent technology for a range of more or less utopian purposes. The fourth and final part maps out ethics approaches to wearables (Smith et al. this volume). A brief conclusion follows.

6.2 The Technologization of Humanitarian Space

The chapter takes a set of interlinked developments as the starting point for reflecting on the ethics of wearables. The first issue pertains to the ongoing technologization of the humanitarian space, taking place in the context of the humanitarian turn to innovation. The much-touted technologizing of humanitarian space, coupled with improvements in global connectivity and lower costs, has brought many useful innovations to the sector (Sandvik and Lohne 2014; Jacobsen 2017; Jacobsen and Sandvik 2018). The use of cell-phones, social media platforms, satellites, drones, 3D printers, digital cash and biometric technology has changed how things are done, the speed and cost of doing things, as well as where things can be done from and by whom. The topic for this chapter is how the miniaturization and personalization of information and communications technologies (ICT) and a growing interface with bio-technology are co-producing 'intimate humanitarian objects' to be used by individual beneficiaries on or inside their bodies.

While the humanitarian sector has been growing radically the past 20 years, the funding gaps as well as the protection gaps are perceived to grow exponentially, to the extent that the sector is not 'fit for purpose'. In response, humanitarians have embraced 'effectiveness' as a sector wide mantra and have turned to technological innovation and a broader engagement with private sector actors. Whereas this relationship has gone from limited and tolerated via the 'corporate social responsibility-route' of corporate benevolence, the humanitarian sector is today rebranding itself as a marketplace that is at once governed by a particular moral economy and open for business (Redfield 2012; Scott-Smith 2016; Sandvik 2017).

Importantly, the use of digital technologies creates corresponding 'digital bodies'—images, information, biometrics, and other data stored in digital space—that represent the physical bodies of populations affected by conflict and natural hazards, but over which these populations have little say or control. A central part of what these technologies accomplish is to generate data (Burns 2014; Read et al. 2016; Sandvik 2016; Fast 2017; Comes et al. 2018). As such, digitization—the collection, conversion, storage, and sharing of data and the use of digital technologies to collect and manage information about beneficiaries—increasingly shift our understandings of needs—and of responses to emergencies. As noted in the introduction, it also opens up fundamental questions about the nature of resource distribution in humanitarian action and how we constitute certain pathways for resources as 'aid' and certain recipients as 'beneficiaries', but not others. As will be explained in the next section, new types of risks and the proliferation of humanitarian technology experimentation are central to identify and understand the implications of these choices (Sandvik 2019).

6.3 Risks and Humanitarian Experimentation

Specifically, these shifts entail that the risks of humanitarian action are changing. While organizations optimistically proclaim that 'that technology redistributes power (OCHA 2013), and that value-added information in itself constitute relief, it is increasingly evident that new risks and harms stem from the adoption of humanitarian innovation and experimentation processes—particularly in relation to data. This necessitates a more precise articulation of these risks and harms, and the way they emanate from the misuse or failure. The existence of risks and harms is already at least partially recognized with respect to surveillance data, where risks can arise in relation to demographically identifiable information. At the institutional level, the humanitarian cyberspace has become a key area of operations. As noted by critics, as actors such as UNHCR and WFP become data hoarders through their new social protection and digital payment programs, data and cybersecurity challenges affecting the individual digital bodies of beneficiaries will become more frequent.[3] This chapter argues that future mass-distribution of wearables in the context of intensifying engagement with the private sector will compound these systemic challenges while also engendering continuous negotiations over the ethical underpinnings and boundaries of humanitarianism at the local, implementing level.

To capture the experimental dimension of wearables, the chapter builds on the concept of 'humanitarian experimentation' articulated by Sandvik et al. (2017). Their point of departure is that experimental innovation in the testing and application of new technologies and practices in humanitarian contexts can underpin unethical, illegal and ineffective trends that result in increased vulnerability and harm for the implicated humanitarian subjects, and potentially also for the implicated humanitarian actors. These consequences can be direct or indirect. Risk can result from both the failure and the success of such experiments. Risk can result from singular harms, which are exacerbated by their relationship to larger, underlying trends in humanitarian aid. Risk can involve for example the privacy violation of collecting personally identifiable information, commercial gains obtained from suspending restrictions on testing technology products on people, the distribution of resources

[3] See controversies surrounding Red Rose (https://www.devex.com/news/new-security-concerns-raised-for-redrose-digital-payment-systems-91619) and Scope https://www.irinnews.org/investigations/2017/11/27/security-lapses-aid-agency-leave-beneficiary-data-risk; https://www.irinnews.org/news/2018/01/18/exclusive-audit-exposes-un-food-agency-s-poor-data-handling

in ways that serve technologies or private-sector actors over the needs of populations in these unregulated contexts.

When humanitarian organizations build systems to distribute relief, they implicitly influence the distribution of harm. Digitization highlights more clearly than ever before how politicization and relationships of power shape mechanisms for needs assessment and evaluation. Power relationships are crucial in the humanitarian domain broadly speaking—and are so too in relation to practices of experimental humanitarian innovation. Such practices may, for example, reinforce a specific distribution of security/insecurity by implicitly enacting assumptions about humanitarian subjects as 'fit' for more experimental practices of innovation than would be found acceptable outside of these humanitarian contexts. Humanitarian innovations unevenly distribute harm, not only by favoring those that are prioritized by a technology's assumptions, but also by exposing recipients of humanitarian assistance to the new harms posed by the underlying innovation itself. Thus, humanitarian actors –and their partners—need to carefully consider the multiple linkages between datafication and harm distribution (Sandvik and Raymond 2017).

Taking this conceptualization of humanitarian experimentation as the point of departure, the next part of the chapter carves out humanitarian wearables as a form of intimate humanitarian digital goods and provides an initial scoping of the type of ethics challenges arising from the introduction of this type of product in humanitarian aid.

6.4 Intimate Humanitarian Objects

6.4.1 The Evolving Modalities of Wearables

The range of sensing and intervention modalities that go into wearable technology is constantly growing. As noted by Wissinger (2017) our understanding of 'wearable tech' is in flux, and varies considerably between social fields, extending from 'wellness, to cuteness, to science fiction level body machine melding'. Wearables range from 'the eminently practical' to the 'utterly fantastical'. The functions of these digital technologies are not new: Paper maps have been in existence for centuries; the earliest pedometers date back to the eighteenth century; and historically worn wearable technologies include mechanical devices that could measure distances and eyeglasses, prosthetic devices, and wristwatches (Carter et al. 2018). However, digital technologies –the human-computer interfaces, and the networked, biosensing, code-emitting nature of the technology—have evolved and diffused rapidly (mass production, affordability) over the last decade (Wissinger 2017). We must interrogate what wearables can do (including the intensification of surveillance of everyday practices), how its capabilities are framed (including through problem re-framing) and *who* does the framing. To that end, we must also consider the *nonhuman* elements that mediate the dynamics and experiences of wearables:

device parameters and affordances, analytical algorithms, data infrastructure, and data itself, as well as the processes and practices around them (Carter et al. 2018).

A key question is what role digital technologies have played in the transformations and commodification of social fields and the bodies that occupy these fields. Wearables can be passive applications (apps) that can be downloaded to smartphones, tablets, and smartwatches that aid wayfinding; dedicated wearables that record activity; and more sophisticated multifunction wearables that also record multiple data streams including biomarkers (such as heart rate). Wearables can also combine data collection and drug delivery. Many of these apps and devices are designed to allow users to directly keep a record of their activities online or communicate with third-party websites used to track and analyze activity (Ruckenstein and Schüll 2017; Carter et al. 2018) Operating upon the emergent interfaces between bio and sensor technology, wearables provide measuring, selection, screening, legibility, calculability and visibility—increasingly they are also becoming vehicles for physical delivery of medicine or reproductive control. Tracking operates through and upon multiple bodily layers: general biodata, such as height, weight, gender, age and race; through bodily fluids, including blood, sweat, sperm and tears; and through the capture of individual characteristics, including but not limited to DNA, fingerprints, iris scans, and voice and face recognition.

Moreover, wearables are constituted through regulation and legalities: we must explore the ethical and legal norms and rules underlying, shaping and constraining the emergence of wearables and its affordances. Among the central regulatory frames for wearables are data protection and privacy law, consumer regulation, and human rights law—these are the frames that constrain and enable the research, development, deployment and integration of wearables in and across social fields. This includes different stakeholder groups (such as designers, data scientists, experts in cybersecurity and intellectual property law) with different approaches and objectives to datafication. In the context of fashion, Wissinger explains that her interviews have revealed that ‘a laissez-faire culture is the environment in which wearable biotech is being developed and will be deployed’: ‘a range of approaches to data protection in design and the producer’s responsibility … when consumers ‘opt in’ to data sharing, they should be cognizant of the risks’ (Wissinger 2018). While this insight is likely relevant across a number of social fields, it should be noted that humanitarian beneficiaries are usually in much more precarious position than the populations considered consumers in the global market economy.

An important aspect of what wearables do is the reframing of practices to become problems, with respect to tracking physical activity for health and ‘wellness’/lifestyle purposes: everyday mobility has been reframed as a public health problem, requiring ‘interventions’ to increase the amount of activity’. Users activities can be monitored and uploaded to the internet, in the process transforming social practices- and contributing to ‘processes of biomedicalisation.’ Thus, the concern is not just with interventions that attempt to promote health but also with those sociotechnical ventures that seek to enhance the biological self (Carter et al. 2018).

6.4.2 The Digital Body: An Initial Scoping

Important critical issues in the wearable technology are how such technology can augment the human body, how it affects human relationships to self and other, and whether wearable technology can promote human autonomy, when it is locked into commercial and power relationships that don't necessarily have the users' best interests at heart (Wissinger 2017). These questions are especially significant for the humanitarian sector—where the risks are greater and the power of intended beneficiaries much smaller (Sandvik 2020).

The literature on datafied self-care focus almost exclusively on wealthy, educated, cosmopolitan citizens and themes relevant to their everyday life and perceptions of citizenship. Thus, the commonly used binary between 'data rich' governments, institutions, and commercial enterprises collecting, storing and mining data; and data poor' individual citizens targeted by such efforts, has been criticized for obscuring global inequities (Ruckenstein and Schüll 2017). This bias is very relevant to the idea of humanitarian wearables, because it shifts our point of departure: it skews our perceptions of the technologies and how they are socially situated, and frames attributes, costs and tradeoffs in a way far removed from the everyday reality of the emergency field (Sandvik 2020).

Concepts such as 'data-double' speaks to the concerns of Global north subjects, focusing for example on identity theft (Whitson and Haggerty 2008). The 'self as laboratory- approach' is concerned with how users experience tracking as restricting their lives. When users report negative attitudes regarding devices, part of the disenchantment was caused by arriving at 'dead ends': devices break, and batteries die, they no longer had fun when playing with the gadgets or the data visualizations, they failed to see progress or had achieved their primary goals, rendering tracking burdensome and restricting (Kristensen and Ruckenstein 2018). In contrast, in the research on the Global South, the focus is typically on connectivity and communication rather than on datafication and digitized self-care (Ruckenstein and Schüll 2017).

Nonetheless, research engaging with the power aspects of tracking in the Global North offers valuable insights about how tracking devices constitute the digital body: Lupton situates individual 'quantifications of the self' within a neoliberal context of coercion and control where intimate biodigital knowledges are converted into biocapital: 'as physical and virtual units of human value to be bought and sold in the digital data economy' (Lupton 2016). Important gender-implications arise from how surveillance technologies focused on bodies and personal lives intersect with identity-based discrimination, particularly gender-based violence (tracking/stalking/honor) and societal power-relation constructs (Woodlock 2017). The intensification of surveillance with self-tracking devices is significant, and the term 'dataveillance', which characterizes the networked, continuous tracking of digital information processing and algorithmic analysis (Ruckenstein and Schüll 2017), is useful for getting at the modalities of surveillance emanating from wearables. Rather than originating from a singular source positioned 'above,' dataveillance is distributed across multiple interested parties—in the case of health, including caregivers, insurance payers, pharmacies, data aggregator and analytics companies, and individuals who provide information (either wittingly or unwittingly). Another feature that distinguishes dataveillance from surveillance is its predictive telos; its aim is not to 'see' a specific behavior so much as to continuously track for emergent patterns (Ruckenstein and Schüll 2017).

6.4.3 Imagining Wearables as 'Game Changers' and Problem Cases

As noted by Craig Calhoun (2010), as cultural constructs, 'humanitarian' and 'emergency' shape understandings of what happens in the world, who is supposed to act and what is supposed to be done.[4] These understandings are usually accompanied by the notion that something *should* be done as a matter of principle. As noted above, with respect to the question of *what* must be done, as a concept and as an emergent practice of humanitarian governance, humanitarian wearables must be understood in the broader context of the technology-focused humanitarian innovation agenda, as well as the general trends of the humanitarian sector. As argued by Sandvik et al. (2017), attention must be paid not only to how humanitarian technology shapes perceptions of what counts as resources, but also to the method of distribution of those resources, in terms of factors that determine access, distribution rights, prioritization of resources and the transparency of the underlying reasoning.

An interesting aspect of wearables is that their use in a humanitarian context would represent a continuation of the traditional practice of counting and monitoring refugees, malnourished children etc. (Sandvik 2020). However, a culture shift has taken place with respect to the permissibility and *necessity* of private sector collaboration as a requirement for success, and this culture shift makes humanitarian wearables 'new'. The technology optimism and sometimes utopianism permeating the sector is articulated in the routine proclamations of digital humanitarian goods as 'game changers' or 'revolutions in humanitarian affairs' (Sandvik and Lohne 2014).

Wearables are often presented in much the same way: 'From wearable gadgets to sophisticated implantable medical devices, the information extracted with mobile technology has the potential to revolutionize the manner in which clinical research is conducted and care is delivered" (Barick et al. 2016, 44). Wearables are often presented as the better data collection alternative, being less invasive: 'Further, by providing patients with sensors, wearable gadgets and apps, data is captured in an unobtrusive way' (Barick et al. 2016, 44), as well as more complete: 'The information assimilated via mHealth allows physicians or investigators to work with more complete data sets and they can identify digital biomarkers that set the path for more

[4] This section draws significantly on the insights presented in Sandvik (2020).

intricate research' (Barick et al. 2016, 44). Finally, wearables are presented as affordable, in an intimate version of the 'boots on the ground' cost saving argument used in the drone-discourse: 'The motto goes 'Healthcare is expensive; health is affordable'—it as to be seen how mHealth will help prevent diseases as well as reduce cost of disease management with remote tracking and management in the evolving future' (Barick et al. 2016, 47; Casselman et al. 2017).

At the same time, the optics of being seen to engage in humanitarian activities has acquired its own commercial logic by creating a marketable moral economy of good intentions. While this has served the objective of creating societal acceptance (in the case of drones), the promoters of humanitarian wearables would potentially be more interested in achieving mass distribution to enable the technology to fulfill its mission as a vehicle for large-scale data collection. Within the panoply of 'tech for good' items intended to provide technical fixes for world poverty, human suffering and similar, in the emergent discourse on wearables, it is the specific characteristics of wearables—mass-scalable, multi-functional, small—that make them uniquely suitable. In a recent article called '*The Application of Wearable Technologies to Improve Healthcare in the World's Poorest People. Technology and Investment*', the author suggests that

> Advantages of wearable technologies include that they are mass scalable, possess many functions, deliver high volumes of potentially high quality data and can be disseminated wherever there are people. For these reasons, one large potential opportunity is for wearable sensing systems to improve the lives of the world's poor. (Levine 2017, 92)

Here, the language of humanitarian wearables is both technical and aspirational: Wearables are partly enabled by the assumed functional integration of other types of 'new' technology:

> Modern distribution and tracking systems enable thousands of units to be tracked in the field with relative ease. Advancement of drone-based systems will further facilitate the distribution of wearable biosensors into remote and potentially dangerous regions. (Levine 2017, 86)[5]

Notably, descriptions of specific devices focus on the technical attributes of wearables, in a manner disconnected from the physical conditions of the emergency context—unstable and insecure, with potentially vulnerable, traumatized and suspicious users with limited data literacy—as well as the specific risks that come with the uses of data-generating and emitting devices in this context. An important concern here relates to the rapidly evolving technological capacity for merging digital and drug therapies. In a press statement from 2015, the start-up Microchip Biotech describes an implantable drug delivery device:

> A self-contained hermetically-sealed drug delivery device that is easy to implant and remove in a physician's office setting that can store hundreds of therapeutic doses over

[5] See also Barick et al. 2016, 47: 'There is a promise that in the near future there will be a time when personalized healthcare and prevention become a possibility on a mHealth platform combined with other technology innovations like drones'.

> months and years and releases each dose at precise times…The system is fully programmable via wireless communications to adjust dosing by physician and/or patient.[6]

The promise of this device is being explicitly linked to the expansion of family planning globally –and to the understanding of implant technology as appropriate for hard-to-reach areas. According to developers 'where access to traditional contraceptives is limited… 'That's a humanitarian application as opposed to satisfying a first-world need'.[7] New partnerships on digital technologies boast that they take the 'drug delivery industry into connected therapeutics': as illustrated by the collaboration between a company called SHL Group and a healthcare startup called QuiO, the idea is of the device as '*not only a drug carrier, but as a data collector*' which '*opens up unique opportunities to support patient adherence and monitor treatment outcomes.*'[8] Yet, identifying individuals in need of rescue does not entail the presence of a political will or logistical or economic capacity to provide the rescue (Sandvik et al. 2014). In tandem with this, 'problems' are often presented as technical problems, concerning bandwidth location issues, interoperability and standardization.

In addition, this chapter suggests that the affordances of the emergency context—underpinned by implicit and explicit moral orientations about agency, suffering and rights of intervention—are important in the case of wearables by way of how they are understood by the public as well as stakeholders in the humanitarian innovation field. The affordances of the emergency context coupled with what is as well as what's *imagined to be* technologically possible engender a permissive imaginary where intrusive uses for intimate tracking devices in the Global South are conceptualized and legitimated. In a remarkable passage of the article '*The Application of Wearable Technologies to Improve Healthcare in the World's Poorest People. Technology and Investment*' referred to above, as part of a case study on the Kibera slum in Nairobi, the author imagines a series of uses for wearables in aid,

> At the simplest level skin patches worn for weeks at a time could detect when a person develops fever. The same skin patch can detect hydration status. Wearable sensors have multiple potential roles in infectious disease. One example, fever patches, has already been discussed. Other skin patches have transdermal capability and can include ELISA technology that could enable early TB or HIV infections to be detected and treated early-this might prevent people from becoming chronically infected… At the very least, these types of wearables would enable disease outbreak clusters to be identified and quarantined. New wearable technologies can be incorporated into intra-vaginal rings that not only incorporate sensors but also can potentially deliver interventions against infectious agents and vaccines. The application of wearable technology to infectious disease in manifold spanning surveillance through treatment. (Levine 2017, 88)

[6] Teva and Microchips Biotech Announce Partnership to Enhance Patient Outcomes through Digital Drug Delivery Technology, June 18, 2015. http://microchipsbiotech.com/news-pr-item.php?news=10. This website is no longer updated.

[7] David Lee, 'Remote control' contraceptive chip available 'by 2018', July 7, 2014. https://www.bbc.com/news/technology-28193720

[8] http://www.shl-group.com/News_SHLQuiO.html

The composite of these suggestions—that users should wear tracking devices—including as a form of contraception—that could be a tool for quarantining individuals deemed to be part of 'disease outbreak clusters' raises multiple ethical questions and invites additional critical scrutiny of discourses surrounding the ability (and acceptability) of certain technologies to 'fix' structural problems while producing economic benefits for data recipients. The final part of this chapter outlines three approaches to these ethical quandaries.

6.5 Humanitarian Ethics as a Bundle of Approaches?

6.5.1 Humanitarian Imperatives and Principles: Top-Down Approaches to Accountability

The first approach takes an applied ethics approach to the possible uses of wearables and comparing them with the normative ideals of the humanitarian imperatives 'do no harm' and 'according to need' and the principles of humanity, neutrality and impartiality.[9]

The imperative of 'doing no harm' compels humanitarian organizations to define and evaluate the potential of an intervention to cause harm, and proof of impact is a necessary component of that analysis. The potential for harm increases significantly when experimental methodologies influence the execution of humanitarian assistance—both in terms of efficiency and distribution. Partnership with private-sector actors combines the extraordinary operational license afforded to humanitarian organizations and the exceptional freedom given to the private sector to commercially trial unregulated technologies. In effect, these partnerships give the least tested interventions the greatest license to operate in contexts where the population has the least recourse. These partnerships bear significantly more legal, operational and principled scrutiny than they currently receive. It is difficult to prove that an untested, experimental intervention will not cause absolute or relative harm, but the onus of proof is on the implementing humanitarian organization and should be a required component of any publicly funded intervention. Even in the absence of ill intentions or negligence, the collection and use of sensitive data creates practical dynamics that inherently question, if not violate, humanitarian principles and the imperative to do no harm.

The imperative of assisting according to need sets out significant constraints on resource use: Resources are notoriously scarce during a humanitarian crisis. Specific practices of humanitarian assistance should be evaluated not only against their individual likelihood of success, but also against their potential impact relative to other forms of humanitarian assistance. Because contemporary humanitarian experimentation is increasingly extractive, there is a need to draw attention to the range of

[9] This section builds directly on Sandvik et al. 2017.

consequences resulting from how the humanitarian sector now sees data as both a means and an end of relief, in programming and policy terms. The humanitarian community's willingness to include commercial application and acquired data as impact metrics is a derogation of its traditional priorities, and a distraction from critical analysis of positive beneficiary impact. Additionally, this chapter suggests that it's not only the beneficiary impact that must be problematized, but the extent to which wearables in fact will amount to a complete reversal of resource flows.

The principle of humanity aligns particularly with the practical consideration of resource scarcity, in that it requires the prioritization of alleviating human suffering and preserving dignity. Humanitarian experimentation, in order to appeal to the principle of humanity, implies a need for both assessment of relative impact on human suffering and, uniquely, a need for mechanisms that give the affected a meaningful ability to hold implementers to account. As seen in the example of the potential future repressive uses of fever patches and intra-vaginal rings to ensure quarantining; wearables can potentially raise significant problems for this principle. The principles of neutrality and impartiality, though distinct, combine to highlight the importance of transparency in core components of humanitarian experimentation, including the priorities of needs assessment, the selection criteria for interventions, and the predictable outcomes or impact of using an intervention. A humanitarian organization that uses wearables to fundraise will no longer be neutral or impartial, but partial to policies, interventions and discourses that support the objective of monetizing beneficiary data.

While the humanitarian imperatives and principles constitute the central moral framework of the humanitarian enterprise, it appears that this framework does no grapple sufficiently with the twin problems of intimate humanitarian digital objects: the exceptionally intrusive nature of wearables and the manner in which this intrusion forms the basis for data flows which can be monetized and securitized.

6.5.2 RBA Revisited: A Bottom-Up Approach to Data and Digital Bodies in Humanitarian Aid?

Recently, commentators have argued for a rights-based approach (RBA) to data, meaning an RBA to information or the way data collection is carried out (Scarnecchia et al. 2017; OHCR 2018). The 2018 Signal code is premised on a right to access, generate, communicate and benefit from information during crisis, and to be protected from threats and harms resulting from the use of information and communications technologies and data during crisis, including a right to data privacy and security, data agency, redress and rectification (Scarnecchia et al. 2017). While a rights-based approach would amend some of the challenges of the top-down imperatives and principles approach, in the following, the chapter will briefly outline the context for an RBA approach, and offer some critical observations.

While there is no scholarly consensus on the initial manifestations of RBA, reference is often made to the 1993 World Conference on Human Rights, the 1995 Women's Conference in Beijing and the 1997 call by Kofi Annan for the UN to 'fully integrate' human rights into its peace and security, development, and humanitarian programming (Cornwall and Nyamu-Musembi 2004). In the late 1990s, the move from needs to rights was conceived by some commentators as a way of formally improving the conceptual framework of humanitarianism (Minear 2002; Slim 2002), while others saw it as a way of allaying public relation concerns, particularly after Rwanda. By the early 2000s, human rights gradually became mainstreamed as a staple of humanitarian rhetoric and numerous handbooks, manuals and codes of conduct (Sandvik 2010).

The RBA turn in humanitarian action was initially subject to significant contestation later to be followed by critical indifference as RBA lost its buzzword-status (Borchgrevink and Sandvik forthcoming). Critics argued that rather than changing what aid agencies did, RBA was linked to the need to reinvent a new identity periodically in an increasingly competitive and skeptical world (Duffield 2001, 223). RBA led to tokenism, instrumentalization and co-option by political actors and inflated and unrealistic claims (Darcy 2004). Reviews of RBA practice noted endemic problems with respect to the conceptualization of rights, and to the exact role these rights would play – but also the absence of reflection on the specific difficulties of applying RBA in situations of conflict (Leebaw 2007; Benelli 2013). The 1998 Sphere Project was an attempt to put humanitarian aid on 'a rights footing,[10] but was criticized for a technicalized RBA and an artificial notion of a 'right to assistance' (Fox 2001; Dufour et al. 2004).

This chapter argues that despite its emancipatory promise, the problem with RBA to data, and consequently for wearables, is twofold: first, that the erstwhile criticisms of RBA still applies in the context of data. RBA to humanitarian action is a problematic concept in itself—and these problems do not go away if we apply RBA to data. The basic dilemma that affects communities in a humanitarian setting can be articulated as the following: 'No longer entitled to rights, they can only have security when embraced by humanitarian non-governmental organisations who have already been enfranchised and contracted by powerful state actors to manage them' (Narkunas 2015). Legalistic versions of RBA are premised on the notion that rights holders are entitled to hold the duty bearer accountable, but according to international law and the view of international humanitarian organisations, the rights are directed principally at the state and its agents. Humanitarian organisations suggest that they must consider 'rights-holders with legal entitlements' but do not see themselves as accountable for the fulfilment of those rights. Organisations sometimes operate with competing definitions of RBA, where humanitarian organisations seek to strengthen the capacities of the rights holders to make claims, and

---

[10] Sphere Project Evaluation, (2004) http://www.odihpn.org/humanitarian-exchange-magazine/issue-53/the-sphere-projecttaking-stock

of duty bearers to satisfy those claims, but are not themselves directly accountable to persons of concern (Lohne and Sandvik 2017).

The second problem is that using an RBA approach as the 'solution' disengaged from any kind of articulated social justice agenda is deeply problematic, and risks shrinking and minimizing the site of struggle. In light of the challenges identified with previous RBA framings, there are costs to frame the struggle for responsible data as a question of rights-based approaches. In particular, this pertains to reducing the scope of claims from transformative to data as focus of struggles. For example, how does participation look where harvesting is the ultimate humanitarian *and* commercial objective?

> Where groups have participated in data collection processes, data collectors should ensure that the resulting data is shared appropriately with these groups. This 'return' of data should be meaningful to the population of interest and delivered in culturally appropriate ways. This demonstrates the impact of their inputs and encourages their ongoing use of data and engagement with the activities of the data collector.[11]

Practically and politically, this is a far cry from the erstwhile transformative impetus of RBA.

6.5.3 Data Justice: New Approaches to New Challenges

A third way of framing the discussion of ethics and humanitarian wearables is to zero in on the resource at stake—the data—and to draw out aspects of the notion of 'data justice' of relevance for humanitarian wearables. To that end, this section draws on recent contributions to data justice in development aid, where framings relate to data collection and use, negative impact of collection and use, how data is handled and framings with respect to more theories of social and moral justice.

As the starting point for discussions, notions of data *in*justice speak to the concept of dataveillance and include particular harms e.g. surveillance, reinforcement of monopoly, loss of privacy, and algorithmic profiling (Heeks and Renken 2018) as well as unequal access to data distribution/non-distribution or how the burden of dataveillance (surveillance using digital methods) has always been borne by the marginalized (Taylor 2017). Taylor suggests that 'markets are a central factor in establishing and amplifying power asymmetries to do with digital data, and that new strategies and framings may be needed that can address the public–private interface as an important site for determining whether data technologies serve us or control us.'[12]

---

[11] https://www.ohchr.org/Documents/Issues/HRIndicators/GuidanceNoteonApproachtoData.pdf

[12] Observing the difficulties with 'the global North freedoms and needs with regard to data technologies have been approached through a fundamental rights framework that includes data protection, framings of informational privacy and the right to free speech and communication.' And the infeasibility of this approach in the global data market.

Heeks and Renken (2018) propose three framings for data justice: 'instrumental, procedural, and distributive/rights-based'. As articulated by Taylor (2017), data justice, conceptualized as fairness in the way people are made visible, represented and treated as a result of their production of digital data, is necessary to determine ethical paths through a datafying world. Data rights determine distribution of data, including data privacy and ownership rights including the right of data portability, representation and inclusion, the right to make decisions on data preservation—and the right to be forgotten. Her strategies for data justice include (in)visibility, (dis) engagement with technology and antidiscrimination.

The problem with adopting a data justice approach, is that humanitarian aid has never purported to be about justice. Whereas RBA has operated with the fallacy of an imagined duty holder, 'humanitarian justice' has never been a stated or implicit objective. Still, three aspects of data justice are of particular interest for an ethics analysis of wearables and needs further analytical attention beyond the scope of this chapter:

First, the central question of how much visibility do citizens owe the state? This concern is particularly relevant for the humanitarian sector: how much visibility do beneficiaries owe humanitarian organizations? Using the example of UN Global Pulse, Taylor (2017) is specifically critical about the notion of a 'collective good' duty of participation, that 'implies that development agencies have a claim to people's data on a utilitarian basis, and that opting out should not be an option because it will impact on the rights of the collective'.

Second, Heeks and Renken (2018) usefully distinguish between 'big data justice' and 'small data justice'. They suggest that this distinction 'reorients to the livelihood needs of individual citizens as data users more than data producers; driven especially by the needs of those in developing countries'. The livelihoods issue is particularly pertinent in a context where data has become as commercial resource extracted 'for free' (potentially free from reciprocal as well as legal constraints) from beneficiaries. This also points to the underlying relationship of duty and obligation in the aid-beneficiary relationship. In the general marketplace the re-selling of consumer data has become big business, and user agreements are increasingly designed to facilitate the possibilities for bundling and re-selling such data.In the humanitarian sector, there are no 'user-agreements'. Moreover, with respect to humanitarian data, the line between commercial and security objectives is increasingly thin, if not already non-existent. How do we think about the freedom to not engage with the data market, not to be represented in commercial databases?

This point links to the third issue, which concerns the formats through which humanitarians articulate their concerns about ethics and experimentation: When Taylor talks about anti-discrimination she talks about hard legal obligations. In lieu of recent debates about the dangers of AI and technology 'ethics washing'(Wagner 2018), the calls for regulation and hard legal obligations are making a comeback.

6.6 Conclusion

This chapter has attempted to carve out the humanitarian wearable as an emergent type of intimate humanitarian goods, and to identify ethical challenges raised by humanitarian wearables as a form of humanitarian technology experimentation. The chapter has been concerned with the extraordinarily intrusive nature of wearables when used in the aid context, and how the specific properties of the relationship between humanitarian actors and beneficiaries is the precondition for the data flow. In addition to the general scholarly contribution to the humanitarian-military innovation nexus, the ambition of the chapter is to initiate a more specific discussion about the ethics of wearables in aid and to provide a conceptual starting point for empirical research. If we are to gauge the 'macro' and 'micro' harms of wearables properly, theoretical and empirical work must be done to study both how wearables are ideologically, discursively and technically constituted by vendors and humanitarian agencies, and how they are deployed, mediated—and resisted—in the field.

With respect to the discussion of ethics approaches, the chapter concludes that at this preliminary stage, it is helpful to work with a bundle approach. Humanitarian actors can not opt out of the imperatives and principles framework: regardless of how this ethical framework is (not) applied in practice, it does constitute the sectors' moral underpinnings and contribute to much of its discursive, symbolic and political legitimacy. However, it is clear that in the context of the specific dynamics engendered by intrusive wearables and one-way data flows, a shift in perspective is needed to incorporate a comprehensive notion of beneficiaries' participation and renumeration as situated within a global data economy. To achieve this, we need practical and applied conceptualizations of data justice that can impose hard legal obligations on humanitarian and private sector actors.FundingResearch for this chapter was funded by the PRIO-hosted project 'Do No Harm: Ethical Humanitarian Innovation' (EtHumIn), and the UiO-hosted project 'Vulnerability in the Robot Society (VIROS), both funded by the Research Council of Norway.

References

Barick, Uttam, et al. 2016. Harnessing real world data from wearables and self-monitoring devices: Feasibility, confounders and ethical considerations. *MEFANET Journal* 4 (1): 44–49.

Benelli, P. 2013. *Human rights in humanitarian action and development cooperation and the implications of rights-based approaches in the field*, ATHA.se.

Borchgrevink, Kaja, and Kristin Bergtora Sandvik forthcoming. *The afterlife of buzzwords: The journey of right-based approaches through the humanitarian sector* (manuscript on file with authors).

Burns, Ryan. 2014. Moments of closure in the knowledge politics of digital humanitarianism. *Geoforum* 53: 51–62.

Calhoun, Craig. 2010. *The idea of emergency: Humanitarian action and global (dis) order*. New York: Zone Books.

Carter, Simon, Judith Green, and Ewen Speed. 2018. Digital technologies and the biomedicalisation of everyday activities: The case of walking and cycling. *Sociology Compass* 12 (4): e12572.
Casselman, Jamin, Nicholas Onopa, and Lara Khansa. 2017. Wearable healthcare: Lessons from the past and a peek into the future. *Telematics and Informatics* 34 (7): 1011–1023.
Comes, Tina, Kristin Bergtora Sandvik, and Bartel Van de Walle. 2018. Cold chains, interrupted: The use of technology and information for decisions that keep humanitarian vaccines cool. *Journal of Humanitarian Logistics and Supply Chain Management* 8 (1): 49–69.
Cornwall, Andrea, and Celestine Nyamu-Musembi. 2004. Putting the 'rights-based approach' to development into perspective. *Third World Quarterly* 25 (8): 1415–1437.
Darcy, J. 2004. *Human rights and humanitarian action: A review of the issues*, HPG background paper 12. London: Humanitarian Policy Group Overseas Development Institute.
Duffield, M. 2001. *Global governance and the new wars: The merging of development and security*. London: ZED Books.
Dufour, C., V. de Geoffroy, H. Maury, and F. Grünewald. 2004. Rights, standards and quality in a complex humanitarian space: Is sphere the right tool? *Disasters* 28 (2): 124–141.
Fast, Larissa. 2017. Diverging data: Exploring the epistemologies of data collection and use among those working on and in conflict. *International Peacekeeping* 24 (5): 706–732.
Fox, F. 2001. New humanitarianism: Does it provide a moral banner for the 21st century? *Disasters* 25 (4): 275–289.
Givoni, Michal. 2011. Beyond the humanitarian/political divide: Witnessing and the making of humanitarian ethics. *Journal of Human Rights* 10 (1): 55–75.
Heeks, Richard, and Jaco Renken. 2018. *Data justice for development: What would it mean?* Information development 2018. Manchester: Global Development Institute, SEED, University of Manchester.
Hunt, Matthew, et al. 2016. Ethics of emergent information and communication technology applications in humanitarian medical assistance. *International Health* 8 (4): 239–245.
Jacobsen, Katja Lindskov. 2017. On humanitarian refugee biometrics and new forms of intervention. *Journal of Intervention and Statebuilding* 11 (4): 529–551.
Jacobsen, Katja Lindskov, and Kristin Bergtora Sandvik. 2018. UNHCR and the pursuit of international protection: Accountability through technology? *Third World Quarterly* 39 (8): 1508–1524.
Kaplan, Josiah, and Evan Easton-Calabria. 2015. Military medical innovation and the Ebola response: A unique space for humanitarian civil–military engagement. *Humanitarian Exchange Magazine* 64: 7–9.
Kristensen, Dorthe Brogård, and Minna Ruckenstein. 2018. Co-evolving with self-tracking technologies. *New Media & Society* 2018: 1461444818755650.
Leebaw, B. 2007. The politics of impartial activism: Humanitarianism and human rights. *Perspectives on Politics* 5 (2): 223.
Levine, J.A. 2017. The application of wearable technologies to improve healthcare in the world's poorest people. *Technology and Investment* 8: 83–95.
Lohne, Kjersti, and Kristin Bergtora Sandvik. 2017. Bringing law into the political sociology of humanitarianism. *Oslo Law Review* 4 (1): 4–27.
Lupton, Deborah. 2016. *The quantified self*. Cambridge: Wiley.
Minear, Larry. 2002. *The humanitarian enterprise, dilemmas & discoveries*. Bloomfield: Kumarian Press.
Narkunas, J.Paul. 2015. Human rights and states of emergency: Humanitarians and governmentality. *Culture, Theory and Critique* 56 (2): 208–227.
OCHA. 2013. *Humanitarianism in the network age (HINA)*, OCHA policy and studies series. Available at https://www.unocha.org/publication/policy-briefs-studies/humanitarianism-network-age. Accessed 30 Sept 2019.
OHCHR. 2018. *A human rights-based approach to data*. Geneva: Office of the United Nations High Commissioner for Human Rights. https://www.ohchr.org/Documents/Issues/HRIndicators/GuidanceNoteonApproachtoData.pdf.

Read, Róisín, Bertrand Taithe, and Roger Mac Ginty. 2016. Data hubris? Humanitarian information systems and the mirage of technology. *Third World Quarterly* 37 (8): 1314–1331.

Redfield, Peter. 2012. Bioexpectations: Life technologies as humanitarian goods. *Public Culture* 24.1 (66): 157–184.

Ruckenstein, Minna, and Natasha Dow Schüll. 2017. The datafication of health. *Annual Review of Anthropology* 46 (2017): 261–278.

Sandvik, Kristin Bergtora. 2010. Rapprochement and misrecognition: Humanitarianism as human rights practice. *The New International Law*: 139–157.

———. 2016. The humanitarian cyberspace: Shrinking space or an expanding frontier? *Third World Quarterly* 37 (1): 17–32.

———. 2017. Now is the time to deliver: Looking for humanitarian innovation's theory of change. *Journal of International Humanitarian Action* 2 (1): 8.

———. 2019. Technologizing the fight against sexual violence: A critical scoping. *PRIO paper*. https://gps.prio.org/Publications/Publication/?x=1274. Accessed 30 Sept 2019.

———. 2020. Making wearables in aid: Bodies, data and gifts. *The Journal of Humanitarian Affairs*.

Sandvik, Kristin Bergtora, and Kjersti Lohne. 2014. The rise of the humanitarian drone: Giving content to an emerging concept. *Millennium* 43 (1): 145–164.

Sandvik, Kristin Bergtora, and N. Raymond. 2017. Beyond the protective effect: Towards a theory of harm for information communication technologies in mass atrocity response. *Genocide Studies and Prevention* 11 (1): 9–24.

Sandvik, Kristin Bergtora, et al. 2014. Humanitarian technology: A critical research agenda. *International Review of the Red Cross* 96 (893): 219–242.

Sandvik, Kristin Bergtora, Katja Lindskov Jacobsen, and Sean Martin McDonald. 2017. Do no harm: A taxonomy of the challenges of humanitarian experimentation. *International Review of the Red Cross* 99 (1): 319–344.

Scarnecchia, Daniel P., et al. 2017. A rights-based approach to information in humanitarian assistance. *PLoS Currents* 9. https://doi.org/10.1371/currents.dis.dd709e442c659e97e2583e0a9986b668.

Scott-Smith, Tom. 2016. Humanitarian neophilia: The 'innovation turn'and its implications. *Third World Quarterly* 37 (12): 2229–2251.

Slim, Hugo. 2002. Not philanthropy but rights: The proper politicisation of humanitarian philosophy. *The International Journal of Human Rights* 6 (2): 1–22.

———. 2015. *Humanitarian ethics: A guide to morality of aid in war and disaster*. Oxford: Oxford University Press.

Smith, Alister, John Pringle, and Matthew Hunt. this volume. Value-sensitive design for humanitarian action: Integrating ethical analysis in the development and implementation of information and communication technology innovations. In *Ethics of medical innovation, experimentation, and enhancement in military and humanitarian contexts*, ed. Daniel Messelken and David Winkler. Dordrecht: Springer.

Taylor, Linnet. 2017. What is data justice? The case for connecting digital rights and freedoms globally. *Big Data & Society* 4 (2): 2053951717736335.

Wagner, Ben. 2018. Ethics as an escape from regulation: From ethics-washing to ethics-shopping. In *Being profiled: Cogitas ergo*, 84–90. Amsterdam: Sum Amsterdam University Press.

Whitson, Jennifer R., and Kevin D. Haggerty. 2008. Identity theft and the care of the virtual self. *Economy and Society* 37 (4): 572–594.

Wissinger, Elizabeth. 2017. Wearable tech, bodies, and gender. *Sociology Compass* 11 (11): e12514.

———. 2018. Blood, sweat, and tears: Navigating creepy versus cool in wearable biotech. *Information, Communication & Society* 21 (5): 779–785.

Woodlock, Delanie. 2017. The abuse of technology in domestic violence and stalking. *Violence Against Women* 23 (5): 584–602.

Chapter 7
Value-Sensitive Design for Humanitarian Action: Integrating Ethical Analysis for Information and Communication Technology Innovations

Allister Smith, John Pringle, and Matthew Hunt

7.1 Introduction

The adoption of new information and communication technologies (ICTs) is leading to many changes in how humanitarian assistance is delivered, with the goal of increasing its effectiveness and efficiency. For example, the World Food Programme trialled iris scanning as a means of confirming identity when providing food allowances in refugee camps (Maron 2013), UNICEF launched an unmanned aerial vehicle (UAV) corridor in Malawi (Matonga et al. 2017) and MSF, Google, and Harvard University collaborated to create an Ebola Electronic Medical Records (EMR) system to capture patient information during the 2014–15 West Africa Ebola epidemic (Jobanputra et al. 2016). While these and other innovative ICT applications are increasingly prominent in humanitarian action," many forms of humanitarian technology and humanitarian action based on the use of digital data … are commonly framed in a humanitarian innovation language in which the possibility that humanitarian principles could be compromised is omitted" (Sandvik et al. 2017). This observation points to the importance for stakeholders involved in ICT innovation in humanitarian action to be attentive to ethical questions that are raised across the cycle of designing, developing, rolling out, and evaluating humanitarian ICTs. As Sandvik discusses in her chapter in this

A. Smith
Faculty of Medicine, McGill University, Montreal, QC, Canada

J. Pringle
Ingram School of Nursing, McGill University, Montreal, QC, Canada

M. Hunt (✉)
School of Physical and Occupational Therapy, McGill University, Montreal, QC, Canada
e-mail: matthew.hunt1@mcgill.ca

D. Messelken, D. Winkler (eds.), *Ethics of Medical Innovation, Experimentation, and Enhancement in Military and Humanitarian Contexts*, Military and Humanitarian Health Ethics,
https://doi.org/10.1007/978-3-030-36319-2_7

volume, these concerns are amplified given the experimental nature of humanitarian innovation and the elevated vulnerability of communities affected by crises.

7.1.1 Scoping the Problem and Existing Literature

Ethical questions associated with the implementation of ICTs in a humanitarian setting include ensuring accuracy of the technology, protecting privacy and security, responding to inequality, respecting individuals and communities, protecting relationships and addressing expectations that cannot be met (Hunt et al. 2016). By design, ICTs are enabled to collect vast quantities of data, which can include sensitive demographically identifiable information that can be a vector for harm (Sandvik and Raymond 2017). The growth of big data from humanitarian settings highlights ethical challenges related to the context sensitivity of collected data, validation of analytic methods, and legitimacy and standardization requirements (Vayena et al. 2015). These concerns are compounded by power imbalances among humanitarian organizations, ICT providers and populations requiring aid (Betts and Bloom 2014), and a lack of clarity regarding ethical responsibilities of groups developing and using innovative ICTs, including for protecting collected data. Moreover, a narrow focus on technological solutions may displace attention from other approaches to addressing humanitarian problems, resulting in missed opportunities (Sandvik et al. 2017).

Existing frameworks and codes of conduct seek to make the innovation process sensitive to the humanitarian principles of humanity, impartiality, neutrality, independence (Slim 2015), as well as linking it to other ethical commitments of humanitarians, such as the principles of do-no-harm, inclusion, respect for persons, and justice. These articulations of ethics guidance for humanitarian innoation, such as the MSF Framework for Medical Humanitarian Innovation (Sheather et al. 2016) or the Framework for Analyzing Ethical Principles in Humanitairan Innovation (Betts and Bloom 2014), can help guide ICT application decisions by outlining value requirements for development and end-use of the innovation. Rights-based and data justice approaches are also being applied to humanitarian ICT innovation (see Sandvik, this volume).

Ethical considerations are not limited to an ICT's end-use; Ethical attentiveness is required throughout the innovation process in humanitarian action. Obrecht and Warner identify the following steps of the innovation cycle: problem recognition, ideation, development, implementation, and diffusion (2016). All of these steps are critical for innovations in humanitarian settings to meet environmental challenges and enhance the capacity of aid organizations to address the needs of populations affected by crisis, including saving lives, alleviating suffering and promoting dignity.

7.1.2 Knowledge Gap

To date, less attention has been given to clarifying how normative commitments can be addressed at different stages of the innovation cycle. Doing so could help support ethically robust innovation, including identifying and minimizing ethical risks that might arise when an innovation is later applied 'in the field.'

One avenue to advance this goal is value-sensitive design, "a theoretically grounded approach to the design of technology that accounts for human values in a principled and comprehensive manner throughout the design process" (Himma and Tavani 2008). While product innovations found in humanitarian action often "begin outside the humanitarian environment" (Betts and Bloom 2014), value sensitive design applied to ICT innovation presents an opportunity to support customized and successful applications with reduced risk of harms for populations affected by crises.

The contribution of this chapter is to consider how value sensitive design can be operationalized in the humanitarian context. We present a framework for Value Sensitive Humanitarian Innovation (VSHI) that can be used by stakeholders in humanitarian ICT innovation, including policy-makers, practitioners, and technology developers to map ethically salient features across the innovation process. The VSHI framework incorporates ethical questions at each stage of the innovation cycle. Underlying the VSHI is a recognition that applying humanitarian principles and other ethical commitments of humanitarian organizations and their innovation partners, along with associated codes of conduct and ethical frameworks, can serve to ground the ethical analysis that value sensitive design requires. In turn, a value sensitive design such as the VSHI provides a structured approach that can complement existing attempts to articulate values for humanitarian innovation and enhance their operationalization across the innovation cycle, including sustained engagement with affected populations.

7.1.3 Framework Development

We developed the framework based on (1) a review of literature on humanitarian ICTs, ethics in innovation, and the application of value sensitive design, (2) our team's experiences in software design and engineering, humanitarian action, and ethics, and (3) discussions with technology developers, humanitarian workers and policy-makers. At the intersection of ethics in ICTs and humanitarian innovation, the framework incorporates value sensitive design in relation to humanitarian principles and other core ethical commitments of aid organizations (Fig. 7.1).

A primary motivation for the development of value sensitive design was the "increasing impact and visibility that computer technologies have had on human lives. Technology may support/enhance and/or undermine/corrupt human values"

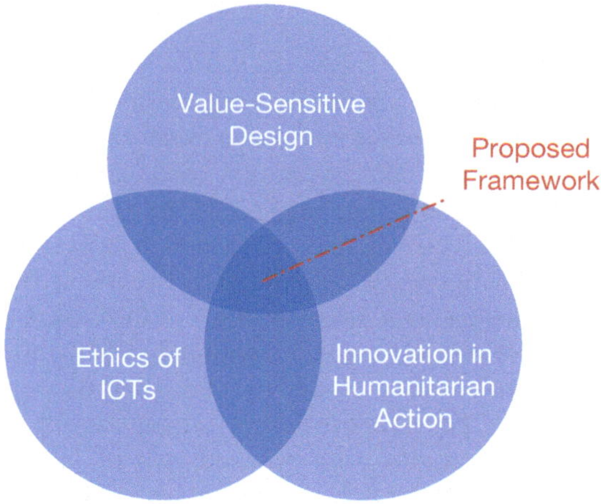

Fig. 7.1 Contextualizing the VSHI framework in existing domains

(Friedman 2004). This increasing influence presents a challenge since traditional design focuses on functionality characteristics, such as "usability, efficiency, reliability, and affordability of (new) technologies" (Manders-Huits 2011). What had been missing was attention to the ethical consequences of the design for stakeholders. Turning attention to functionality in terms of human lives is in keeping with the concept of humanitarian design, which has been described simply as "the creation of objects and systems that improve the lives of poor people" (Schwittay 2014) or as "generative of new ways of addressing poverty that present both continuities and ruptures with previous development regimes" (Schwittay 2014) According to Schwittay:

> Rather than presuming to know what people need, humanitarian designers wonder if they are even asking the right questions. They seek to learn what people value through collaborative processes that account for power and knowledge dynamics, and that at their best embrace indigenous and collective ways of knowing and living. Once problems have been unearthed, they are often redefined in ways that align with an understanding of how differently-situated people are affected by design interventions.

While innovation in business systems focuses on products whose design favour optimization of costs and establishment of longer-term profit, "effective humanitarian logistics are means-centered; operating according to a shared moral code, in pursuit of strategic obsolescence, and within maximum uncertainty" (Mays et al.). Humanitarian design then, seeking to aid populations affected by crisis in contexts of uncertainty, also encompasses multiple stakeholders. Nielson has recently introduced the concept of "agenda spacing mapping": demonstrating it is "the 'needs' or agendas of multiple stakeholders that determine the impact and selection of a design" to be used in humanitarian action (Nielsen 2017).

In line with humanitarian design, value sensitive design incorporates attention to values throughout the design process. To accomplish this, it incorporates the following steps (Himma and Tavani 2008):

- "Start with a value, technology or context of use
- Identify direct and indirect stakeholders
- Identify benefits and harms for each stakeholder group
- Map benefits onto corresponding values
- Conduct a conceptual investigation of key issues
- Identify potential value conflicts
- Integrate value considerations into one's organizational structure"

Drawbacks of value sensitive design include ambiguity in identifying stakeholders and the underpinning ethical theory that will ground the evaluation of ethical trade-offs (Manders-Huits 2011). Applying humanitarian principles and other ethical commitments of aid agencies and their innovation partners, along with associated codes of conduct and ethical frameworks, can serve to ground the ethical analysis that value sensitive design requires. In turn, value sensitive design provides a structured approach that can complement existing attempts to articulate values for humanitarian innovation and enhance their operationalization across the innovation cycle.

7.2 VSHI Framework

The VSHI framework acts as a tool to aid in the design, development, and implementation of emergent ICTs in humanitarian settings, and in ways that are responsive to relevant ethical commitments. The framework supports humanitarian innovators and their partners to flag, track, and account for ethical considerations associated with an ICT innovation. To do so, it identifies ethical questions pertinent to each phase of the innovation cycle. Before examining these questions, however, the framework directs attention to context and values, including contextualizing the innovation being examined, identifying values and ethical commitments of ICT developers, humanitarian organizations, and crisis-affected communities, as well as the collaborative environment amongst partners. Depending on the particular innovation and its context of application, additional stakeholders may also be engaged in the VSHI process (e.g. governmental or intergovernmental organizations). The VSHI framework (see Fig. 7.2) may be applied to technologies that are either adapted for humanitarian use or *de novo* applications created to address an identified humanitarian need.

Before outlining the stages of the VSHI, we will briefly introduce the topic of refugee biometrics. We then use this example to illustrate how the VSHI might function to examine a humanitarian ICT innovation.

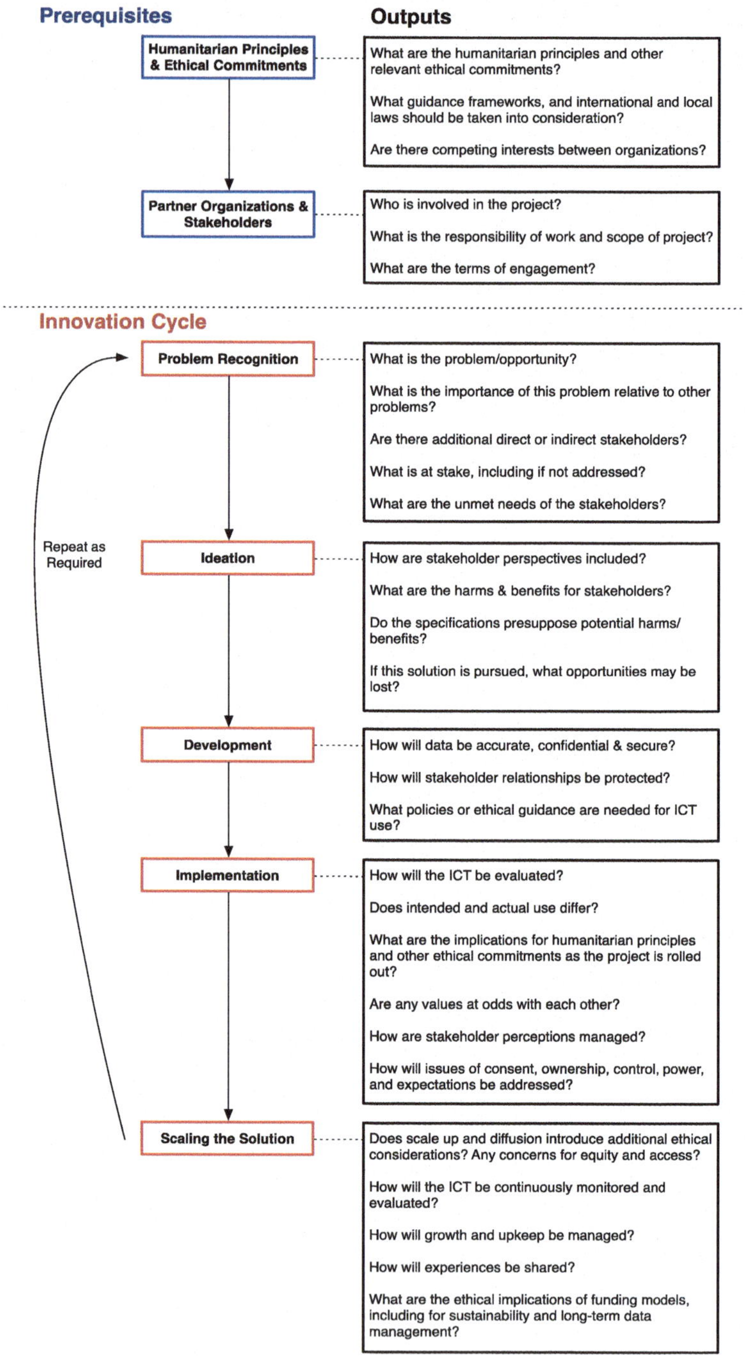

Fig. 7.2 Framework for applying value-sensitive design to innovation in humanitarian action

7.2.1 *Illustrative Example: Refugee Biometrics*

Generally speaking, biometrics encompass "biological or physiological characteristics which can be used for automatic recognition, or it refers to the automated process of recognizing individuals based on such characteristics" (Lodinova 2016). In a humanitarian context, the inability to prove one's identity limits access to basic services and rights, and is most likely to be experienced by the most vulnerable members of a population, including displaced persons and refugees (Group 2016). Biometrics thus presents a method for providing individuals with international identification and access to services (Reyes 2016). It is also used for functions such as financial transactions and medical records.

Applying biometrics in a humanitarian context has been described as leading to a new category of persons: the digital refugee who is "a refugee whose safety has become inseparable from a now digitalized body part (e.g. iris patterns) that has been made 'machine-readable' (Van der Ploeg 2005), and thus appears in a form that humanitarian actors trust a biometric machine to make authoritative judgement about" (Jacobsen 2014). In contrast to other forms of authentication, such as physical- or knowledge-based tokens (e.g. passports, passwords), once collected, biometric measurements cannot be "revoked, changed or reissued" (Abernathy and Tien 2017).

Biometrics represent a strong candidate for analysis given the variety of ethical considerations associated with its development, implementation, and current societal interest. Our framework can be applied to a fictional scenario examining both point-of-use instances (such as iris scanning for food allowance (Staton 2016)) and the use of global databases (such as the Biometric Identity Management System (United Nations High Commissioner for Refugees 2015)).

In the analysis that follows, we discuss biometrics as an illustrative example: trialing of biometrics software by a major intergovernmental organization for allocation of food allowances in a refugee camp. The software in question was built by a software vendor using technology stemming from military-backed research. A local non-governmental organization and the software vendor will be partners in testing the solution for deployment in a newly established refugee camp.

7.3 Prerequisites

A. *Humanitarian principles and ethical commitments*

Humanitarian action aims to save lives, alleviate suffering and promote dignity of people affected by crises. The provision of humanitarian action, and its foundations in international humanitarian law, are closely tied to humanitarian principles which function as a "cornerstone of aid effectiveness" (DARA 2014) and "guide the work of relief agencies in conflict" (Mackintosh 2000). ICT innovations can be examined in the light of these principles (Sandvik et al. 2017). For example, the following considerations for upholding specific humanitarian principles have been identified in regard to big data-driven crisis analytics, such as mapping crisis response during a natural disaster (Ali 2016):

- **Humanity**: Data collected from or about populations affected by humanitarian crises require a restrictive data policy based on a 'need-to-know' approach
- **Neutrality**: Data should be collected in ways that will not be perceived as taking sides in a conflict or in situations of political instability or tension
- **Impartiality**: Data collection should be designed in an unbiased manner and be sensitive to the needs of marginalized groups
- **Independence**: Humanitarian data collection should not be used to advance political, economic, military or other objectives

Value-sensitive design necessitates explicit delineation of values. While humanitarian principles serve as overarching points of reference, humanitarian principles are not the only relevant values. It is important for those involved in an innovation process to identify additional ethical commitments relevant to the project. To do so, they may turn to articulations of values that are specific to the organizations involved (i.e. values identified in an organization's mission statement) or that have been identified across humanitarian organizations (e.g. the Humanitarian Charter of the Sphere Project or the NGO/Red Cross Code of Conduct). Wider sets of normative considerations may also be considered such as the medical ethics principles of justice, non-maleficence (do no harm) and beneficence (do good), and respect for autonomy (these values will be particularly relevant for humanitarian medical organizations), and risk mitigation strategies, such as collecting data on a 'need-to-know' basis (Dette and Streets 2017).

Expressly framing an innovation project in terms of humanitarian principles and partners' ethical commitments is important for foregrounding these values, and so that all partners have confidence in the project outlook. It also clarifies that the responsibility to attend to values is not the responsibility of funding or operational organizations only, but of all involved in humanitarian innovation (Betts and Bloom 2014). As partners may have different commitments (e.g. among humanitarian organizations, for-profit companies and networks of volunteers) or interpret and prioritize common values differently (e.g. how the humanitarian principles of neutrality or impartiality are operationalized), value conflicts may arise (these aspects are discussed further below). Beginning the innovation process by highlighting these commitments as design considerations will provide opportunities for discussion and exchange.

The importance of delineating relevant values at the outset is due to their impact on the entire design process. As noted in a 2015 review of ICTs, "application of core humanitarian principles and values to the potential operational uses of the technologies with affected populations, and an analysis of their possible impacts on affected populations and humanitarian actors, should be the preceding requirement of ICT-supported humanitarian action" (Raymond and Card 2015). At this stage, stakeholders to a refugee biometrics innovation process would consider the interplay between their planned project and humanitarian principles and other ethical commitments. For example, they would examine concerns for upholding impartiality (collection of data is not biased and is sensitive to the needs of vulnerable groups) and independence (autonomous from political, economic or military objectives), as well as how to avoid harm and promote (or, at least, not undermine) dignity of refugees. They would also need to map national and international legal standards for management of confidential data and standards of information management for refugee populations.

Framework Outputs
Identify humanitarian principles and relevant ethical commitments
Identify relevant guidance frameworks, and international and local law
Identify potential value conflicts

B. *Partner Organizations & Stakeholders*

The humanitarian innovation ecosystem involves many actors, including for-profit organizations, local and international non-governmental organizations, inter-governmental organizations, governments, philanthropic foundations, networks of volunteers, militaries, researchers, and populations affected by crises (Betts and Bloom 2014). Multiple actors may partner in an ICT innovation process and it is common for them to differ in their values, priorities, desired outcomes, and preferred implementation strategies. Funders and funding models can also significantly influence how project aims and outputs are understood. Such diversity can present opportunities for complementarity of strengths, but also present the potential for value conflicts that will impact stakeholders.

Given differences of outlook between humanitarian actors and their innovation partners, terms of engagement are needed (Leader and Macrae 2000). Such agreements should be put in place prior to beginning a project. The content of the agreement would be tailored to the innovation, context and partners involved. It is important to address such terms of engagement early on (e.g. addressing responsibilities regarding data), especially as data collection can already take place as product, legal, intellectual, and business objectives are designed (not just when the final ICT is 'launched'). As the private sector engages and forms deeper connections with humanitarian actors, agreements should also consider the concept of 'data philanthropy' (donation of data from both individuals and for-profit companies) and how this may impact impartiality (Taddeo 2016). In the illustrative example, stakeholders include refugees, local communities, the software vendor, and inter and non-governmental organizations. The scope of the project is to field test a computerized iris scanning biometric software for allocation of food allowances by the intergovernmental organization. The partners would then develop and document specific terms of engagement and collaboration, and clarify the roles and responsibilities.

Framework Outputs
Outline responsibility of work & scope for project
Document terms of engagement/collaboration
Formalize project partnership agreement

1. **Problem Recognition**

Successful innovation, whether applied to a market, government or society, requires a thorough understanding of the problem and opportunity, as well as the needs of stakeholders. Innovation in humanitarian action broadly follows a similar innovation pathway to other settings; however, the stakes associated with innovation are elevated given the involvement of populations affected by war or disaster, and the context of innovation in acute, chronic or recurrent crisis settings. Thus, careful attention to problem recognition is critical to guide all subsequent development and, in particular, who participates in identifying and defining problems.

When considering problem recognition, what may come to mind is an emergency solution, such as a sanitizable EMR system during an Ebola epidemic (Jobanputra et al. 2016). However, innovation can also lead to a stage-wise process improvement, such as decreasing costs in aid delivery via mobile phone banking (Kikulwe et al. 2014). What should be consistent in both problem and opportunity recognition is the care taken to consider the ethical considerations associated with the proposed solution.

Specifically related to ICTs, it is also necessary to beware of what is *not* the solution. Technological solutionism, or an exaggerated 'techno-optimism', is the act of looking for places to add technology (a solution in search of a problem) or assuming technology is the sole solution (Sandvik 2017). As political problems are commonly the root of humanitarian crises, solutions are also likely to be political – even if they involve technology (Leader and Macrae 2000). An ICT may not be the most suitable response in a given context or to a particular problem, and it is important to consider alternatives to ICTs and what might be displaced by focusing on technological solutions. Therefore, Innovators should "study the context before choosing the tools: understand who influences and spreads information and can impact it" (Dette and Streets 2017). ICTs can also "have a self-reinforcing logic whereby one set of technologies (for example, information gathering) leads to another set of technologies (for example, information processing). This becomes potentially problematic if technologies become naturalised and mainstreamed to the extent that they are not subject to fundamental questioning, or they exclude other methodologies" (Read et al. 2016).

Value-sensitive design stipulates identifying both direct (such as the local population and humanitarian practitioners) and indirect stakeholders (e.g., funding agencies). Stakeholder identification is followed by the development of a detailed list of risks and potential benefits. Such a list can be used to track harms and benefits throughout the innovation cycle and be updated as the project unfolds.

A commonly identified problem within humanitarian action is "a longstanding and unjustifiable lack of engagement with recipients of aid" (Ramalingam et al. 2015). Lack of engagement will severely limit understanding of the context of the problem or solution that is being developed. Building relationships among stakeholders that will be strengthened throughout the stages of the innovation cycle will help keep benefits and harms for stakeholders at the forefront of the design considerations.

Unlike innovation in some other sectors, humanitarian action has particularly pronounced and "inherent power asymmetries between those providing protection or aid and those in need of that assistance" (Betts and Bloom 2014). People in need of assistance may also have few other options for getting help that they need. In this context, they may have little choice but to agree to participate in biometric identification to access food allocation in a refugee camp. With this in mind, ICT innovation should consider unmet needs of the local population and be cognizant of constraints on autonomy, and look for opportunities to support meaningful engagement and participation from the outset of the innovation process.

How affected populations will be included in the project development should be considered early on and be formalized. What are their rights? How will they be engaged? What will permission or consent look like? A refugee from Myanmar reported the following dynamic during the rollout of a biometrics program in 2006: "I don't know what it is for, but I do what UNHCR wants me to do" (Ismail and McKinsey 2006). This description highlights challenges for the agency of people caught up in a humanitarian crisis. Planning the innovation project in ways that support robust consent processes and meaningful participation of stakeholders, especially end users, is important (Betts and Bloom 2014; Sheather et al. 2016). As evidenced in a recent privacy affront, biometrics data collection in Xinjiang, China occurred under the guise of medical physicals, but also served to bank population biodata (Haas 2017). Such situations reinforce the importance of transparency and engagement with communities. The problem that the developers of a refugee biometric system seek to address is that physical tokens for receiving food allowances can be stolen or lost, leading to allocation inefficiencies and potentially to system abuse (affecting equitable distribution and leading to higher costs). This problem is made more acute due to the context of insecurity, strained trust and high mobility. In this setting, potential challenges include (but are not limited to): insecure and inequitable resource allocation, being forced to protect physical tokens in an environment with few resources, and encountering delays or experiencing deprivation when physical tokens are lost. As the partners involved in the innovation process clarify the nature of the problem in relation to the unmet needs of the local population, they will be better placed to consider potential solutions.

Framework Outputs

Determine the problem or opportunity

Contextualize the problem in terms of other issues and what is at stake (including if not addressed)

Identify direct & indirect stakeholders

Outline the unmet needs of populations affected by crises and other stakeholders

2. **Ideation**

Ideation in humanitarian innovation can be split into *de novo* inventions or adaptation of existing technology (Warner 2017). The contexts of humanitarian innovation development has also been categorized as grants and finance, research and development, and collaborations and networks (Betts and Bloom 2014). It is important to recognize that different innovation types and contexts will raise a partially overlapping set of ethical questions. In the ideation phase, innovators develop the project's scope of what will be included and what will not. This process sets boundaries on the extent of the solution and, by association, guides how ethical considerations will be examined and which ones will be particularly salient.

While a project's scope sets the parameters for ethical analysis, technical or design specifications guide the focus toward particular values. As an example, consider an Electronic Health Records (EHR) app where usability considerations (such as the use of optical character recognition to read and store data from a photo ID card) are given priority over 'need-to-know' (only collecting the ID number). These usability specifications imply greater risk of harm were there to be a data breach. Such specifications should therefore be given careful consideration in regard to the harms and benefits they present, as well as questions of consent (what information the person gives permission to collect) and accountability (who should provide oversight for such decisions). The importance of doing so is underlined by Read et al.*'s* description of the implications of increased data collection and processing power:

> Our analysis suggests that declarations of emancipation via a data revolution are premature. There is a danger that much of what we see is the same information being processed more quickly. Content analysis of data, even if that data is collected by local actors on the ground, is rarely conducted in local languages. The data revolution risks reinforcing technocratic specialists who are often based in headquarters. Greater connectivity has produced greater demands from humanitarians to support their own connectivity. (Read et al. 2016)

At the ideation stage, a range of avenues exist for engaging with communities in areas where the ICT is likely to be deployed. This can include working directly with communities through a 'bottom-up' approach that understands local populations to be integral contributors to the innovation process. In this way, innovation activities are more likely to be relevant to and accepted by local communities, contributing to self-reliance, better understanding of the problem, and greater accountability (Betts et al. 2013). As is the case for any software development project, primary stakeholders are best positioned to predict and understand how new technologies will impact them when they are involved in the design process.

The ideation stage should also include non-traditional design metrics, such as cultural, religious, and social beliefs of the local population. A 2016 review on the impact of technology on health of populations affected by humanitarian crises found that "a lack of evaluation of these technologies, a paternalistic approach to their development, and issues of privacy and equity constituted major challenges" (Mesmar et al. 2016). A paternalistic approach misses the mark on being sensitive to the needs and context of the local population and their values. Acknowledging that many populations are heterogenous with regard to culture, religion and normative commitments, VSHI guides its users to carefully consider how an ICT implementation can best be aligned with the (diverse) values that are important to local communities.

When considering the ideation of a refugee biometrics application, the project scope should clarify the uses of the biometric information and any limits on its use (i.e. who can use the data and for what purposes). Also relevant for consideration is whether focusing on the biometrics application will displace attention from another more promising approach. To ensure stakeholder perspectives are heard and an inclusive approach is adopted, the project team could also establish a system of feedback mechanisms and actively solicit user experience, including the impact of biometrics on the everyday life of refugees. Careful attention to potential harms and benefits is essential. They might consider features such as gender inequalities, potential for harms due to data breach, impact on relationships between humanitarian workers and refugees, and social and cultural considerations specific to the refugee communities.

Framework Outputs
Delineate the project scope and what is not included
Outline how stakeholder perspectives are included
Examine alignment with and challenges to cultural, religious, and social beliefs
Consider how the specifications presuppose potential for stakeholder harms and benefits
Start a list of potential harms & benefits to stakeholders

3. **Development**

Currently, there exists a 'doctrine gap' in humanitarian action, where "technology adoption by humanitarian actors [happens] prior to the creation of standards for how and how not to apply a specific tool" (Raymond and Harity 2017). As tools, the spectrum of proper to improper use of ICTs completely changes the ethical considerations. Policies (such as terms of service or privacy policy) and codes of conduct should thus be considered a core component of an ICT and be created with the goal of minimizing risk of harm to local populations.

ICT development should make use of more general principles, such as 'do no harm' and collecting data on a 'need-to-know' basis (Dette and Streets 2017). Like the prototypical values sensitive design approach 'privacy-by-design', inherently including these principles as ethical guidelines will reduce the downstream potential for value conflicts. Identifying actionable information for decision-making and only recording what you need to know mitigates harm in over-collection of data. Collected data must also be accurate, confidential, and secure. Minimum technical standards (encryption, backups, failovers) should be incorporated with relevant legal standards, such as the EU's General Data Protection Regulation (GDPR). As a regulation and not a directive, GDPR enshrines rights of data subjects, including the rights of access, rectification, restriction of processing, data portability, objection to certain processing and the right to be forgotten/erasure (Official Journal of the European Union 2016).

Stakeholder relationships and rights ought to be protected in pilot testing. Pilots are experimental in nature: they may result in harm, and any data collected should be responsibly handled for its lifespan. Considerations of individual consent and collective permission should be carefully considered. Transparency is crucial for accountability, and the nature of the pilot should be clearly communicated to local populations. Depending on the features of the pilot, it may constitute research and require research ethics oversight.

A range of considerations should be accounted for at this stage. In this context, fear of information leaks may severely curtail implementation of the project (Mesmar et al. 2016). ICTs also influence relationships with local populations and their trust in humanitarian actors. For example, drones are widely associated with their military use, which can raise suspicion, fear and re-traumatization when autonomous vehicles are used in humanitarian action. In line with these concerns, refugee biometrics require careful and rigorous strategies to ensure data protection, akin to research data protection standards. Attention is also needed toward the influence on relationships. For example, humanitarian innovators should be attentive to refugee perceptions of the technology and how it will affect trust. In carrying out this analysis, relevant codes of conduct should be considered, and a tailored set of guidelines developed to support ethically robust rollout of the program.

Framework Outputs
Detail how data will be accurate, confidential, and secure
Explain how stakeholder relationships will be protected
Identify or develop policies and/or codes of conduct that are needed for ICT use
Update list of harms and benefits to stakeholders

4. **Implementation**

ICT testing in parallel with theoretical use and existing (or alternative) solutions can provide measurement of change produced. A common shortcoming of ICT development is that evaluation models are not integrated within the limited evaluative frameworks that are employed (Mesmar et al. 2016). This gap leads to a lack of evidence-based decision making. Similarly, evaluation methodologies can be used to identify inequalities or bias, such as gender gaps in access to and use of ICTs (Anderson 2013).

Lessons from healthcare technology assessment (HTA) illustrate that "ethical analysis is integral to the whole HTA process as it contributes to how HTA is defined, interpreted, and acted upon. It includes equity and distributional considerations but also all value judgments inherently involved in assessments" (Abrishami et al. 2017). Evaluation methodologies have inherent assumptions about definitions and interpretations of how outcome measures should be applied. These evaluation methods should also receive ethical scrutiny to align their outputs with the values and needs of the stakeholders involved.

The differences between intended and actual use can also illuminate ethical shortcomings. Regardless of the operating framework or plan, consideration should be paid to how the technology is actively being used, and any unintended consequences or unanticipated uses of collected data. Open source technologies and information can be used malignly. For example, a descriptive crisis map showing displaced refugees and where aid is needed most could be exploited by warring factions to harm civilians (Kent 2014).

The importance of avoiding harm (Raymond and Card 2015) should be revisited once again at this stage of the cycle: What harms have arisen as the biometrics program was initiated? How might they be avoided in future? In evaluating the implementation, what lessons are learned? These questions should be asked by the partners involved in the refugee biometrics program. Value conflicts might be identified, such as between flexibility of the system vs potential for system abuse, or refugees' capacity to provide meaningful consent given the lack of other options available to them.

Framework Outputs
Define how the ICT will be evaluated
Examine if intended and actual use differ
List any value conflicts
Outline how stakeholder perceptions are managed
Outline how issues of consent, ownership, control, power and expectations are addressed
Update list of harms and benefits to stakeholders

5. **Diffusion**

Adapting an ICT for a new population can present unexpected challenges. Diffusion should be approached similarly to the innovation cycle as a whole: iteratively, with careful rollout planning and extensive monitoring. In preparation for diffusion, research or evaluations should prove innovation was successful, identifying key lessons and what can be applied elsewhere (Warner 2017). This approach is needed for successful diffusion, but also for ethical responsiveness: did harms arise? What benefits were produced and for whom? Are they likely to occur elsewhere? In the 2014–15 Ebola outbreak in West Africa, "effectiveness and uptake of delivered services [were] negatively affected by users' perceptions" and "political factors and poor communication between humanitarian workers and beneficiaries" were cited as failings to be avoided (Colombo and Pavignani 2017). Such analysis is essential in order to improve future ICT implementation in the same context, or elsewhere.

A further concern for diffusion is funding. Humanitarian health funding requires sustainable, long-term investments to counteract annual provisioning models (Spiegel 2017). The concern about sustained investment for humanitarian innovations is highlighted by Neilsen and colleagues who suggest that the "humanitarian market is unable to absorb and harness innovation due to a short-term focus" (Nielsen et al. 2017). Diffusion, and innovation more broadly, should be thought of in a long-term perspective with evaluation strategies that make use of data collected longitudinally. There is the potential for harm to local populations if solutions are implemented without adequate funding or if ICTs are unmaintained and therefore become candidates for abuse by third parties. In this way, attention is needed regarding the role of dedicated funding in sustaining an innovation initiative. To overcome this barrier, new models of collaboration may be needed. Project independence may be enhanced when resources and risks are pooled across multiple organizations (Dette and Streets 2017). As funding might be multi-staged, depending on the trajectory of the project, terms of engagement for these collaborations should be addressed proactively so that different funding rounds do not modify the initial engagement terms and lead to loss of independence.

When scaling a solution, Elrha recommends measured and valid outcomes that are ranked in importance by the affected population (Oleskog 2017). As is common in other aspects of humanitarian action, temporary solutions can often become permanent. Technology cannot exist in a vacuum devoid of proper oversight – considerations need to be given to how upkeep and any growth will be managed, and if these new developments will continue to be aligned with the values and needs of the local population. Ever a transient environment, ICT endpoints should also be applied to humanitarian action when the solution is no longer needed (and before it becomes harmful by not adapting to the new environment).

As mentioned in the prerequisite stage, diffusion of an ICT should still reference the partnership and original terms of engagement. Collected data can be subject to secondary uses, raising additional ethical concerns (Kaplan and Easton-Calabria 2017); similarly, military-derived technology can "raise questions about costs, lobbying and the framing of political agendas" and, potentially, clashes between military and humanitarian identities and goals (Sandvik and Lohne 2014).

Regardless of their origins, static technology does not fare well: even highly siloed applications typically require routine maintenance and confirmation that security is upheld. Data collected throughout the lifecycle of an ICT must therefore be managed until termination. A particular concern for humanitarian actors may be the interplay between technology developed for military applications and possible humanitarian uses, including concern for both collection and end-use bias, and impacts for humanitarian impartiality and perceptions of neutrality. Our biometrics example could raise such concerns given that the program utilizes technology originally designed for military purposes. This association could raise concerns and undermine trust among some refugees. Attention to data storage and maintenance is also crucial. Once collected, data cannot be returned, and terms of use therefore must outline where the data can(not) go, who has access, when it is to be destroyed and how data ownership will be handled. For example, humanitarian street mapping during the 2015 Nepal earthquake meant that mapping information was no longer solely the property of an aid organization, representing a potential social good if

Framework Outputs

Outline how the ICT will be continuously monitored
Explain how growth and upkeep will be managed
Outline how the experience will be shared
List the ethical implications of funding models
Update list of harms and benefits to stakeholders

disseminated (Read et al. 2016). The delegation of such responsibilities is a part of project planning, and should be clearly defined for the biometrics program.

7.4 Conclusion

Humanitarian organizations continue to expand their use of ICTs. New technology applications provide a range of opportunities to improve the coordination and provision of services to affected communities. They have also led to new stakeholder relationships and have raised ethical issues and risks of harm. This reality raises questions about how humanitarian ICTs can be developed, implemented and evaluated in ways that are attentive to values and ethical commitments. Having tools that foreground the place of values and principles across the innovation cycle will help ICT developers, humanitarian actors, and other innovation partners meet this challenge. Sensitizing ICT development and implementation to humanitarian principles is a critical consideration for humanitarianism moving forward. It is important to note, however, that value sensitive design is insufficient to ensure ethical innovation. Steep imbalances of power will continue to exist and require careful attention and vigilance on the part of all involved in innovation processes.

In this chapter, we presented a framework for developing and implementing ICTs in the humanitarian setting. Approaching the innovation process with prerequisites rooted in humanitarian principles, ethical commitments, and partner organization collaboration, the framework provides accountability and a structure for clarifying and documenting the ethical considerations associated with an ICT application. Making use of value-sensitive design, these salient ethical considerations are integrated into the design process, not just as a process of retrospective review.

The VSHI framework can guide innovators to be attentive to ethical considerations across the innovation cycle, and to support careful analysis and appraisal of a technology's ethical merits and any sources of concern. In turn, these insights can be used to enhance the ethical robustness of an innovation process. Having humanitarian actors and innovators use and provide feedback on our framework will yield further refinement and new approaches to effective humanitarian action. As presented by Sandvik et al. (2017), humanitarian action must begin "building systems that actualize and operationalize our values" – it is our hope that VSHI can be incorporated into this approach and support principles-driven and value-sensitive development and use of information and communication technologies in humanitarian settings.

References

Abernathy, William, and Lee Tien. 2017. *Biometrics: Who's watching you?* December 13.

Abrishami, P., W. Oortwijn, and B. Hofmann. 2017. Ethics in HTA: Examining the "need for expansion". *International Journal of Health Policy and Management* 144: 212–217. https://doi.org/10.1016/j.sbspro.2014.07.289.

Ali, Anwaar. 2016. Crisis analytics: Big data driven crisis response. *Journal of International Humanitarian Action* 1: 1–21. https://doi.org/10.1186/s41018-016-0013-9.
Anderson, Jessica. 2013. Policy report on UNHCR's community technology access program: Best practices and lessons learned. *Canada's Journal on Refugees* 29: 21–30.
Betts, Alexander, and Louise Bloom. 2014. Humanitarian innovation: The state of the art. *United Nations Office for the Coordination of Humanitarian Affairs*, OCHA policy and studies series, 1–30.
Betts, Alexander, Louise Bloom, University of Oxford. Refugee Studies Centre. 2013. *The two worlds of humanitarian innovation*.
Colombo, Sandro, and Enrico Pavignani. 2017. Recurrent failings of medical humanitarianism: Intractable, ignored, or just exaggerated? *Lancet* (London, England) 390: 2314–2324. https://doi.org/10.1016/S0140-6736(17)31277-1.
DARA. 2014. *Now or never: Making humanitarian aid more effective*, 1–13.
Dette, Rahel, and Julia Streets. 2017. *Innovating for access: The role of technology in monitoring aid in highly insecure environments*, October 14.
Friedman, Batya. 2004. Value sensitive design. In *Encyclopedia of human-computer interaction*, 769–774. Great Barrington: Berkshire Publishing Group.
Group, World Bank. 2016. *Identification for development: Strategic development*.
Haas, Benjamin. 2017. Chinese authorities collecting DNA from all residents of Xinjiang. *The Guardian*, December 13.
Himma, Kenneth E., and Herman T. Tavani. 2008. *The handbook of information and computer ethics*. Hoboken: Wiley.
Hunt, Matthew, John Pringle, Markus Christen, Lisa Eckenwiler, Lisa Schwartz, and Anushree Davé. 2016. Ethics of emergent information and communication technology applications in humanitarian medical assistance. *International Health* 8: 239–245. https://doi.org/10.1093/inthealth/ihw028.
Ismail, Yante, and Kitty McKinsey. 2006. *Fingerprints mark new direction in refugee registration*, November 30.
Jacobsen, Katja Lindskov. 2014. Experimentation in humanitarian locations: UNHCR and biometric registration of Afghan refugees. *Security Dialogue* 46: 144–164. https://doi.org/10.1177/0967010614552545.
Jobanputra, Kiran, Jane Greig, Ganesh Shankar, Eric Perakslis, Ronald Kremer, Jay Achar, and Ivan Gayton. 2016. Electronic medical records in humanitarian emergencies – The development of an Ebola clinical information and patient management system. *F1000Research* 5: 1477. https://doi.org/10.12688/f1000research.8287.1.
Kaplan, Josiah, and Evan Easton-Calabria. 2017. *Military actors and humanitarian innovation: Questions, risks and opportunities*, October 14.
Kent, Randolph. 2014. *Positive and negative noise in humanitarian action: The open source intelligence dimension. Open source intelligence in the twenty-first century*. London: Palgrave Macmillan UK. https://doi.org/10.1057/9781137353320_7.
Kikulwe, Enoch M., Elisabeth Fischer, and Matin Qaim. 2014. Mobile money, smallholder farmers, and household welfare in Kenya. Edited by Ajay Mahal. *PLoS ONE* 9: e109804–13. https://doi.org/10.1371/journal.pone.0109804.
Leader, N., and J. Macrae. 2000. *Terms of engagement: Conditions and conditionality in humanitarian action*.
Lodinova, Anna. 2016. Application of biometrics as a means of refugee registration: Focusing on UNHCR's strategy. *Development, Environment and Foresight Journal* 2 (2): 1–10.
Mackintosh, Kate. 2000. *Principles of humanitarian action in international humanitarian law: Study 4 in the politics of principle: The principles of humanitarian action in practice*. bases.bireme.br
Manders-Huits, Noëmi. 2011. What values in design? The challenge of incorporating moral values into design. *Science and Engineering Ethics* 17: 271–287. https://doi.org/10.1007/s11948-010-9198-2.

Maron, Diane Fine. 2013. Eye-imaging ID unlocks aid Dollars for Syrian Civil War refugees. *Scientific American*, September 18.

Matonga, Doreen, Patsy Nakell, and Georgina Thompson. 2017. *Africa's first humanitarian drone testing corridor launched in Malawi by Government and UNICEF*, August 26.

Mays, Robin E, Robert Racadio, and Mary Kay Gugerty. *Competing constraints: The operational mismatch between business logistics and humanitarian effectiveness*, 132–137. IEEE. https://doi.org/10.1109/GHTC.2012.29.

Mesmar, Sandra, Reem Talhouk, Chaza Akik, Patrick Olivier, Imad H. Elhajj, Shady Elbassuoni, Sarah Armoush, et al. 2016. The impact of digital technology on health of populations affected by humanitarian crises: Recent innovations and current gaps. *Journal of Public Health Policy* 37. Palgrave Macmillan UK: 167–200. https://doi.org/10.1057/s41271-016-0040-1.

Nielsen, Brita Fladvad. 2017. Framing humanitarian action through design thinking: Integrating vulnerable end-users into complex multi-stakeholder systems through "Agenda Space mapping". *Journal of Design Research* 15: 1. https://doi.org/10.1504/jdr.2017.10005345.

Nielsen, Brita Fladvad, Kristin Bergotora Sandvik, and Maria Gabrielsen Jambert. 2017. *How can innovation deliver humanitarian outcomes?* 1–4.

Obrecht, A., and A.T. Warner. 2016. *More than just luck: Innovation in humanitarian action*. ALNAP.

Official Journal of the European Union. 2016. *Regulation (EU) 2016/679 of the European Parliament and of the Council*.

Oleskog, Nora. 2017. How do you scale humanitarian innovations? A new tool that will help. *Elrha*, October 19.

Project, Humanitarian Innovation. 2015. *Principles for ethical humanitarian innovation*, 1–12.

Ben Ramalingam, Howard Rush, John Bessant, Nick Marshall, Bill Gray, Kurt Hoffman, Simon Bayley, Ian Gray, and Kim Warren. 2015. *Strengthening the humanitarian innovation ecosystem*, 1–52.

Raymond, Nathaniel A., and Brittany L. Card. 2015. *Applying humanitarian principles to current uses of information communication technologies: Gaps in doctrine and challenges to practice*.

Raymond, Nathaniel A., and Casey S. Harity. 2017. *Addressing the "doctrine gap": Professionalising the use of information communication technologies in humanitarian action*, October 14.

Read, Róisín, Bertrand Taithe, and Roger Mac Ginty. 2016. Data hubris? Humanitarian information systems and the mirage of technology. *Third World Quarterly* 37: 1314–1331. https://doi.org/10.1080/01436597.2015.1136208.

Reyes, Duina. 2016. *Identification for development (ID4D)*. The World Bank

Sandvik, Kristin Bergtora. 2017. Now is the time to deliver: Looking for humanitarian innovation's theory of change. *Journal of International Humanitarian Action* 2: 145. https://doi.org/10.1186/s41018-017-0023-2.

Sandvik, Kristin Bergtora, and Kjersti Lohne. 2014. The rise of the humanitarian drone: Giving content to an emerging concept. *Millennium: Journal of International Studies* 43: 145–164. https://doi.org/10.1177/0305829814529470.

Sandvik, Kristin, and Nathaniel Raymond. 2017. Beyond the protective effect: Towards a theory of harm for information communication technologies in mass atrocity response. *Genocide Studies and Prevention* 11: 9–24. https://doi.org/10.5038/1911-9933.11.1.1454.

Sandvik, Kristin Bergtora, Katja Lindskov Jacobsen, and Sean Martin McDonald. 2017. Do no harm: A taxonomy of the challenges of humanitarian experimentation. *International Review of the Red Cross*: 1–26. https://doi.org/10.1017/S181638311700042X.

Schwittay, Anke. 2014. Designing development: Humanitarian design in the financial inclusion assemblage. *PoLAR: Political and Legal Anthropology Review* 37: 29–47. https://doi.org/10.1111/plar.12049.

Sheather, Julian, Kiran Jobanputra, Doris Schopper, John Pringle, Sarah Venis, Sidney Wong, and Robin Vincent-Smith. 2016. A Médecins Sans Frontières ethics framework for humanitarian innovation. *PLoS Medicine* 13: e1002111–e1002110. https://doi.org/10.1371/journal.pmed.1002111.

Slim, Hugo. 2015. *Humanitarian ethics: A guide to the morality of aid in war and disaster*. London: Hurst.

Spiegel, Paul B. 2017. The humanitarian system is not just broke, but broken: Recommendations for future humanitarian action. *Lancet* (London, England). https://doi.org/10.1016/S0140-6736(17)31278-3.

Staton, Bethan. 2016. Eye spy: Biometric aid system trials in Jordan. *IRIN*, May 18.

Taddeo, Mariarosaria. 2016. Data philanthropy and the design of the infraethics for information societies. *Philosophical Transactions. Series A, Mathematical, Physical, and Engineering Sciences* 374. https://doi.org/10.1098/rsta.2016.0113.

United Nations High Commissioner for Refugees. 2015. *Biometric identity management system: Enhancing registration and data management*, 1–2.

Van der Ploeg, Irma. 2005. *The machine-readable body: Essays on biometrics and the informatization of the body*. Maastricht, the Netherlands: Shaker Publishing.

Vayena, Effy, Marcel Salathé, Lawrence C. Madoff, and John S. Brownstein. 2015. Ethical challenges of big data in public health. Edited by Philip E Bourne. *PLoS Computational Biology* 11: e1003904–7. https://doi.org/10.1371/journal.pcbi.1003904.

Warner, A.T. 2017. Working paper: Monitoring humanitarian innovation. *HIFALNAP Working Paper*, 1–24.

Part II
Military Human Enhancement: "Science-Fiction" in the Real World

Chapter 8
Military Enhancement: Technologies, Ethics and Operational Issues

Ioana Maria Puscas

8.1 Introduction

On the morning of the 8th of August 2012, US Task Force Mountain Warrior was on a patrol escort mission in the Asadabad District, Kunar Province, Afghanistan. In this mission, U.S. Army Capt. Florent Groberg served as a commander responsible for the security of 28 personnel from the coalition and the Afghan National Army. Their duty was to go and attend a weekly security meeting at the provincial governor's compound, including covering a portion of the distance on foot. En route to the governor's compound, the patrol reached a choke point at a small bridge over a canal to the Kunar River. Approaching the bridge, the patrol stopped as they saw two motorcyclists heading towards them from the opposite direction, crossing the bridge. Midway, the motorcyclists stopped and turned back, but at that point Capt. Groberg also spotted a lone individual walking in the direction of the patrol. Although there were other civilians in the area, the individual appeared suspicious to Capt. Groberg, and as soon as he made an abrupt turn, Capt. Groberg rushed to him, pushed him away and instantly observed he was wearing a suicide vest. With the help of a fellow soldier, Capt. Groberg grabbed the bomber, and pushed him to the ground, at which point his vest detonated. This caused another suicide bomber, who had stayed hidden all this time, to detonate his vest prematurely. While this ambuscade resulted in four casualties, Capt. Groberg's intervention prevented a more dramatic turn of events

I. M. Puscas (✉)
Geneva Center for Security Policy (GCSP), Geneva, Switzerland
e-mail: ioana-maria.puscas@graduateinstitute.ch

D. Messelken, D. Winkler (eds.), *Ethics of Medical Innovation, Experimentation, and Enhancement in Military and Humanitarian Contexts*, Military and Humanitarian Health Ethics,
https://doi.org/10.1007/978-3-030-36319-2_8

and saved many lives. In recognition for his acts of bravery, medically-retired Capt. Groberg received the Medal of Honour in November 2015.[1]

The age of military human enhancements will bring about changes to military values and confront us with new questions about what courage, resilience, merit and sacrifice mean. Let us imagine a similar script for an enhanced soldier of the near future, who, while having accomplished similar acts of bravery as Capt. Groberg, will have acted while under a type of enhancement, or a combination of enhancement procedures: with the help of transcranial direct current stimulation, his cognitive abilities would have been enhanced for heightened vigilance on the battlefield, despite exhaustion or severe sleep deprivation. The soldier's capacity to detect threats would have been enhanced with electrical stimulation of the brain. Tiny implantable devices, monitoring and controlling nerve activity, could have neuromodulated peripheral nerves, stimulated the immune system, reduced anxiety, and provided quick relief from pain. This may sound purely hypothetical but it is not: such solutions are envisaged as part of DARPA's Electrical Prescriptions (ElectRx) program and "could be automatically and continuously tuned to the needs of warfighters without side effects".[2]

The benefits of these interventions have a psychological effect too, not just a physical one. For example, the soldier would be emboldened to take extreme risks incurring severe physical injury, knowing that even in case of extreme blood loss (which is the number one cause of death in the military), they could be aided to survive past the "golden first hour"[3] and potentially enter a state of suspended animation whereby their biological functions slowed down. DARPA's Biostatis program aims to do just that: "slow life to save life".[4] The five-year research program essentially investigates modalities to control "the speed at which living systems operate" and extend the window of intervention. More specifically, the program investigates biochemical ways of controlling cellular functions at the protein level and the inspiration for this comes from nature, where animals such as wood frogs exhibit what is known as "cryptobiosis": a state in which the metabolic state is slowed down and reduced to an almost undetectable level, yet not altogether, thus allowing life to continue. If the same physiological mechanisms could be replicated in human soldiers, it would reduce significantly the risks of loss of life in the minutes or hour following an incident.

Would the enhanced soldier be less deserving of recognition for their conduct if their endurance and courage were enhanced by technology or drugs? How is that

---

[1] Captain Florent Groberg, US Army, 2015, https://www.army.mil/medalofhonor/groberg/. Accessed 30 September 2019.

[2] Dr. Eric Van Gieson, "Electrical Prescriptions (ElectRx)", DARPA, https://www.darpa.mil/program/electrical-prescriptions. Accessed 30 September 2019.

[3] The US Department of Defense refers to the critical window of time for reaction as the "golden hour" but very often, the necessary time is much more limited.

[4] "Slowing Biological Time to Extend the Golden Hour for Lifesaving Treatment", DARPA, March 1st 2018, https://www.darpa.mil/news-events/2018-03-01. Accessed 30 September 2019.

qualitatively different from the sophisticated technology that many troops already use, which places them far ahead in terms of capabilities compared to their enemies?

Such questions merely scrap the surface of what promises to be a challenging and re-defining era in military ethics. To better engage with some of these questions, it is firstly critical to define human enhancement – and understand what it means in a military context. The chapter then proceeds to provide examples of enhancement technologies and account for their evolution in a historical continuum. The following sections delve more explicitly into some of the ethical conundrums expected from soldier enhancement – with the important caveat that in the limited space of a chapter, this coverage is not exhaustive.

8.2 Human Enhancement in the Military

8.2.1 Definition

The most concise definition of military human enhancement was encapsulated in the description of a DARPA program entitled "Targeted Neuroplasticity Training". This program aims to facilitate learning through precise activation of peripheral nerves and by strengthening neuronal connections in the brain, and it explicitly and clearly distinguishes such efforts from medical treatment: "unlike many of DARPA's previous neuroscience and neurotechnology endeavors, it will aim not just to restore lost function but *to advance capabilities beyond normal levels*".[5]

Human enhancement refers to the suite of interventions that endow individuals with physical and cognitive abilities that *go beyond the statistically normal range of functions for humans* (Dinniss and Kleffner 2016, 434). Conceptually, enhancement is opposed to treatment, preventive and restorative medicine (although this distinction has its limitations). Enhancements are therefore augmentations to human functions beyond what is necessary to sustain or restore health, and thus, they do not respond to legitimate medical needs (Juengst 1998, 29–31).

8.2.2 Controversies

The definition typically takes "normal functioning" as a benchmark against which to assess if an intervention is treatment or enhancement: bringing lost or damaged functions back to "normal" is treatment, whereas taking the point of departure at "normal functioning" and boosting physiological or cognitive functions above that line is enhancement.

---

[5] Boosting Synaptic Plasticity to Accelerate Learning, DARPA press release, 16 March 2016, https://www.darpa.mil/news-events/2016-03-16. Accessed 30 September 2019. Emphasis added.

However, what qualifies as statistically "normal" for our species is not always clear-cut, even in medical terms, which is why, looking at individual physiology is more appropriate. Enhancement, by that logic, would be any kind of *improvement to a healthy individual, which provides them with physical or cognitive skills above anything that treatment or training alone could achieve*. Even so, the concept leaves some room for ambiguity. What is qualitatively different between a soldier whose brain has been zapped with electricity for enhanced ability to detect threats versus a soldier 'merely trained' with Virtual Reality technology? The latter has also been demonstrated to augment perceptual abilities (Wright 2014, 2). If the effect of an enhancement is comparable to that of training or equipment, then what makes enhancement stand out at all? This question will, however, only apply to those situations where the effects of enhancement vs. training are comparable. When the effects of a procedure or pharmacological agent on the cognitive and physical skills of the soldiers are radical, then enhancement is self-evident. However, before a radical alteration of human physiology or cognitive skills has been achieved, most attempts to clarify the concept will be fraught with some difficulties due to the possible parallel that can be drawn to treatment.

Patrick Lin et al. propose a set of dichotomies to better explore the concept of human enhancement as well as identify the difficulties in pinning it down:

1. "natural vs. artificial": the idea that we can easily draw a distinction between activities that can be considered "natural" (meaning activities that humans supposedly have always done throughout history) or "artificial" (those activities that take us beyond 'natural limitations' of human functioning) is simply not tenable. Here, the temptation may be to easily slip into extreme categorizations: either consider everything created by humans as 'natural', which would also include any form of modern 'enhancement', or consider everything created by humans as artificial and a form of enhancement – this means that the definition becomes so broad it is finally all-encompassing and therefore meaningless (Lin et al. 2013, 12).
2. "external vs. internal": the distinction between an external tool or technology versus an internal one provides a great qualitative difference and could create a line between enhancement and 'mere tools'. The fundamental added value of a technology or device implanted in the body is that it delivers an *always-on* access to that tool. For example, a smartphone connected to the Internet is just a tool that can be located in one's hand, or on a piece of furniture etc. A computer chip implanted in the head, which would allow the same features as a smartphone is an enhancement precisely due to its unlimited accessibility and comparative advantage. This proximity to the device can create a difference in degree and kind. A person looking up information on the internet via an implanted "Google chip" is at a far greater comparative advantage to a person searching that same information on a laptop, not because the person becomes smarter, but because of the unlimited access (ibid, 13). In a military context, we can imagine such an advantage to be further useful in that the enhancement of the soldiers could be entirely concealed and thus further confuse and disorient the enemy.

Of course, the notion of "always-on" accessibility comes with a host of nuances. As the authors point out, some tools or technologies may be close to the body and be considered as either enhancements or tools depending on their practical use. Exoskeletons are tools (and not enhancements) not only because they are worn outside of the body, but also because they are bulky structures and difficult to wear and remove. Should exoskeletons become significantly more lightweight and unobtrusive, they could perhaps be considered enhancements. This demonstrates that some nuances are critical in the debate: a miniaturized technology that can be integrated within the body or attached to it, and become somehow intimately connected to the wearer, or become part of their identity, is more likely to be considered 'enhancement' – even though a similar structure in a larger format may already exist and is now considered a 'tool'.

While the inside/outside distinction can be generally useful in establishing the line between enhancement and tools, it is, however, not clear-cut. Furthermore, as the authors notice, the inside/outside distinction does not appropriately account for dual-use technologies that are internal-only. The use of anabolic steroids is a pertinent example: this is a case of a pharmacological intervention that is always internal to the body and which can serve for treatment, for patients with muscular dystrophy, but it can also be taken by healthy athletes.

3. "enhancement vs. therapy" is the most common distinction attempted to delimit enhancement. The difference between the two is on many occasions very clear, as described in the definition of enhancement provided above. This distinction also accounts for the dual-use aspect highlighted above, in the example of anabolic steroids, and is in line with Eric Juengst's definition: an enhancement does not respond to legitimate medical needs: while the muscular dystrophy patient would be justifiably entitled to the dose of steroids, the athlete would not be as there is no medical necessity involved and his/her motivations would serve other purposes (ibid, 14). The distinction between enhancement and therapy is usually sensible and clear, although it runs into some theoretical and philosophical quagmires, the most critical of which is the meaning of 'normal' and 'species-specific'.[6] Enhancement usually means to go beyond the normal limits of our species but what is normal may not always be self-evident, not to mention that medical understandings of normal and 'healthy', 'disease' evolve in time.

Nick Bostrom and Anders Sandberg (2009) demonstrate some of the practical challenges of differentiating treatment from enhancement. A person with *naturally poor* memory may receive an enhancement that at the end, still leaves them with a memory that is below the level of that of a patient with early-stage Alzheimer's disease. Technically speaking, the distinction between enhancement and treatment

[6] The distinction treatment/enhancement very often looks into whether vaccinations fall into the first or second category. Vaccinations, to the extent that they aim to prevent a diseased condition, are not considered enhancements, although, again, the distinction may be more controversial at times.

is useful most of the times but it lends itself to an individual, case-by-case assessment. Indeed, a critical reminder in the debate on human enhancement is that context is everything.

However, while it is important to flag these 'conceptual gray areas', the remainder of the chapter will deal more concretely with the impacts of enhancement in the military and will focus on examples of enhancement technologies, which do not reside in the blurry lines between enhancement/therapy, but are rather evident cases of enhancement.

8.3 Military Physical and Cognitive Enhancement

8.3.1 Historical Perspective

The rationale underpinning soldier enhancement has a history that accompanies the history of warfare. At every point in history, armies attempted to give soldiers not only the best weapons, but also whatever available mood enhancers afforded by the folk knowledge or science of the time. Indigenous peoples in the Orinoco basin between Venezuela and Brazil duelled under the influence of hallucinogenic drugs; one group, the Otomac Indians, would take *yupa*, a powdered snuff extracted from the seeds of *Piptadenia peregrina*, which is a tree that contains hallucinogenic alkaloids. This type of intoxication had an important ritualistic role in combat (*Kamieński* 2016, 2). Later on, English sailors and infantrymen were given generous ratios of rum, reaching a peak in 1875, when the British armed forces consumed around 5386 million gallons of rum (*ibid, 9)*. Hindu warriors often drank *bhang*, an infusion of dried flowers from cannabis, in order to control their fears and increase their energy. Similarly, when Napoleon arrived in Egypt, his soldiers started replacing alcohol (which was banned in Egypt) with hashish as the 'military intoxicant of choice'. However, the behavioral problems following rampant use were finally recognized because the drug undermined the soldiers' fighting spirit and morale. In October 1800, Napoleon issued a prohibitive order (*ibid, 52–53*).

In modern history, the war in Vietnam was probably the most glaring example of the widespread use of intoxicants in warfare and their devastating effects. During World War II, Nazi soldiers heavily relied on drugs too, with millions of pills of methamphetamines distributed to soldiers before the invasion of France, Belgium and Holland. However, the Vietnam War is regarded as "the first true pharmacological war" as by 1973, as much as 70% of US soldiers were using intoxicants (*ibid, 187*). A veteran explained that amphetamines or some kind of uppers were available "like candy" – with little or no control over prescribed doses. Psychostimulants were given to boost endurance and strength and combat fatigue, boredom and trauma (Cornum et al. 1997, 55).

A more recent and safer alternative of 'stay awake' pills is Modafinil, which, although an amphetamine-like drug, is biochemically and pharmacologically distinct. It improves daytime wakefulness for individuals with narcolepsy and has (unlike amphetamines) minimal peripheral side effects in therapeutic doses. In healthy individuals, it significantly enhances performance in visual pattern recognition memory, spatial planning, and overall alertness, attentiveness and energy levels (Eliyahu et al. 2007, 385).[7] These benefits of the drug were not lost on the military, especially for flight operations where sleep deprivation can be a decisive factor in the outcome of the battle. A study with 18 helicopter pilots during two 40-h periods of sustained wakefulness when they were administered moderate doses of 100 mg of Modafinil at 4-h intervals showed that Modafinil maintained alertness, cognitive functions, risk perception, a feeling of well-being and situation awareness (Estrada et al. 2012). Before it was fully authorized, in 1991, during the Gulf War, the French army purchased 18,000 tablets of Modafinil (sold under the name Virgyl), which promised to keep the troops awake for up to 72 h (Bordenave and Prieur 2005). The UK Ministry of Defense started purchasing Modafinil in 1998. For example, in 2001, the same year when Allied forces started the offensive in Afghanistan, the MoD ordered 5000 pills, and in 2002, the year before the invasion of Iraq, the MoD ordered 4000 pills (Sample and Evans 2004).[8]

8.3.2 Contemporary Military Human Enhancement

8.3.2.1 The Role of DARPA

Psychostimulants target specific parts of the circadian rhythm (the human biological clock) such as by releasing excitatory neurotransmitters during prolonged sleep deprivation, and they have rather limited effects or they help compensate for some of soldiers' impaired functions, not necessarily alter or add new functions. In the past two decades, DARPA has tinkered with bolder ideas and with a more diverse set of technological tools. In the 1990s, DARPA started to take a more active interest in biology, and in 1999, the Agency created the Defense Sciences Office (DSO), and appointed a biologist and venture capitalist as its director.

In the following years, a new vision was integrated in DARPA's goals, which was to transform soldiers "from the inside out" and reduce or eliminate the physiological, psychological or cognitive limitations that impaired their "operational dominance" (Jacobsen 2015, 307–308). This objective was reconfirmed in a declassified statement release from 2003, which justified DARPA's thrust into life science: the "Bio-Revolution" was a "comprehensive effort to harness the insights and power of biology to make US warfighters and their equipment safer, stronger, and more effective." One of the pillars of this effort was "Enhanced Human Performance", which

[7] Uri Eliyahu et al., "Psychostimulants and Military Operations", *Military Medicine* Vol. 172 (April 2007), 385.

[8] Between 1998 and 2004, the period the two journalists covered at the date of the publication of their article, the MoD had purchased more than 24,000 pills of Provigil (the brand name of the Modafinil pills).

aimed at preventing the human soldiers from "becoming the weakest link in the US military" and to exploit life sciences to make the individual warfighter stronger, more alert, more endurant, and better able to heal" (Tether 2003, 12).

Some of the initiatives and programs created in the following years included projects to explore how to get humans to enter a hibernation-like state of suspended animation, devise ways of rapid healing, accelerate learning, enhance attention, enhance hearing and the olfactory system, or change metabolism (Lin et al. 2013, 23–26). For example, the DARPA-commissioned Crystalline Cellulose Conversion to Glucose was premised on the goal to have warfighters be able to eat materials otherwise indigestible to humans, such as grass (ibid; 25). The "Wound Stasis System" program worked to develop a "stabilizing treatment that would keep warfighters alive until they could be delivered to a surgical setting" and identify the mechanisms for controlling bleeding.[9] A program with a similar scope, "Surviving Blood Loss", aimed to develop strategies to extend the time that injured warfighters could survive critical blood loss. This interdisciplinary effort engaged scientists working towards a comprehensive exploration of energy production, metabolism, oxygen use. The research yielded some important findings, suggesting that exposure to low levels of hydrogen sulphide could induce a hibernation-like state in mammals by reducing cellular oxygen consumption.[10] Following almost two decades of research, another important finding was that coping with trauma and blood loss could also be achieved with a single dose of the female hormone 17β estradiol (E2).[11] Blood loss is a major area of research in the military because it is the number one cause of preventable deaths. For example, studies for a 10-year period (between 2001 and 2011), showed that 80% of the potentially preventable deaths resulted from blood loss. The DARPA-funded project at the University of Alabama at Birmingham further discovered that, following massive blood loss, E2 significantly improved heart performance, heart output and liver function and it could do so for 3 h without any fluid resuscitation and longer if fluid resuscitation was provided after the initial first 3 h (Hansen 2015).

Sleep is one of the key areas of interest and research in the military. Fighting against sleep deprivation, or against other hostile conditions such as extreme temperatures, for instance, is one area where the soldier cannot be trained (Ozanne 2015, 13). At the same time, sleep deprivation has debilitating effects and incurs dramatic risks on the battlefield, both for the safety of the soldier and the military operation. Lack of sleep leads to stress and impacts spatial memory, with cumulative effects as more sleep deprivation leads to further decrease of cognitive abilities. During sleep, certain brain regions (such as the prefrontal cortex, the occipital cortex, the medial parietal cortex and the thalamus) are metabolically 'deactivated'. These regions are critical in cognitive performance, taking decisions, initiative, and in planning abilities. It is believed that sleep helps replenish the glycogen stores in

[9] DARPA, Wound Stasis System, https://www.darpa.mil/program/wound-stasis-system . Accessed 30 September 2019.

[10] DARPA, "Surviving Blood Loss", https://www.darpa.mil/program/surviving-blood-loss . Accessed 30 September 2019.

[11] Ibid.

the brain, which had been depleted during wakefulness, meaning that after sleep, the brain is enabled to restart its ability to conduct such cognitive roles (Cotting and Belenky 2015, 15).

In the early 2000s, the Continually Assisted Performance program at the DSO explored ways to create a "24/7" soldier, a soldier who can stay awake for many days at a time – up to seven (Jacobsen 2015, 310. See also Tether 2003). A class of medications studied extensively, as part of this programme, was the ampakines, normally used for the treatment of Alzheimer's disease. Ampakines are a group of pharmaceuticals used for low-risk cognitive enhancement, improvement in memory and alertness, but not without counter-indications. In fact, ampakines are believed to have serious effects on neural communication in the normal brain, eliciting brain plasticity in the regions that are associated with emotive and affective functions. Brain plasticity may seem as a big gain initially – implying faster and more efficient learning and heightened cognitive functions; however, excessive plasticity also means high activity in *all* synapses and thus a reduction in synaptic pruning, which was identified as one of the symptoms associated with autism disorders (Urban and Gao 2014, 6). Synaptic pruning is crucial for the developing brain and it occurs until the mid to late 20s. Its role is to weed out the weaker neural connections while allowing others to strengthen (Zukerman and Purcell 2011). The use of certain cognitive enhancers could potentially pose risks for the brain development of young servicemen and women.

8.3.2.2 Brain Stimulation

In recent years, non-pharmacological methods have been increasingly tested for neuro-stimulation. Brain stimulation with electricity promises to enhance cognitive functions by targeting specific brain regions or nerves to accelerate learning and skills acquisition. Non-invasive brain stimulation is premised on running small doses of electrical currents through specific parts of the brain to improve alertness and acuity. Initial experiments at the Air Force Research Lab in Ohio showed that the volunteers performed consistently – a useful feat especially for drone pilots who spend hours monitoring surveillance footage (Sample 2016).

Neuroenhancement, unlike other tools for cognitive enhancement, acts directly in the brain and in the nervous system. Neuroenhancement can enhance "fluid intelligence", which is the ability to learn flexibly, to adapt and to understand abstract notions. Different methods of brain stimulation have been tested in recent decades for clinical purposes, such as for helping patients with attention deficit disorder or schizophrenia. When used for military enhancement purposes, brain stimulation can be employed for enhancing learning, as well as for modifying functions 'on demand'. For example, stimulation of the prefrontal cortex can enhance or inhibit the tendency to lie. The dorsolateral prefrontal cortex is considered to be the brain area involved in deceptive behaviour; with brain stimulation techniques it has been shown that stimulation of the right hemisphere decreases the tendency to lie, while stimulation of the left hemisphere increases lying (Karton and Bachmann 2011).

Two main approaches are transcranial magnetic stimulation, which works by placing a magnetic coil above parts of the skull to deliver magnetic pulses, and transcranial electrical stimulation, which works by applying low electrical currents to the skull through one or more electrodes to modulate neuronal excitability in the brain – the latter approach offers the more diverse kinds of cognitive enhancement (Academy of Medical Sciences et al. 2012). One way of passing electrical current via the electrodes is a method called transcranial direct current stimulation (tDCS). tDCS enhances attention, learning, memory, and can double performance accuracy, with effects that last for up to 24 h after the stimulation ended (Clark and Parasuraman 2014, 890). In the military, experiments with tDCS showed significant improvement in processing information and enhancing performance in multitasking. tDCS applied over the left dorsolateral prefrontal cortex also showed improvement in visual search via stimulation of the frontal eye fields (Nelson et al. 2016, 3).

The aforementioned DARPA program "Targeted Neuroplasticity Training", announced in 2016, focuses on harnessing the potential of peripheral nerve stimulation in order to enhance learning processes in the brain. More specifically, the program explores ways to increase both the pace and effectiveness of cognitive skills training with the goal to reduce the costs and duration of the DoD's trainings in foreign languages, intelligence, cryptography and other fields.[12] In concrete terms, it means that warfighters could be aided to learn a foreign language significantly faster, to memorize extensive and complex details or perhaps even entire maps or location details. As mentioned in the first pages of this chapter, programs such as this have an avowed goal to enhance human capacities beyond what is normally attainable with typical human abilities.

Another research track has focused on ways of bypassing human consciousness in order to harness the potential of the brain area that processes information *before* the brain is consciously aware of it. DARPA's Neurotechnology for Intelligence Analysts sponsored research that explores how the P300 signal can be used in a way that would cut human consciousness out of the loop in searches for specific weaponry such as rocket launchers in satellite or drone imagery. The P300 signal is a fleeting electrical signal produced by the brain when it recognizes an object it was seeking, and it is detectable by electrodes on the scalp before the individual becomes consciously aware of having recognized that object. Experiments with skull enclosing caps fitted with 32 electrodes, which recorded the signal emitted by the brain in response to stimuli, found that the speed at which the objects were found tripled (*The Economist* 2017). Brain stimulation can take this further, as demonstrated by an experiment with over 1000 volunteers who had their brains stimulated with a 9 V battery connected to electrodes on the scalp. The result was that after 30 min, the participants in the experiments reported a 13% increase in the amount of snipers, bombs and other weaponry they were able to spot (ibid.)

[12] Boosting Synaptic Plasticity to Accelerate Learning, DARPA press release, 16 March 2016, https://www.darpa.mil/news-events/2016-03-16. Accessed 30 September 2019.

In addition to neurostimulation, which uses external devices, another approach has centred on implantable technology. One prominent example is "neural dust", which is a very small electronic microchip that could be implanted in the body, in individual nerves, and is capable of detecting electrical activity deep in the muscles and nerves. It is activated with ultrasound and implanted by needle injection or other non-surgical procedures.[13] The smallest neural dust mote available in 2018 was the StimDust, developed at Berkley: measuring only 6.5 cubic mm (which is 3–4 times smaller than a grain of rice), it senses neural activity and stimulates peripheral nerves (Saracco 2018). It is a sensory that does not work with batteries and the ultrasound is used both to power the mote and to read out the measurements – more effective than radio waves because it can penetrate anywhere in the body.

The envisaged uses for this stimulator include the treatment of heart irregularities, chronic pain, asthma, epilepsy. The neural dust can also be used for a range of other functions, such as to stimulate the immune system, to tamp down inflammation, or suppress the appetite. However, its long-term prospects are much broader and in the future this tiny stimulator could be used in the central nervous system, for the next generation of brain-computer interfaces, or, as its developers hope, anywhere in the body (Sanders 2016). In the military, in addition to its clinical uses, 'neurograins' could push the limits of soldiers' bodies, such as, for example, by detecting biomarkers early at the onset of diseases and trigger neural responses, or by helping the body heal itself faster. DARPA's Electrical Prescriptions (ElectRx) program supports several projects for 'non-pharmacological' treatments and interventions in order to "exploit and supplement the body's natural ability to quickly and effectively heal itself", including technologies to enable artificial neuromodulation of peripheral nerves, and delivery of therapeutic signals to peripheral nerve targets.[14]

The military forays into physical and cognitive enhancement will generate an abundance of ethical questions. These will concern several domains of ethics, such as bioethics and medical ethics (involving debates about 'informed consent', 'testing and experimentation on human subjects' etc) as well as military ethics, for example concerning the legitimacy of the superior's orders (in cases when the enhancement comes as an order from the superior to the soldier) and fundamental values the military is built upon, such as courage, merit, sacrifice.

The use of psychostimulants in the twentieth century, especially amphetamines, raised significant ethical questions in the military and society, in addition to tarnishing the reputation of the army, and affecting the lives of countless veterans upon their return home. Contemporary examples of enhancement technologies include much more sophisticated and invasive type of interventions and some of the ethical concerns will be entirely new. Addiction of the type created by amphetamines and the resultant social anxiety spurred by tens of thousands of addicts returning home will be replaced by other kinds of concerns. If some types of enhancements are

---

[13] DARPA, "Implantable 'Neural Dust' Enables Precise Wireless Recording of Nerve Activity", 3 August 2016, https://www.darpa.mil/news-events/2016-08-03. Accessed 30 September 2019.

[14] Dr. Eric Van Gieson, "Electrical Prescriptions (ElectRx), https://www.darpa.mil/program/electrical-prescriptions

irreversible and permanently alter the physiology of the soldier, the reintegration of the 'enhanced soldier' back into society as an 'enhanced civilian' will be problematic in its own way. For example, a cognitively enhanced veteran may even have a comparative advantage in the job market compared to non-enhanced civilians. However, as it is beyond the scope of this chapter to cover all these ethical issues exhaustively, the remaining parts of this chapter will focus on a range of narrow and specific ethical dilemmas and consequences of soldier enhancements.

8.4 Enhancements, Ethics and Operational Dilemmas

In exploring the ethical implications of soldier enhancement, the initial questions to consider are: "do all enhancements in the military deserve the same ethical scrutiny?", and "do enhancements everywhere in the military bear the same ethical and operational costs?". The short answer is no. Multiple enhancement procedures and pharmacological agents, such as Modafinil or neurostimulation with electricity, have been tested for drone pilots working remotely from the battlefield. They too need to stay focused for many hours at a time, to be hyper-vigilant and able to detect threats and suspicious elements, so from a military perspective, their enhancement is entirely justifiable. However, given the specific context in which they operate, such pilots face different dilemmas, stressors and physical and cognitive strains compared to soldiers fighting on the battlefield, and especially in urban combat. The system of rewards and decorations in the military has so far only acknowledged the merits of servicemen and servicewomen physically present on the battlefield and it is the enhancements to those members of the army that will be critically assessed here.

8.4.1 Challenges to Army Values

Members of the military community share a number of values that are fundamental to the sense of community and the self-identity of the soldier. These values are timeless and, in their absence, the cohesion that binds soldiers together in a community would simply collapse. Courage, honour, loyalty, discipline, and sacrifice are some of the most frequently mentioned virtues in military life. For example, the Australian Army lists four values as 'the bedrock to everything it does': courage, initiative, respect, and teamwork (Beard et al. 2016, 13). The US lists similar principles: courage, honour, selfless service, loyalty, respect, integrity. In addition, the US Law of War Manual (2015) lists honour as one of the foundational principles of most of the treaty and customary rules of war. Honour incorporates requirements of "a certain amount of fairness in offense and defense", "a certain mutual respect between opposing forces", and also recognition of the fact that "combatants are a common class of professionals who have undertaken to comport themselves honourably" (US Department of Defense 2015, 66–68).

Irrespective of the particular phrasing, courage, the ability to take risks, own up to one's responsibilities and demonstrate personal sacrifice reign supreme among military virtues. That is why the highest distinctions across militaries recognize precisely these merits. For example, in the US, the Medal of Honour, the foremost military decoration, is awarded for acts of bravery "at the risk of life, above and beyond the call of duty" – as the brave acts accomplished by Capt Groberg, mentioned in the beginning of this chapter. Victoria Cross is the highest military distinction in the British Armed Forces and is awarded for extreme bravery in the face of the enemy. Israel's Medal of Valor is awarded to soldiers of the Israeli Defense Forces who perform acts of valor and supreme heroism in the face of the enemy, at the risk of one's life.

An enhancement can significantly reduce the risk of losing one's life in several ways: it can boost vigilance and focus in the face of sleep deprivation and fatigue, and thus reduce the likelihood of committing errors; it can augment acuity, perception and ability to detect threats or spot the enemy; in the event of injury and loss of blood, therapies such as those investigated through the "Surviving Blood Loss" programme can mobilize physiological functions to compensate for blood loss and thus increase one's chances of survival. An enhanced warfighter will feel less vulnerable and will enjoy a physical and/or cognitive advantage over the enemy. An intrinsic aspect of combat virtue and honour is thus diminished (Beard et al. 2016, 14). This would be even more the case if the enhancement procedure were to target precisely the neuronal mechanisms underpinning risk-taking.

In many instances, as a result, the personal courage and merit of enhanced warfighters be assessed by typical standards because these virtues appear less authentic, being an upshot of the enhancement or the combination of enhancements. A similar claim could be made that a warfighter equipped with sophisticated weaponry could feel more courageous as a result of the weapon they are carrying, and which allows them to outfight the enemy. This is, strictly speaking, correct. Nevertheless, in conditions of long and exhausting hours in urban combat (where fighting is particularly difficult and taxing), fatigue, injuries or basic human limitations can still make the same soldier commit fatal errors. An enhancement to the soldier themselves, therefore, overrides many of the risks originating from the inherent frailties and limitations of the human body, and which can make the soldier inefficient or under-performing even with advanced weaponry at their disposal.

Enhancements, however, do not render courage and sacrifice altogether obsolete. In the context of drone warfare, Christian Enemark noted that warfare entered a post-heroic age, characterized by aversion to physical risk and casualties (Enemark 2014, 9). However, unlike drone operators, enhanced soldiers are physically present on the battlefield (although arguably at times enduring less physical strains) and still face the risk of violent death. It would, therefore, be incorrect to conclude that enhanced soldiers are 'post-heroic soldiers'. (Puscas 2019, 212) If they are different from non-enhanced soldiers, yet not 'post-heroic', what best describes them? The answer to this question will vary depending on the type and nature of enhancements, as well as context. For example, let us consider the following situations:

(a) A soldier that underwent neurostimulation for augmented cognitive abilities may be readily deployable, they may be better and faster at recognizing targets, remembering geospatial coordinates, or communicating with the local population, having mastered the basics of the language. Combat is a relatively easier experience to them now, compared to not having been enhanced. However, the same warfighter is engaged in a battle where it is none of these enhanced abilities that distinguish him, but, say, his selfless conduct and willingness to stand fully exposed to enemy fire, or shield his teammates. If no distinct connection can be identified between the enhancements as enablers of the brave act, there is no reason why the warfighter's actions should not be considered worthy of utmost recognition.
(b) The expectations in combat may be commensurate to the enhancements: if a soldier is enhanced to have stronger muscles, to be more vigilant for long hours, to withstand high levels of pain and to have heightened threat-detection abilities, they may be held to different standards of courage and merit. The fact that they would go 'above and beyond the call of duty' compared to their non-enhanced soldiers may come as an obvious, if not minimal, requirement given that they were enhanced beforehand, and that special resources were spent on them.

Even so, the result of their actions can still be extraordinary, and the warfighter can demonstrate selfless dedication and an ethos for sacrifice for their unit, and for their country. For example, their swiftness and skills may help spare the lives of tens of civilians around them and of fellow soldiers, quickly neutralizing the enemy while, at the same time, acting under an imminent threat of death or irrecoverable injury. Should artificially enhanced efficiency annul altogether the demonstration of courage and devotion to duty, especially if the risk of loss of life (of the soldier) was obvious? It would appear unfair to deny enhanced soldiers any special recognition for their behaviour in combat, especially if their lives were on the line. At the same time, an unenhanced soldier committing similar acts of bravery may consider it unfair to them that they are recognized on par with a soldier that benefitted from performance enhancers. The former could claim they are in the right since the ethics and codes of warfare emerged and always only accounted for 'normal', unenhanced individuals, not for soldiers enhanced with technology. Why should, therefore, the enhanced soldiers' actions be judged against a set of values that traditionally applied to a category less privileged than them?

This point leads to another ethical conundrum as enhancements may not be regarded as a privilege but rather as a token of sacrifice. As research into biotechnologies progresses, the available methods of enhancement may be increasingly safer, and with minimal side effects. However, some procedures, such as some types of neurostimulation with electricity, may have unforeseen consequences with effects that will take longer to manifest, such as speech impairments, brain plasticity compromise etc. A soldier that volunteers to undergo complex and dangerous forms of enhancements, under full and informed consent, may in fact demonstrate selflessness (and this may potentially trigger a race to sign up for enhancements so as to prove one's dedication and patriotism). Rather than eliminate sacrifice from the battlefield, biotechnological enhancements may change and update the meaning of personal sacrifice in warfare. (Puscas 2019, 213)

With these considerations in mind, it appears increasingly challenging to parse the connotations of courage, merit and sacrifice because it is very easy to reach a dead end. For the time being, and in the next few decades, the range of enhancement technologies used on the battlefield will unlikely call for a complete separation of rewards and decorations for enhanced and unenhanced soldiers. However, the questions of 'real/authentic' courage vs. enhanced/artificial neuromodulation of courage cannot be avoided either. A more likely scenario will be that a more complex, at times contentious, investigation will be carried out to determine the circumstances of a military action, and the validation of a medal-worthy act could require input and assessment from a wide range of experts, including biomedical engineers or doctors who administered the enhancement procedure.

These questions are theoretical for now but will need to be resolved through a thoughtful and thorough methodology of implementing enhancements. Patrick Lin correctly noted that the precise manner in which military human enhancements are rolled out is going to be critical in how those enhancements will affect the army and societies at large. He flags several issues that will need to be clarified: should enhancements be restricted to special or elite teams, and if so, which enhancements and why? Should enhancements be routinely reversed upon discharge? (Lin 2010, 322) Additionally, if enhancements are not adopted by all warfighters at once, this may increase dissension among them, affecting morale and unit cohesion. The asymmetry of needs and capabilities could finally lead to resentment, also because enhanced soldiers, being better-abled, may be sent to more complex operations, rather than be asked to complete 'mundane' tasks. Furthermore, could it be conceivable at all that soldiers are enhanced but their superiors not? The command structure may be greatly affected if commanders are unenhanced and thus seen as physically or intellectually inferior to those they command (Lin et al. 2013, 39–40).

Dismissing these concerns as merely hypothetical is not an option. These idiosyncratic factors matter, not in the least because they ultimately have a bearing on operational success, on how warfighters accept the legitimacy of their mission and the orders they receive. To avoid situations that may damage the sense of community, as well as the hierarchical disposition of the army, it will be crucial to establish guidelines for the implementation of enhancements and to ensure that they are accepted as legitimate.

8.4.2 Ethics in the Military Profession and Combat

By challenging well-established norms in the military and warfare, enhancement technologies remind us why ethics is present in the military to begin with, both at the micro level, of the unit, and at the macro level, on the battlefield. In addition, soldier enhancement re-emphasizes the moral foundation of the military profession.

If winning the battle at any cost were the sole guiding principle in warfare, there would be little space left for the application of humanitarian principles and few checks on the amount of brutality permissible in war. It would also not matter if and how soldiers are enhanced, or how they relate to their community, since a very crude notion of 'military necessity' would be the sole guiding principle. In that case, the moral component of the military community would not rest on anything but the objective to kill. However, as poignantly underlined by Christian Enemark (2014, 4), "if war is to remain [...] a pursuit more virtuous than organized murder, it is vital also to think in terms of what is permissible"; "ethics is thus constitutive of the practice of war as a form of violence that is (or is held to be) morally distinguishable from other forms".

At the same time, militaries equally must acknowledge the contextual, relative nature of ethics in warfare, where the context may dictate interpretations of what is permissible or not.

Traditionally, most Western militaries employ hybrid ethical codes, which draw a lot of inspiration from Aristotelian virtue ethics, and deontological ethics. Contemporary codes of military ethics have been impacted by three elements of virtue ethics in particular:

1. the *contextual* relativity of virtues,
2. the *actor* relativity of virtues,
3. the emphasis on *character formation*.

Aristotle argued that moral virtues are defined as 'the mean between extremes of deprivation and excess' – the particular elements of the context vary a lot, and so does what qualifies as 'extreme' from a context to another, the character and role of the actor (Schulzke 2016, 189). This relative nature of virtues distinguishes them from standard ethical rules, which are more rigid moral imperatives that urge the same type of action for every person and in every situation. Since that cannot work in a military context, virtue ethics provides a practical guidance, especially for those soldiers who find themselves confronted with unexpected moral challenges, soliciting quick responses. Aristotle had called the ability to cultivate the wisdom to apply the appropriate virtue at the right time "phronesis", a skill that can only be acquired through practice. That is why the military has a transformative role in character formation (ibid, 190). In addition to deontologist ethics, which imposes absolute rules in combat and very clear limits (such as through the laws of war), soldiers must be able to apply their own virtue-guided judgement and initiative, especially those soldiers who act with relative autonomy on the battlefield (ibid, 191–192). For all of these reasons, the military has a role in developing and training soldiers to see themselves as members of a unique profession, with a special identity. Courage, the ability to act honourably and 'to do the right thing' is therefore something that does not manifest only in one-off situations on the battlefield but is an ability cultivated continuously. In other words, soldiers are habituated to have good character through training, and this is in line with Artistole's observation that virtues are carved through repetition (ibid, 191. See also McMaster 2009).

Enhancement technologies can impact these timeless military values. For example, if it is possible to shortcut one's way to courage with the appropriate neurostimulation, one might argue the intrinsic value of courage is diminished. That is because courage in a military context does not manifest only as an isolated and intrinsic quality of an individual, but it is also nurtured and fostered through social cohesion and military education (Olsthoorn 2007, 275–276). Additionally, if a soldier's capacity for multitasking and decision-making is enhanced artificially, the warfighter may find himself alienated or less esteemed by his peers in the military community, even though that skill could be sorely needed when acting without direct instructions. Nevertheless, just as they become (or as perceived as) less virtuous within their own community, the enhanced soldier may have an improved ability to act ethically on the battlefield. As H.R. McMaster argues, very often, unethical behaviour in war is not due to a lack of ethical training but simply to combat stress, even more so in irregular warfare, which creates abnormal pressures. To counter this, he advises realistic and tough educational packages, including language and cultural training (which is one of the objectives of the "Targeted Neuroplasticity Training" programme, mentioned above), in order to steel soldiers against the stress of combat (McMaster 2009, 15). The challenge is that enhancements may help fill this 'ethical gap' but at the expense of the communal and professional values the armed forces rely upon – an option that is hardly desirable or practical in the long run.

Nevertheless, ethicists may be at odds attempting to build a strong moral case even against the most extreme forms of enhancement if the enhanced soldier is, due to that enhancement, able to spare or save lives. In effect, this perfectly resonates with the objectives of deontological and utilitarian ethics (including International Humanitarian Law), and key principles such as the principle of proportionality and distinction, or the principle of prohibition of superfluous injury and unnecessary suffering. These latter principles made their way into law because military ethics developed to exclude certain forms of behavior as morally unacceptable, even in the harsh and gut-wrenching conditions of the battlefield (Lucas 2016, 21). In other words, enhancements could well enable soldiers to better implement the laws of war, so often disregarded today. But this observation confirms the perplexing nature of enhancements: they may help soldiers save their own lives and apply humanitarian doctrines, and simultaneously clash with the ethics of the military profession, and with values that have historically weaved the military's social fabric.

There is an additional consideration worth signalling here. Military ethics is generally founded on the basic premise of ensuring the "greater good" and the best possible outcome in a limited timeframe, which is the duration of the military operation. This would imply that military ethics is exclusively framed in 'short-termism'. But what if, despite the overall positive gains of a given action in combat, the long-term effects offset the short-term gains? What if the known or anticipated long-term impacts of enhancement technologies reveal enormous risks for the armed forces as a community, for individual units, for the health and integrity of the soldiers, and more globally, for the future of humanity (considering some technologies will inevitably transfer into the civilian sphere)?

The laws of war are overwhelmingly concerned with conduct in warfare, but on several occasions take 'future generations' into account. Long-term considerations are part of IHL insofar as, for example, considerations about environmental degradation are enshrined in the Geneva Conventions and Additional Protocol I, or the 1976 ENMOD Convention (ICRC 2010). Therefore, the principle of military necessity must also be considered vis-à-vis potential damage and harm outlasting the military operation and extending into the future or impacting "generations unborn".[15] Utilitarianism in military ethics has limitations that cover *long-term consequences* that can be disastrous for humanity.

8.5 Conclusions

The long-term implications of soldier enhancement will be mooted extensively in the coming decades, both in legal and ethical circles. This chapter offered a preview into a range of ethical and practical dilemmas raised by enhancement technologies. In order to stimulate a realistic discussion of the ethical implications and operational consequences of increased use of enhancement technologies, it was first and foremost important to understand what military human enhancement means, beyond exaggerated misrepresentations from Sci-Fi or popular culture, and the kind of interventions and technologies under research and development today.

The dilemmas prompted by enhancements will cut across ethics, law, and military planning and operations. While it is unrealistic to call for a halt on enhancement technologies, a minimal requirement would be to ensure they are used for legitimate military purposes, and that their legitimacy, overall, is accepted by service members. As it was emphasized on several occasions, the expectation of legitimacy cannot be dismissed as naïve or idealistic. Ensuring that enhancements are accepted both as necessary and legitimate will be a critical task going forward, and ultimately, the only way they will bring an added value to militaries.

References

Academy of Medical Sciences, et al. 2012. *Human enhancement and the future of work*. Report from a joint workshop hosted by the Academy of Medical Sciences, the British Academy, the Royal Academy of Engineering and the Royal Society, November 2012. https://royalsociety.org/~/media/policy/projects/human-enhancement/2012-11-06-human-enhancement.pdf. Accessed 30 Sept 2019.

Beard, Matthew, Jai Galliott, and Sandra Lynch. 2016. Soldier enhancement. Ethical risks and opportunities. *Australian Army Journal* 13 (1): 13.

[15] The reference to "generations unborn" comes from a 1996 ICJ Advisory Opinion on *Legality of the Threat or Use of Nuclear Weapons*, where the environment is described as not an abstraction, but "the living space, the quality of life and the very health of human beings, including generations unborn". Full text here: http://www.icj-cij.org/files/case-related/95/095-19960708-ADV-01-00-EN.pdf (page 19)

Bordenave, Yves, and Cécile Prieur.2005. Les cobayes de la guerre du Golfe, *Le Monde*, December 18. https://www.lemonde.fr/societe/article/2005/12/18/les-cobayes-de-la-guerre-du-golfe_722462_3224.html. Accessed 30 Sept 2019.

Bostrom, Nick, and Andreas Sandberg. 2009. Cognitive enhancement: Methods, ethics, regulatory challenges. *Science and Engineering Ethics* 15 (3): 311–341.

Clark, Vincent, and Raja Parasuraman. 2014. Neuroenhancement: Enhancing brain and mind in health and in disease. *NeuroImage* 85: 889–894.

Cornum, Rhonda, et al. 1997. Stimulant use in extended flight operations. *Airpower Journal* 11 (Spring): 55.

Cotting, Dave I., and Gregory Belenky. December 2015–January 2016. "Le sommeil et la performance opérationnelle", *Défense & Sécurité Internationale*, 15.

DARPA. *Surviving blood loss*. https://www.darpa.mil/program/surviving-blood-loss

DARPA. *Wound stasis system*. https://www.darpa.mil/program/wound-stasis-system

DARPA. 2016a. Boosting synaptic plasticity to accelerate learning. *DARPA Press Release*, March 16. https://www.darpa.mil/news-events/2016-03-16. Accessed 30 Sept 2019.

———. 2016b. *Implantable 'neural dust' enables precise wireless recording of nerve activity*, August 3. https://www.darpa.mil/news-events/2016-08-03. Accessed 30 Sept 2019.

———. 2018. *Slowing biological time to extend the golden hour for lifesaving treatment*, March 1. https://www.darpa.mil/news-events/2018-03-01

Dinniss, Heather A. Harrison, and Jann K. Kleffner. 2016. Soldier 2.0: Military human enhancement and international law. *International Law Studies* 92: 432–482.

Eliyahu, Uri, et al. 2007. Psychostimulants and military operations. *Military Medicine* 172: 383–387.

Enemark, Christian. 2014. *Armed drones and the ethics of war: Military virtues in a post-heroic age*. New York: Routledge.

Estrada, Arthur, et al. 2012. Modafinil as a replacement for dextroamphetamine for sustaining alertness in military helicopter pilots. *Aviation, Space, and Environmental Medicine* 83 (6): 556–567. Accessed 30 Sept 2019.

Hansen, Jeff. 2015. Female sex hormone may save injured soldiers on the battlefield. *UAB The Mix*, October 20. https://www.uab.edu/news/research/item/6625-female-sex-hormone-may-save-injured-soldiers-on-the-battlefield . Accessed 01 Oct 2019.

International Committee of the Red Cross. 2010. *Environment and international humanitarian law*. https://www.icrc.org/eng/war-and-law/conduct-hostilities/environment-warfare/overview-environment-and-warfare.htm. Accessed 30 Sept 2019.

Jacobsen, Annie. 2015. *The Pentagon's brain. An uncensored history of DARPA, America's Top Secret Military Research Agency*. New York: Back Bay Books.

Juengst, Eric T. 1998. What does enhancement mean? In *Enhancing human traits. Ethical and social implications*, ed. Erik Parens, 29–47. Washington D.C.: Georgetown University Press.

Kamieński, Łukasz. 2016. *Shooting up. A short history of drugs and war*. New York: Oxford University Press.

Karton, Inga, and Talis Bachmann. 2011. Effects of prefrontal transcranial magnetic stimulation on spontaneous truth-telling. *Behavioural Brain Research* 225 (1): 209–214.

Lin, Patrick. 2010. Ethical blowback from emerging technologies. *Journal of Military Ethics* 9 (4): 313–331.

Lin, Patrick, et al. 2013. Enhanced warfighters: Risks, ethics and policy. *The Greenwall Foundation*, January 1, 12.

Lucas, George. 2016. *Military ethics. What everyone needs to know*. New York: Oxford University Press.

McMaster, H.R. 2009. Preserving soldiers' moral character in counter-insurgency operations. In *Ethics education for irregular warfare*, ed. Don Carrick, James Connelly, and Paul Robinson, 15–26. Burlington: Ashgate.

Nelson, Justin, et al. 2016. The effects of transcranial direct current stimulation (tDCS) on multitasking throughput capacity. *Frontiers in Human Neuroscience* 10: 589. https://www.frontiersin.org/articles/10.3389/fnhum.2016.00589/full. Accessed 30 Sept 2019.

Olsthoorn, Peter. 2007. Courage in the military: Physical and moral. *Journal of Military Ethics* 6 (4): 270–279.

Ozanne, Eric. 2015. Les forces et les faiblesses du soldat sur le champ de bataille. *Défense & Sécurité Internationale*, 45 Hors-série: Le soldat augmenté, 2015: 12–13.

Puscas, Ioana. 2019. Military human enhancement. In *New technologies and the law in war and peace*, ed. William H. Boothby, 182–229. Cambridge: Cambridge University Press.

Sample, Ian. 2016. US military successfully tests electrical brain stimulation to enhance staff skills. *The Guardian*, November 7. https://www.theguardian.com/science/2016/nov/07/us-military-successfully-tests-electrical-brain-stimulation-to-enhance-staff-skills. Accessed 30 Sept 2019.

Sample, Ian, and Rob Evans. 2004. MoD bought thousands of stay awake pills in advance of the war in Iraq. *The Guardian*, July 29. https://www.theguardian.com/society/2004/jul/29/health.sciencenews. Accessed 30 Sept 2019.

Sanders, Robert. 2016. Sprinkling of neural dust opens door to electroceuticals. *Berkley News*, August 3. http://news.berkeley.edu/2016/08/03/sprinkling-of-neural-dust-opens-door-to-electroceuticals/. Accessed 30 Sept 2019.

Saracco, Roberto. 2018. Neural dust is getting ready for your brain. *IEEE*, May 21. http://sites.ieee.org/futuredirections/2018/05/21/neural-dust-is-getting-ready-for-your-brain/. Accessed 30 Sept 2019.

Schulzke, Marcus. 2016. Rethinking military virtue ethics in an age of unmanned weapons. *Journal of Military Ethics* 15 (3): 187–204.

Tether, Tony. 2003. *Statement to the Subcommittee on Terrorism, Unconventional Threats and Capabilities, House Armed Services Committee, US House of Representatives*, March 27. https://www.darpa.mil/attachments/TestimonyArchived(March%2027%202003).pdf. Accessed 30 Sept 2019.

The Economist. 2017. How to make soldiers' brains better at noticing threats. *The Economist*, July 27. https://www.economist.com/science-and-technology/2017/07/27/how-to-make-soldiers-brains-better-at-noticing-threats. Accessed 30 Sept 2019.

Urban, Kimberly R., and Wen-Jun Gao. 2014. Performance enhancement at the cost of potential brain plasticity: Neural ramifications of nootropic drugs in the healthy developing brain. *Frontiers in Systems Neuroscience* 8: 38.

US Department of Defense. 2015. *Law of war manual*. http://archive.defense.gov/pubs/law-of-war-manual-june-2015.pdf. Accessed 30 Sept 2019.

Van Gieson, Eric. Electrical prescriptions (ElectRx). *DARPA program information*. https://www.darpa.mil/program/electrical-prescriptions. Accessed 29 Sept 2019.

Wright, W.Geoffrey. 2014. Using virtual reality to augment perception, enhance sensorimotor adaptation, and change our minds. *Frontiers in Systems Neuroscience* 8: 2.

Zukerman, Wendy, and Andrew Purcell. 2011. Brain's synaptic pruning continues into your 20s. *New Scientist*, August 17. https://www.newscientist.com/article/dn20803-brains-synaptic-pruning-continues-into-your-20s/. Accessed 30 Sept 2019.

Chapter 9
Human Enhancement, Transhuman Warfare and the Question: What Does It Mean to Be Human?

Dirk Fischer

9.1 Introduction

The possibility of enhancing a human being leads inevitably to the question *What does it mean to be human?* In a military context, a huge number of conceivable applications for enhancement techniques are the subject of rigorous research in natural and technical sciences as well as humanities. Throughout the twentieth century, it seemed to be appropriate to confine human enhancement scenarios to the field of science fiction. By the beginning of the twenty-first century, most of the ideas about human enhancement have lost a great deal of their fictional character and could be classified as future realities, which, if they are not yet realized, will be within a foreseeable period of time.

First, the case of the German philosopher and university professor Helmut Dubiel (1946–2015), who was treated with deep brain stimulation because of Parkinson's disease, will demonstrate the effects which invasive techniques have on human beings. Although, at first sight, it might seem that this has nothing to do with the debate on human enhancement in a military context, it is very thought provoking from a medical ethics perspective and helps to understand the particular challenges which the people involved might have to face, when it comes to the application of human enhancement techniques.

Second, the problem of defining the term *human enhancement,* will be addressed. None of the various attempts to define the term, so far, have received full acceptance by both the philosophical and the natural scientific community. Therefore, in depth terminological work is necessary, alongside the discourse about human enhancement.

D. Fischer (✉)
Teaching and Research Unit for Military Medical Ethics, Bundeswehr Medical Academy Munich, Munich, Germany
e-mail: LFWME@bundeswehr.org

D. Messelken, D. Winkler (eds.), *Ethics of Medical Innovation, Experimentation, and Enhancement in Military and Humanitarian Contexts*, Military and Humanitarian Health Ethics,
https://doi.org/10.1007/978-3-030-36319-2_9

To support this effort, the Teaching and Research Unit for Military Medical Ethics in Munich, has developed a rather wide, working definition which identifies the central elements which are at stake if human enhancement techniques are employed.

The idea of the posthuman is a central element in transhumanism. Based on the enhancement-mediated transformation of humans into posthumans, some light will be shed on its possible implications for military conflicts (transhuman warfare) and their underlying ethical codes of conduct. As with debates, the question, what does it mean to be human, is of particular interest at this point. Although it is impossible to give an ultimate answer to that question, its crucial importance must be pointed out, particularly when human enhancement is being discussed in a military medical ethics context.

9.2 A Reluctant Cyborg: Helmut Dubiel's Struggle with Deep Brain Stimulation

At the age of forty-six Helmut Dubiel, a professor of sociology at the University of Gießen in Germany, was diagnosed with Parkinson's disease. Fearing the reactions of the people around him, he at first did his utmost to conceal his condition. But when his symptoms became too obvious to camouflage so that he could no longer hide his illness, he had to take some serious decisions about his further treatment. Dubiel decided to undergo deep brain stimulation surgery. In 2006, he published a detailed report describing both his life with Parkinson's disease and his experience of deep brain stimulation. The English version, *Deep in the Brain,* followed in 2009 (see Dubiel 2006, 2009).

Surgical therapy may be considered in the course of Parkinson's disease; Surgical intervention can reduce symptoms particularly if medication alone no longer helps. For this purpose, thin electrodes, which are connected to a pacemaker situated under the skin below the clavicle, are placed into specific areas of the human brain. After adapting the neurostimulator to the individual patient's situation, the electronic impulses help to control the patient's neuro-muscular status and reduce the amount of Parkinson's drugs. Although it normally takes some time to achieve the necessary equilibrium between medication and neurostimulation, in most cases deep brain stimulation helps to achieve an acceptable balance between remaining symptoms, medication needed and possible side effects.

In Helmut Dubiel's case, deep brain stimulation was considered because he suffered from an increasing number of periods when medication was inadequate, symptoms returned (off-times) and uncontrolled, involuntary movements (dyskinesia) appeared. Although the surgery went well, Dubiel developed various symptoms of posttraumatic stress and he processed the experience of stereotactic surgery in very impressive images of violence: "A friend of mine told me of a telephone conversation that I apparently began while still in the best of moods. He asked me about the operation and I described the drilling in my skull. Apparently, in order to amuse the caller and demonstrate my detachment from the event, I compared myself to a dog whose house is being attacked with a chainsaw. My speech then became unintelligible, I suddenly began to cry and simply aborted the call" (Dubiel 2009, 84–85). He also experienced an excruciating dream which he believed was an attempt to cope with the atrocious experience of having brain surgery while being conscious: "It was a dream about a massacre in a supermarket. Even though the violence was boundless, its portrayal in the dream was still stylized. It was an ultramodern ballet. Not a movement, not a gesture, seemed coincidental. Everything served a secret choreography that had an even more nightmarish effect because there was no sound in the dream. Yet despite all this visible choreography, the silent mutual butchering didn't come across like media content; it wasn't that I was dreaming a film. I numbered among the dramatis personae in the dream, although I was not directly involved in the depicted actions themselves. I *suffered* this dream. The dream sequences themselves offered no comment or reflection on the violence they depicted. Experiencing the horror of the images was my part alone. [...] It was obvious that the dream represented an attempt by my psyche to come to grips with the operation" (Dubiel 2009, 87–88).

In addition to the posttraumatic stress symptoms, Dubiel also had to face various side effects. Although the symptoms of Parkinson's disease became much better due to the deep brain stimulation, it also led to the effect that he lost his ability to speak and think properly under neurostimulation: "My overall mobility and endurance were markedly better than before. In contrast, I now suffered from a series of very weird and stressful symptoms [...] My worst post-operative symptom [...] is a speech disturbance: my volume is too low, and my articulation is poor, slurred. Often, I can't even command twenty percent of my normal speech volume" (Dubiel 2009, 94). This is certainly a very serious side effect, especially for a university professor. It vanished immediately when the neurostimulator was switched off however, on the other hand, doing this aggravated the Parkinson's symptoms again. Dubiel describes the effect of switching of the pacemaker for the first time very impressively: "One year after the operation, a light appeared in the tunnel of uncertainty when a neurologist at a different clinic suggested simply turning off the pacemaker as an experiment. It was as if I were channeling a spirit. That very second, my voice returned, sonorous and clearly enunciated, only slightly hoarse. Interestingly, not only was my speech immediately functional in a technical sense, but my intellectual activity and cognitive faculties were quite literally switched on again. During the fifteen minutes that we turned off the device, it was as if a PC were booting up in my head, and its clicking and whirring were signaling that my brain was working" (Dubiel 2009, 118).

Once he had found out this, Dubiel saw himself in possession of a peculiar power. By using a remote control, he was able to influence his personal state via neurostimulation and to choose between having the ability to think and speak clearly or reducing the symptoms of Parkinson's. This ability caused his family and friends to regularly ask him, whether he was switched on or off. The fact that his ability to speak and to communicate with others was based on a technological device, forced Dubiel to think about the effects this might have on the people around him. He was clearly aware of the social consequences attached to the application of invasive techniques like brain-pacemakers. This is a challenge, not only to a person's self-understanding but also for the different forms of human interaction and communication: "The worst part of my new condition is being ashamed in society that my human communications are being mediated by a piece of equipment. More and more often it's being pointed out to me how creepy [...] the technology used in my case must seem to the naïve mind. Long-term medication use will turn a person with a neurological illness into a zombie, a pacemaker turns him into a Frankenstein" (Dubiel 2009, 126). That Dubiel called himself a reluctant cyborg (see Kutter 2015), is more than understandable.

Dubiel's report on deep brain stimulation as a medical treatment, is also of particular value to the discussion about human enhancement because he experienced an invasive technique and was capable of reflecting on this experience philosophically. Although, in his case, deep brain stimulation was a part of medical treatment for Parkinson's disease, Dubiel clearly sees its importance even in a non-medical context. As a sociologist, he emphasizes the social consequences of human enhancement. He points out the shift in the demarcation line between sickness and health and raises the question of what consequences for our understanding of ambition-based society and justice could result from it (see Dubiel 2009, 97–100).

Dubiel also reflects on the consequences which the application of invasive techniques has on the physician-patient-relationship: "Scientifically oriented human medicine is based on the acceptance of mute natural laws that apply without respect of the person. According to this theory of science, a doctor's actions are similar to those of an engineer. Doctors do not view the body within its organic context but see it as an isolated entity. In fact, the practice of medicine consists of a silent intervention in an array of objects, of things. The fact that human beings are endowed with intellect isn't a consideration in this approach. Enlightened physicians have long been aware that ignorance of this dimension can itself lead to disease" (Dubiel 2009, 33).

9.3 Human Enhancement: A Terminological Problem

The non-medical usage of pharmacological or surgical techniques, has serious consequences for our understanding of fundamental medical ethics concepts. It leads to the need to define human enhancement, which is quite a terminological challenge. The problem of giving a definition of the phenomenon, arises right at the beginning of the ethical discourse. There are several approaches to be found, illustrating different aspects. It is quite obvious, that some of these definitions are formulated in the light of the aim which human enhancement should help to achieve, rather than considering the phenomenon itself and the question of its possible consequences for human beings.

Apart from the fact that human enhancement is realized in different fields, such as performance, appearance and capabilities, it follows the aim of accentuating different human traits. In 2008, Nick Bostrom and Rebecca Roache (2008) identified a group of different human enhancement fields, such as physical enhancement, mood enhancement, personality enhancement, cognitive enhancement and genetic enhancement. The field of methods and techniques which are subsumed under the term *human enhancement,* is commonly extremely wide and extends from ordinary, everyday practices, like drinking coffee, to neurosurgical interventions, like setting up a brain-machine-interface. This inflationary usage of the term *human enhancement* is quite problematic, as it eliminates the methods and techniques' different characters and the resulting ethical challenges. Not all of them need to be subsumed under human enhancement (see Bostrom and Savulescu 2009, 2–3).

Though man uses all sorts of techniques to improve his personal skills, the techniques invented by human enhancement have a special character: The term should be used without exception, for those methods and techniques which can be described as invasive. Based on the experiences reported by Helmut Dubiel, the German philosopher Gernot Böhme emphasizes that invasive techniques, like deep brain stimulation, have an effect on the unity of body and mind and fundamentally change both our understanding of being human and living a human life. This is true in a personal and social context (see Böhme 2008, 12–13).

Applying Gernot Böhme's term, invasive techniques might be defined in the following way:

> The term *invasive techniques* refers to technical methods and tools, which modify the set of traits which human beings as members of the species homo sapiens, usually possess. As such, invasive techniques have an impact on what it means to be human and lead to a shift in human self-understanding, both from a personal and social perspective.

Based on these considerations, human enhancement can be defined in the following way:

> Human enhancement means a medically non-indicated invention and the application of invasive technical methods and tools to overcome human being's natural, given limits, who thereby enter a new stage of existence. After a method or tool of human enhancement has been used, being human means something different than before.

In view of the given definition, many methods or techniques which are currently characterized as human enhancement, are wrongly attributed to this term. Indeed, in most of these cases, it appears preferable to speak of optimization rather than enhancement. Any non-invasive optimization techniques therefore do not cause a fundamental shift in our self-understanding as human beings. Based on the fact that modifications can take place in different spheres, such as appearance, performance

Table 9.1 Optimization and enhancement

	Optimization	Enhancement
Appearance	***H**uman **A**ppearance **O**ptimization*	***H**uman **A**ppearance **E**nhancement*
Performance	***H**uman **P**erformance **O**ptimization*	***H**uman **P**erformance **E**nhancement*
Capacity	***H**uman **C**apacity **O**ptimization*	***H**uman **C**apacity **E**nhancement*

and capacities, both human optimization and human performance can be subdivided into three different categories (see Table 9.1).

Most forms of human optimization and enhancement can be applied in a military context. The amount of research, carried out today to enhance different aspects of human appearance, performance and capacities, is immense. Though most methods are still not applied in quotidian military practice, they already, today, play a fundamental role in strategic foresight scenarios (see Fischer 2018). To differentiate between optimization and enhancement by pointing out the effect of invasive techniques on the human being, underlines the special character of human enhancement, when it comes to ethical decision taking. This becomes even clearer, as human enhancement is for many transhumanists, the core element in view of the creation of the posthuman.

9.4 Transhuman Warfare

The declared aim of transhumanism is to overcome human beings' naturally given limits by the invention and application of technical methods and tools: "Humanity stands to be profoundly affected by science and technology in the future. We envision the possibility of broadening human potential by overcoming aging, cognitive shortcomings, involuntary suffering, and our confinement to planet Earth. [...] We favor morphological freedom – the right to modify and enhance one's own body, cognition, and emotions. This freedom includes the right to use or not to use, techniques and technologies to extend life, preserve the self through cryonics, uploading, and other means, and choose further modifications and enhancements" (Transhumanist Declaration 2013). Whether the enhanced individual can still be called a member of the species homo sapiens, is a controversy at the center of the transhumanism debate (see Sorgner 2016, 17–18). Whatever the answer to this question may be, it clearly demonstrates the tremendous consequences of human enhancement techniques. Nothing less than the self-understanding of man as a human being, is at stake.

The fundamental ideas of transhumanism and human enhancement were originally expressed by thinkers like John Burdon Sanderson Haldane (1892–1964), John Desmond Bernal (1901–1975) and Julian Sorel Huxley (1887–1975) in the 1920s and 1930s (see Heil 2010). Their publications are still very thought provoking today, when it comes to thinking about the future existence of man. Haldane's *Daedalus or Science and the Future* (1924), Bernal's *The World, the Flesh and the*

Devil (1929) and Huxley's *What dare I think?* (1931) are classical reference texts, which demonstrate impressively that in the first part of the twentieth century, scientists were already aware of the tremendous impacts that sciences would one day have on the human being. All of them foresee profound changes. Many of the topics which they presented are still determining the philosophical discourse on transhumanism and the posthuman. Although the term *transhuman* has a long history in theological and philosophical thinking (see Cole-Turner 2011), it was Julian Sorel Huxley who first developed a definition of the term *transhumanism* in his book *New Bottles for New Wine,* published in 1957: "The human species can, if it wishes, transcend itself – not just sporadically, an individual here in one way, an individual there in another way, but in its entirety, as humanity. We need a name for this new belief. Perhaps *transhumanism* will serve: man remaining man, but transcending himself, by realizing new possibilities of and for his human nature. 'I believe in transhumanism.': once there are enough people who can truly say that, the human species will be on the threshold of a new kind of existence" (Huxley 1957, 17).

As well as the ever-increasing interest of natural sciences and philosophy in the subject, human enhancement and the development of the posthuman were central elements of science fictional literature. The importance of science fiction as a literary genre can be seen, if we consider how these future visions were spread in a popular scientific way. Most science fiction texts refer to philosophical ideas and it is worth making the effort to discover the latter's impact on science fiction literature. One very interesting example is the similarity of content between Julien Sorel Huxley's future visions as described in *What dare I think?* and his brother Aldous Leonhard Huxley's (1894–1963) novel, *Brave new world* (Huxley 1932). Both books were written in 1931 and offer a wide range of examples of human enhancement (see Heil 2010, 59).

One does not have to be a transhumanist to recognize the influence of the concepts related to transhumanism, when discussing future scenarios in philosophy and particularly in ethics. The aim of creating a posthuman being is no longer unattainable. When the term *cyborg* was originally introduced by Manfred Clynes (born 1925) and Nathan Kline (1916–1983) in their article *Cyborgs and Space* (1960, 27), it was used to describe a technologically improved human organism, who would be adapted to the conditions in space: "For the exogenously extended organizational complex functioning as integrated homeostatic system unconsciously, we propose the term 'Cyborg'. The Cyborg deliberately incorporates exogenous components extending the self-regulatory control function of the organism in order to adapt it to new environments. If man in space, in addition to flying his vehicle, must continuously be checking on things and making adjustments merely in order to keep himself alive, he becomes a slave to the machine. The purpose of the Cyborg, as well as his own homeostatic systems, is to provide an organizational system in which such robot-like problems are taken care of automatically and unconsciously, leaving man free to explore, to create, to think, and to feel." Today, the term *cyborg* is of course applicable to a huge number of technically altered men and women, who use all sorts of human enhancements (see Spreen 2015, 27–38).

The fact, that humans might transform themselves into posthumans by overcoming naturally given limits in the way stated above, clearly also has implications on military ethics, military medical ethics and the law of armed conflict. As soon as somebody who must be declared as posthuman becomes involved in military action, most of the concepts of humanitarian law and related ethics may need to be revised.

The term *transhuman warfare*, which will be proposed in this context, takes into account the deployment of posthuman soldiers in military conflicts. Transhuman warfare could be defined in the following way:

The term *transhuman warfare* relates to military conflicts involving soldiers who, for reasons of military necessity, have been altered by human enhancement, by the invention of technical methods and tools to overcome various physical and psychological limits and the implementation of different traits beyond those of technically unaltered members of the species homo sapiens.

Both in civil and in military life, the creation of a posthuman being by inventing human enhancement techniques will obviously have consequences for man's self-realization as an autonomous moral subject. To clarify what a posthuman autonomous moral subject might be like, is a very thought-provoking challenge; it is important to realize that each term (autonomy, morality, subject) only makes sense in relation to the understanding of human beings as human beings. This leads to the next point and the crucial question: What does it mean to be human?

9.5 What Does It Mean to Be Human?

In the light of these possibilities, the question, what does it mean to be human, takes on a new significance. This very old, anthropological question (anthropological in the sense of continental philosophy) achieves a topicality which when compared to the past, cannot fail to consider both the essence and the existence of man. The urgent need to answer this question in the light of today's developments, affects both natural-scientific and humanities thinking.

Not least ethical thinking is called to provide a service for humanity here. Ethics must offer support for decision-making, which helps morality to assert itself in the face of the posthuman challenge. To be able to cope with future realities in this field, whatever they may be, there is a strong need to adopt a forward-looking visionary view in moral philosophy. There is a number of bioethicists who proclaim the need for a so-called moral enhancement (see Persson and Savulescu 2012). This demand is due to the fact that they do not believe that today's humans are intellectually capable of handling the outstanding moral challenges. Aside from the call for moral enhancement and the question of what shape a moral enhancement might take, human enhancement leading to posthumanity is certain to be connected with a paradigm shift in moral philosophy.

In the past, military medical ethics emphasized the importance of helping humanity achieve a succeed even in times of war. The asymmetric conflicts of our time represent an extraordinary threat to this aim (see Fischer 2015). Nevertheless, a transformation of military conflicts into transhuman warfare, will intensify these challenges: Which component of the posthuman is a legitimate target to fight: the technical or the human part? Concerning military medical care, the question, whether physicians or members of the military medical service are obliged to help the posthuman in the same way as they currently help the human, must be answered. How should we handle the demand for human enhancement both in our own society and the military, in view of the development generally and the military opponent's commitment to human enhancement techniques in particular?

The questions, what does it mean to be human and how can humanity be served, are of crucial importance when physicians are obliged to face the possibility of human enhancement. Should the application of human enhancement techniques be part of a physician's duty? In fact, there is no absolute necessity for making it part of their responsibility.

If physicians find themselves at the center of the human enhancement movement, their core concepts of self-understanding will be shaken. Terms like *health*, *illness*, *physician*, *patient* or *physician-patient-relationship* are unlikely to be defined in the same way as they are now. Even today, they are critically questioned because of the involvement of physicians in the field of performance, appearance and capacity optimization. The fact that these treatments promise an enormous economic profit, does not justify a commitment to pushing every demarcation line aside. The above-mentioned must to take into consideration that whether medical treatment is due to medical necessity or not, appears to be a crucial point in this context.

To restrict one's own commitment strictly to what is medically indicated, is a way to protect both the physician's profession and its underlying ethical concepts. Even if human optimization is close to prevention, diagnostics, therapy and rehabilitation, human enhancement as defined above, is not.

Pointing out these aspects does not mean condemning human optimization and human enhancement. Regardless of whether or not it is desirable to transform humans into posthumans, the development of human enhancement techniques has already begun and will provide a huge number of benefits to mankind. The physician's role within this development will need to be defined. Whether human enhancement has to be part of a physician's practice is questionable. As an alternative, a new profession, of an enhancer, could be established, leaving the profession of physician, as someone who takes care of the sick and wounded in a particular way, untouched. Of course, anyone who is involved in enhancing human beings must have a fundamental knowledge e. g. of anatomy, physiology, neurology or pharmacology, as well as technical skills, but there is no need for the enhancer to be a physician.

9.6 Conclusion

In conclusion, the main elements of the deliberations above, will be summarized. Helmut Dubiel's case offered a very impressive insight into the effects on the existential dimension which invasive techniques might have, both on our self-understanding and the way we are seen by others. The development of invasive techniques to treat an illness like Parkinson's disease, is just one step towards the employment of similar techniques in non-medical cases to improve human traits and skills. The difference between medical and non-medical necessity plays an important role when it comes to handling human enhancement techniques.

The term *human enhancement* is being used to describe a wide range of methods and techniques. For most of them, the term *human optimization* would appear to be more appropriate. This is particularly true if the term *human enhancement* is used to describe those invasive methods and techniques which are used to go beyond human beings' natural, given limits, who thereby enter a new stage of existence. It is crucial to realize that by employing a method or tool for human enhancement, being *human* means something different to before.

The consequences of this development can hardly be foreseen today. Nevertheless, developing future scenarios seems to be unavoidable if moral philosophy is to offer decision support for dealing with human enhancement in general and the application of invasive techniques in particular. The term *transhuman warfare* has been introduced in this context. It takes into account that warfare as it is known today, will be changed completely by the invention of means of human enhancement.

The role which the military medical service has to play in this context, still has to be defined. As in civilian contexts, human enhancement will lead to a paradigm shift in medicine. Therefore, the question, whether human enhancement must be part of the physician's profession, has to be answered. It might possibly be better to develop a new kind of professional, called an enhancer, who clearly must have medical knowledge (as well as knowledge of engineering, chemistry, biology etc.), but who does not necessarily have to be a physician.

Medical ethics are one thing, the ethics of enhancement are something different, even if both fields of applied ethics are closely linked to each other. Mixing up the two would have severe consequences for medical ethical concepts. The latter, which are based on our understanding of physician, patient, health, illness etc., could be preserved by distinguishing medical ethics from the ethics of enhancement. The demarcation line between the two fields of conduct could be that medical personnel react solely on the basis of medical necessity. The practical guideline of medical necessity is medical indication, a term which has been brought to renewed importance in the light of the enhancement debate.

Whatever the future developments in this field might be, medical personnel today are already being confronted with the question, what does it mean to be human? In military medical ethics, the aim of helping humanity to succeed, even in times of war, has always been a crucial element. In the light of the future opportunities offered by human enhancement, this will be a particular challenge.

References

Bernal, John Desmond. 1929. *The world, the flesh and the devil*. London: Kegan Paul.

Böhme, Gernot. 2008. *Invasive Technisierung. Technikphilosophie und Technikkritik*. Kusterdingen: Die Graue Edition.

Bostrom, Nick, and Rebecca Roache. 2008. Ethical issues in human enhancement. In *New waves in applied ethics*, ed. Jesper Ryberg et al., 120–152. Basingstoke: Pelgrave Macmillian.

Bostrom, Nick, and Julian Savulescu. 2009. Human enhancement ethics: The state of the debate. In *Human enhancement*, ed. Julian Savulescu and Nick Bostrom, 1–22. Oxford: Oxford University Press.

Clynes, Manfred, and Nathan Kline. 1960. Cyborgs and space. *Astronautics* 1960 (26–27): 74–76.

Cole-Turner, Ronald. 2011. The transhumanist challenge. In *Transhumanism and transcendence. Christian Hope in an age of technological enhancement*, ed. Ronald Cole-Turner, 1–18. Washington, DC: Georgetown University Press.

Dubiel, Helmut. 2006. *Tief im Hirn*. München: Kunstmann.

———. 2009. *Deep in the brain. Living with Parkinson's disease*. New York: Europa Editions.

Fischer, Dirk. 2015. The threat to humanitas in asymmetric conflict. *Medical Corps International Forum* 1: 42–45.

———. 2018. Wir sollten menschliche Existenz nicht unreflektiert verändern. *Ethik und Militär. Kontroversen der Militärethik und Sicherheitskultur* 1: 57–58.

Haldane, John Burdon Sanderson. 1924. *Daedalus or the science and the future*. London: Kegan Paul.

Heil, Reinhard. 2010. Human Enhancement – Eine Motivsuche bei J. D. Bernal, J. B. S. Haldane und J. S. Huxley. In *Die Debatte über Human Enhancement. Historische, philosophische und ethische Aspekte der technologischen Verbesserung des Menschen*, ed. Christopher Coenen et al., 41–62. Transcript: Bielefeld.

Huxley, Aldous Leonhard. 1932. *Brave new world*. London: Chatto and Windus.

Huxley, Julian Sorel. 1931. *What dare I think? The challenge of modern science to human action and belief*. London: Chatto and Windus.

———. 1957. *New bottles for new wine*. London: Chatto and Windus.

Kutter, Susanne. 2015. "Ich kam mir vor wie ein Versuchskanienchen…" – Interview mit Helmut Dubiel. *Wirtschaftswoche*, 11. März.

Persson, Ingmar, and Julian Savulescu. 2012. *Unfit for the future. The need for moral enhancement*. Oxford: Oxford University Press.

Sorgner, Stefan Lorenz. 2016. *Transhumanismus – »Die gefährlichste Idee der Welt«!?* Freiburg: Herder.

Spreen, Dierk. 2015. *Upgrade Kultur. Der Körper in der Enhancement-Gesellschaft*. Bielefeld: Transcript.

Transhumanist Declaration. 2013. In *The Transhumanist reader*, ed. Max More and Natasha Vita-More, 54–55. Chichester: Wiley-Blackwell.

Chapter 10
Genetic Science and the Future of American War-Fighters

Sheena M. Eagan

10.1 Introduction

In 2017, the United States Defense Advanced Research Projects Agency (DARPA) budgeted 100-million-dollars to fund gene-editing technology. Recent reports speculate that the military's interest in this technology focuses on enhancement (JASON 2010; Committee on Opportunities in Biotechnology for Future Army Applications, Board on Army Science and Technology, National Research Council 2001). Specifically, they hope to use genetic manipulation to enable war-fighters to run at super-human speeds, carry enormous weight, live off their fat stores, and go without sleep. While this enhancement would inevitably lead to increased survivability in war, there are significant and warranted ethical concerns.

This paper will provide a brief overview of genetic enhancement technology, focusing on the Department of Defense (DoD) priorities and interests in the field. Relevant ethical issues will be explored, discussing topics such as autonomy, informed consent, and risks of coercion. This analysis will also discuss issues around transparency, privacy, secrecy, and harm by way of psychological injury. While much discussion has focused on the dilemma of dual-use (Patrone et al. 2012; Resnik et al. 2011; Selgelid 2009; Drew and Mueller-Doblies 2017), or research ethics (Azarow 2003), little has been said about the long-term ramifications of enhancing service-members who must ultimately rejoin civil society. A specific concern highlighted by this research is that these genetically enhanced service-members may be unable to re-integrate into civilian society. These issues must be discussed, as a plethora of reintegration issues already exist (Shay 2010), and the

S. M. Eagan (✉)
Department of Bioethics and Interdisciplinary Studies, Brody School of Medicine, East Carolina University, Greenville, NC, USA
e-mail: EAGANSH17@ecu.edu

D. Messelken, D. Winkler (eds.), *Ethics of Medical Innovation, Experimentation, and Enhancement in Military and Humanitarian Contexts*, Military and Humanitarian Health Ethics,
https://doi.org/10.1007/978-3-030-36319-2_10

enhancement of service-members to be ideal warriors may only serve to widen the divide between soldier and civilian. This paper will specifically focus on this under-discussed element within the body of literature on this topic, examining genetic enhancement's precarious position at the intersection of civilian and military spheres, as well as the possible ramifications for service-members as they separate from service. While this analysis will focus on the United States military, the ethical issues and dilemmas that this research highlights are meant to be generalizable and can be extrapolated to most professional militaries in similarly situated cultures. Importantly, much of this analysis is speculative addressing concerns for the near-future application of genetic enhancement technologies within the military. Ongoing ethical analysis will be necessary as research regarding the military application of this new technology continues.

10.2 Enhancement in the Military

It is important to note that military attempts to enhance war-fighters are not new. Instead, it is the technological approach (genetic enhancement) that is novel and uniquely invasive within the history of military enhancement research. Historically, militaries have always sought to increase both their fighters' chances of survival in conflict as well as their ability to successfully complete a mission (or win a war). This history of enhancements has included the introduction of personal protective equipment such as shields, helmets, gas masks and eventually Kevlar body armour; weapons development; and the development of military vehicles such as tanks and armoured personnel carriers (APCs). Of course, other important enhancements that have had positive effects on service-member survivability and mission success were less technological. The introduction of sanitation and hygiene can be understood as a widely successful enhancement in early warfare where more soldiers died of epidemic diseases and infections than battle wounds. Similarly, ongoing advancements in vaccine development, front-line aid, trauma surgery, and evacuation systems continue to enhance war-fighter success. These enhancements represent essentially morally neutral interventions into the lives and bodies of service-members, setting them apart from more recently proposed enhancements. While some of the above-listed enhancements brought with them ethical dilemmas, their benefits were evident and negative ramifications relatively low. The proposed near-future enhancements of war-fighters—the focus of this research—will involve interventions on the genetic level and thereby challenge this history of military enhancement while raising new ethical issues. How can we ensure responsible research conduct when using gene-editing technology? Can soldiers consent to permanent biological enhancement or manipulation? If there is a permanent genetic alteration that affects the germline, can they consent for their future children? How can we ensure that this enhancement is not coerced or forced? Will civilians have access to the same types of genetic enhancement, or will it be limited to military use? Moreover, who owns the enhancements—the military or the individual service-member? Additionally, if these enhancements are permanent, how will the service-member function in society after separation from the military?

The U.S. Department of Defense (DoD) has long demonstrated an interest in implementing genetic technologies within the military (Committee on Opportunities in Biotechnology for Future Army Applications, Board on Army Science and Technology, National Research Council 2001; JASON 2010). In 2010, the JASON defense advisory panel released a report on the opportunities and challenges of genomic technologies in the military context (JASON 2010). The JASON report's primary recommendation was a call to action, advising the DoD to establish policies and practices based on the latest genomic science (JASON 2010). To date, the DoD has successfully implemented a DNA registry for identifying human remains and routinely screens service men and women for genetic conditions (Department of Defense 2017). There have also been other large pilot projects and research collaborations in the field of genetic science. Much of this research makes use of the DNA registry, and other voluntary biobanks to examine genotypes and phenotypes relevant to military readiness (Pence 2014). While a considerable amount of the research in this field has focused on therapeutic innovation, disease prevention, and genetic screening, the military is also interested in how this technology can be used to create better soldiers (Illing 2018; Mehlman and Li 2014).

There are arguably two ways that the military aims to use genomic science to make better soldiers and improve its overall mission readiness (Mehlman and Li 2014). First, it may be possible to identify genotypes with predictive power concerning ability and military performance. Then, by way of genetic screening, it is possible to screen incoming recruits and use this genetic information as inclusion/exclusion criteria for military service (Lázaro-Muñoz and Juengst 2015). Second, it may soon be possible to use gene-editing technologies to enhance those already in service. Gene editing would enable the military to use genomic technologies that may enable soldiers to sleep less, eat less, and increase overall physical ability. Each of these methods of enhancement will be discussed in turn, focusing on any morally relevant components and highlighting ethical issues.

10.3 Genetic Screening

Many may not consider genetic screening to be a form of enhancement. However, when considering the military at a population level, it seems clear that the inclusion/exclusion criteria for service, or combat, has a direct effect on the military's ability to succeed and be mission ready. Additionally, in any discussion of medical (or biological) human enhancement, it is important to note that the line between therapeutic or diagnostic technologies and those aimed at enhancement is not clear-cut. Bioethicists and others in the field have long debated the line between therapy and enhancement. While genetic screening might not initially appear to be a form of enhancement, its purpose within the military context is to enhance mission readiness. The genetic screening tools that will be discussed do not aim to identify and treat disease, but to exclude or include populations based on genetic markers. Aims and intent are a morally relevant component to understanding policy and practice. We will show that the intent behind screening is force health and readiness enhancement, and thus the military's involvement in screening can be assessed as part of the genetic enhancement debate.

Currently, the DoD's largest endeavour in genomic technologies is the biobanking of service-member DNA. The Armed Services Repository of Specimen Samples for the Identification of Remains (AFRSSIR) was first established in 1993 and as of 2007 included over 5 million blood samples (Miles 2007). Under current policy, all service-members are required to participate, with no ability to opt-out (Department of Defense 2017). Additionally, unless destruction is requested, these DNA samples are stored by AFSSIR for up to 50 years—long after their separation from service or retirement (Department of Defense 2017). Of relevance to our discussion, the DoD's biobank has been used for more than the mere identification remains. In line with the recommendations from the 2010 JASON report (JASON 2010), the DNA collected by the military can be decoded or sequenced and connected to the service-members medical records creating a 'geno–phenobank.' In 2009, a $65 million, six-year collaborative project between the U.S. Army and the National Institute of Mental Health aimed to use this biobank to identify risk factors for military suicides (The National Institute of Mental Health 2011). An additional component of this research was the *New Soldier Study*, which asked new enlistees to voluntarily complete surveys and undergo neurological and DNA testing aimed at identifying genetic and other risk factors (The National Institute of Mental Health 2011). Finally, the US Department of Veterans Affairs (VA) has also established its own biobank. The "Million Veteran Program," is a geno–phenobank that collects DNA and health information from veterans and then combines it with their VA medical records for research purposes with over 250,000 veterans enrolled (Office of Research and Development 2018).

One of the ultimate aims of these bio-banks is to create the aforementioned geno–phenobanks, which can then be used to 'determine which phenotypes might reasonably be expected to have a genetic component that has special relevance to military performance and medical cost containment'. (JASON 2010) Thus, military interest in genomic testing extends far beyond the prevention, diagnosis, and treatment of disease. Instead, as the JASON report anticipates, these biobanks can be used to understand genetic profiles for inclusion and exclusion from military service (The National Institute of Mental Health 2011; JASON 2010).

As an example, many speculate that there is a genetic component to Post-Traumatic Stress Disorder (Connelly 2012; Chang 2009). Twenty different genes associated with PTSD have been identified, with continued research aimed at identifying a genotype that predicts susceptibility to PTSD. When identified, service-members with this genetic predisposition could be either excluded from service or given non-combat duties. Of course, identifying genetic predictors of PTSD is complicated and perhaps even impossible, but these geno-phenobanks aim to do just that—and PTSD is not the only target.

These biobanks could also be used to understand the genetic makeup of the so-called *ideal-soldier*. The research aims to identify genetic markers associated with genotypes for exclusion as well as inclusion. Theoretically, it could be possible to assess the genetic makeup of war-heroes or successful leaders and identify commonalities on the genetic level, then selecting for these traits within the population of incoming recruits. Alternatively, the military could identify those whose genetic profiles are more easily edited and enhanced, then prioritizing this population for enlistment and commission to combat roles. In this way, the blurred line between diagnostics, therapy, and enhancement is evident. Biobanks and genetic testing could enhance the military service-member population by weeding out specific populations and creating a military of the identified ideal genotype.

10.3.1 Ethical Issues in Genetic Screening

Military biobanks raise critical ethical concerns surrounding consent and autonomy. Since the military population is one of reduced autonomy and liberty, they are often understood as a vulnerable population warranting specific protections. The hierarchical system upon which the military institution is built relies on obedience and loyalty. Especially within the enlisted ranks, service-members are expected to unreflectively follow orders, and not question them (unless the order is illegal). The recognition of the coercive power of military rank structure has led to the recognition of service-members as a vulnerable population (or special population)—a fact thoroughly discussed in the field of research ethics (McManus et al. 2002, 2005). In the case of biobanks, service-members cannot say no—they cannot opt out and must surrender a blood sample to AFRSSIR. Additionally, the maintenance of this genetic material for a half-century means that the ability to use service-members' DNA for research purposes will likely ebb and flow depending on changes in government and related political shifts. While former service members can now request that their DNA sample be destroyed, this is only possible after separation from service and must be specifically requested using appropriate paperwork and channels. The requirement to specifically request destruction, as opposed to an automatic system of specimen destruction upon separation, means that many service-members will inevitability not take this opportunity as they may not be aware of the option or the administrative burden may be too onerous.

The use of this material for research complicates the moral picture, as service-members become unknowing participants in genomic research. This fact is worthy of concern and highlights a need for two separate biobanks; the first mandatory and used for the identification of remains, and a second that is voluntary and used for research. While these concerns have been recognized, and research is now limited within AFRSSIR, the military maintains the ability to use this information for operational purposes. Additionally, the fact that the military does not automatically destroy samples upon separation also places their intent into question—if the intent of this database is the identification of remains for those killed in action, it seems unnecessary to maintain these samples once the person has separated from service and is no longer active-duty.

The military's aim of identifying connections between desirable phenotypic traits and genetic markers raises additional concerns surrounding discrimination. While American civilians are protected by the Genetic Information Non-Discrimination Act (GINA), this Act does not apply to military personnel (Evans and Moreno 2014). In light of this, service-members provide DNA that could ultimately be used to determine either their role in the military or their ability to serve. Conceivably, if genetic markers for PTSD were identified, those service-members with these markers could be assigned to state-side, desk jobs for the entirety of their career, or alternatively removed from military service entirely. If genetic screening was implemented in the recruitment phase, it could also lead to discrimination against an entire population, who has now been deemed unfit for military service.

This fact highlights the positioning of this technology at the intersection of civilian and military spheres. Genetic screening could have the unintended consequence of marginalizing a population based on their perceived inability to serve. This type of stigmatization is recurrent in the history of Post-Traumatic Stress Disorder, dating back to the First World War when militaries believed that shell-shock had a genetic or familial basis (Eagan Chamberlin 2012). This understanding of shell-shock was tightly bound to concepts of eugenics and racial medicine, and thus inherently discriminatory. The same type of discrimination is possible with genetic material, especially given our increasing understanding of epi-genetics and the social determinants of health. Therefore, we must consider the possible individual and social harms that genetic screening may bring with it and attempt to address these. In fact, the importance of social, legal and ethical considerations in genetic research is well established and mandated by the National Institutes of Health in the United States.

Those that would be excluded or removed from service are not the only ones that may be harmed. Those without these markers could be assigned to high-stress combat roles and sent on repeated deployments possibly causing psychological harm. As mentioned earlier, it is likely that genetic screening would aim to not only exclude specific genotypes from service, but also to select for genotypes identified as best-suited for military roles. With its focus on war-fighters, the military may preferentially select those that possess the traditional traits of the model warrior, or those that are more easily edited (gene-editing will be discussed in the next section). This population would then be deployed to their genetically-determined warrior role. Depending on the state of international affairs and the number of ongoing conflicts, these service-members could face an increased operational tempo with multiple deployments. As will be discussed in detail later (Sect. 10.4.2), this could potentially cause significant harm to this population due to the connection between repeated deployments and psychological trauma. Even if this population is understood as genetically superior and able to psychologically withstand battle better than others, we do not know the psychological harms brought with military dependence on one specific genetic population for repeated deployments.

As discussed, genetic screening is not the only form of genetic enhancement that interests the military. In this next section, the analysis will focus on the use of gene-editing technologies to enhance those already in service, possibly enabling soldiers to require less sleep and less food, while also improving physical ability. This second type of genetic enhancement is arguably more problematic since it is more invasive and involves changes to service-members on the genetic level. These changes, either temporary or permanent, could have far-reaching ramifications in both the service-member's military and post-military life.

10.4 Gene-Editing

Gene-editing is not new. However, the discovery of CRISPR/Cas9 technology at the end of the twentieth century has propelled research and capabilities in this field. CRISPR is an acronym for the name *Clustered Regularly Interspaced Short Palindromic Repeats*. The CRISPR system is a natural one developed by bacteria over the course of evolution; a system the bacteria use as its own gene-editing system. Essentially, CRISPR systems identify foreign genetic material that has been inserted into the bacterial genome by viruses and edits itself by snipping away the invasive genetic material and thus repairing the genome. In the past two decades, scientists have come to recognize that this system is usefully exploited in the laboratory to enable genetic engineering. CRISPR rapidly overtook other methods for gene-editing and became the go-to system as it is easier and cheaper than previous methods, with increased target specificity. CRISPR technology holds promise in the worlds of both therapy and enhancement. The ease with which one can edit genes using CRISPR technology is unrivaled by other types of gene-editing and has led to a type of democratization of science wherein lay-people (non-scientists) are using this technology at home. These so-called 'biohackers' have attempted to use CRISPR to cure disease and change physical attributes byway of self-experimentation that has been widely publicized (Williams 2016). Within the scientific and medical community, CRISPR is being recognized as a way to enhance human performance by editing out detrimental genes and replacing them with a preferred alternative. Gene-editing could be used to make people smarter, less error-prone, increase muscle mass, or alter their metabolism to function on fewer calories, among other abilities. These specific traits would make for a good war-fighter, and the military has taken notice.

In 2001, the Committee on Opportunities in Biotechnology for Future Army Applications of the Board on Army Science and Technology at the National Research Council provided direction to the military in this area. They recommended that the Army should "lead the way in laying ground-work for the open, disciplined use of genomic data to enhance soldiers' health and improve their performance on the battlefield" (Committee on Opportunities in Biotechnology for Future Army Applications, Board on Army Science and Technology, National Research Council 2001). The relevance of gene-editing to the military was re-emphasized by both the

JASON report and a 2002 report by DoD Information Assurance and Analysis Center (JASON 2010). Most recently, DARPA budgeted 100-million-dollars to fund gene-editing technology, granting four-year contracts to seven teams (DARPA 2017). While this type of enhancement is still mostly theoretical and not yet in use, the emphasis on this type of enhancement is undeniable, and the ethical issues should be discussed before the technologies are implemented.

Many of these projects are defensive and seek to make service-members immune to neuro-weapons that would use gene-editing against them in a battlefield scenario (DARPA 2017). This defensive research raises the classic dual-use concerns that are often discussed in military medical ethics. Although most of the projects aim to develop genomic technology and pharmacological interventions that are effective in the prevention of gene-editing, they must first also develop the offensive, or weaponized, gene-editing technology to then be able to develop defenses against it. The dual-use issue highlights this blurring between defensive military developments, and the need to first develop offensive technologies. According to a variety of international doctrine including the Geneva protocol, and the Biologic and Toxin Weapons Convention (BTWC), the use of biological and chemical weapons is forbidden in armed conflict and cannot be developed by militaries (Drew and Mueller-Doblies 2017). They can, however, do research and development in order to provide defensive measures against these types of threats. Importantly, genetic weapons are not explicitly covered by this doctrine. Some have argued for its inclusion, while others have argued for additional and separate guidelines for the use of genetic technologies in a military context. Either way, most agree that genetic weapons should be restricted, or at least limited.

Beyond weaponized gene-drives, some of the funded DARPA projects aim to develop safe, and possibly reversible, gene-editing technologies in animal models. This research is in the early stages, as it aims at "building initial mathematical models of gene editing systems, testing them in insect and animal models to validate hypotheses, and feeding the results back into the simulations to tune parameters" (DARPA 2017). In light of the early stages of research, discussion of ethical issues on this topic are mostly speculative and ongoing analysis will be required as this technology further develops. However, the underlying objective is undeniable and clearly stated in the numerous reports cited earlier (JASON 2010; Committee on Opportunities in Biotechnology for Future Army Applications, Board on Army Science and Technology, National Research Council 2001). This research aims to use gene-editing in the military population to enhance it at both the population and individual level—specifically; the military is interested in creating real-life super soldiers.

10.4.1 Ethical Issues in Responsible Research Conduct

There are complex ethical dilemmas related to these proposed enhancements. Since this work is in the research stage, we must first examine questions concerning responsible research conduct in gene-editing or genetic manipulation.

Inevitably, this research will be done on military installations, involving military populations. As previously mentioned, military service-members should arguably be classified as a 'vulnerable population' (McManus et al. 2002, 2005). Their vulnerability lies in the moral realities of being a service-member; this population is understood to have relinquished their autonomy in service of the state. In light of being a service-member, these men and women inhabit a world of limited autonomy and constrained liberty (Huntington 2008); they do not get to decide how to dress, where to live, or whether or not to go to war. Even given these contextual features, research remains a protected space where this population should be able to exercise unrestricted autonomy.

Genomic research also raises essential questions about the ability to obtain informed consent from a population given the novel nature of the work. Serious doubts have been expressed concerning the ability of a research subject to adequately understand the information given to them about the study, especially in complicated research involving genomics. In light of this population's limited autonomy, and education level (enlisted populations generally hold high-school diplomas, or increasingly Associate's and Bachelor's Degrees), there must be clear expectations of informed-consent in military research. Additionally, informed consent must be obtained with particular attention to the avoidance of coercion. The coercive powers of the military hierarchy cannot be denied, has been well-established, and must be controlled. Currently, there is a vast body of policy governing American research both inside and outside of the military context. Military installations conducting research have Institutional Review Boards composed of military personnel and tasked to evaluate any ethical issues that may arise in research. These boards pay special attention to coercion, particularly of enlisted personnel and require that there are sufficient safeguards for this population. Safeguards often attempt to remove the coercive power of military hierarchy by not permitting commanding officers' presence during information briefings and consenting. Those that are providing information and consenting subjects often do not wear military uniforms during this process. Of course, these same requirements should be maintained in future gene-editing research, with additional concern paid to informed consent.

Many have argued that genetic manipulation, such as gene-editing, requires even stronger protections. War-fighters may feel compelled to participate in research out of patriotic duty, or because of military culture that obliges them to be the best warrior that they can be. According to this cultural narrative of strength and excellence, war-fighters may feel as though they must consent to enhancement as part of this duty to be the best they can be. On a pragmatic level, why wouldn't someone want to be stronger, run faster, and require less food or sleep, especially when heading to a war-zone? These capabilities would improve one's chances of survival, and better contribute to their unit thereby increasing the chance of survival for others. In light of these dimensions, any research protocol involving military personnel must include safeguards against coercion.

It is also important to remember that laws or policies establishing civilian protections may not apply to this population. For instance in the United States, the Genetic

Information Non-Discrimination Act (GINA) does not apply to military personnel (Evans and Moreno 2014). Gene-editing interventions will inevitably include the collection of genetic material and information about all participants, including additional data for inclusion and exclusion decisions. Genetic testing and genome sequencing (that may be involved in this research) have the potential to generate large amounts of data about individuals and their health, which has the potential to challenge current data protection models and cause potential harm. While current policies and regulations do exist, the ever-expanding scope of genetic information available and increased use of electronic medical records may challenge these protocols. Access to this data must be restricted by way of policy.

Data should be accessible only to the individual patient, the research team, health care providers directly involved in treatment or genetic counseling and others only with patient permission (following a specific consenting process regarding appropriate information of risks). While this might be complicated by some militaries' lack of doctor-patient confidentiality for operationally relevant health information, a new standard for genetic information is needed. The specific concern is that the accessibility of this information to military command, insurance companies, and possible future employers could lead to discrimination. Therefore, access rights to the genetic information by insurance companies or employers have to be clearly defined. To date, genetic information remains an inappropriate basis for making employment or insurance decisions. This genetic information should be protected and not widely disseminated. It is also critical that military service-members are not discriminated against based on this information. For instance, should their genetic information reveal pre-disposition for a specific disease they must be counseled according to ethically sound genetic-counseling standards and not victimized for this genetic issue. Genetic information and genetic counseling should always be required in tandem, to ensure adequate information is provided to the patient. Additionally, service-members should not be discriminated against based on their ability to be successfully enhanced using gene-editing. The lack of formalized policy protecting service-members genetic information must be adequately addressed and rectified.

A further complication to informed consent and data protection lies in the unknown or unintended harms. When discussing gene-editing, we must keep in mind that the potential permanence of this genetic manipulation has the potential for long-term and unforeseen harm, or unintended consequences. As with any permanent change, it is possible that gene-editing will be impossible to reverse without further complications, interventions, or harm. While DARPA projects aim to develop safe and reversible technologies, this begs the question as to whether or not 18-year-old recruits can adequately consent to genetic manipulations that may affect them for the rest of their lives. Thusly, the possible permanence of gene-editing also raises additional issues concerning informed consent. Specifically, if gene-editing can potentially affect the germline, how should we understand the potential harm to future offspring, as well as their inability to consent to this genetic modification? Within the military, this is especially relevant given that the majority of combatant service-members are at peak procreative ages—predominantly males age 18–24. Of additional concern within the United States military, is the high rate of unintended pregnancies within this population. The combination of high rates of unplanned or unintended pregnancies and possible germ-line affects from gene-editing could create issues for not only the future off-spring of service-members but also the role of military healthcare to support future dependents.

10.4.2 War-Fighters for Life? Ethical Issues in Veteran Re-integration

The possible permanence of gene-editing brings us to an under-discussed and morally relevant consideration—what happens to enhanced war-fighters when they separate from military service? Whether in research or as part of future military policy, the ethical discussion surrounding genetic enhancement in military service members focuses on them as war-fighters. However, these people are not soldiers for life. At some point, whether upon discharge or retirement, service-members return to civilian life, and must rediscover their civilian selves. Currently, the American military, along with American society as a whole, has failed to adequately reintegrate service-members (Shay 2010). Upon separation from service, these men and women can end up neglected and struggling. Veterans suffer from substantially higher rates of PTSD, and depression; are at increased risk of suicide, addiction, violent acts; and are more likely to end up homeless or in the prison system (VA 2016; Shay 2010). These issues are perhaps most prevalent among combat veterans. Importantly, these realities are not indicative of a personality type that joins the military, but instead of a cultural inability (or unwillingness) to adequately re-integrate our war-fighters into American civilian society. If we fail to re-integrate soldiers now, what will happen when these former-service members have been further distanced from civilian society by way of genetic enhancement?

The military institution has created a sub-culture for its service-members that is purposefully separated from civilian society (Huntington 2008). Upon enlistment or commission, service-members undergo a rigorous, intensive, and emotionally-laden indoctrination process. Put bluntly; this process is meant to take civilians and transform them into warriors. They repeat their service's values, or ethos, ad infinitum—the taboos of civilian life are removed, and a new set of values are instilled. The military also wears uniforms and uses language to separate themselves from their civilian counterparts (Huntington 2008). Beyond that, those in military service have experiences that civilians will never have, nor ever fully understand. In fact, there has been a growing concern that American civilians are becoming increasingly distanced from the military community. A lack of understanding between these two communities (civilian and military) inevitably affects both service-member morale and veteran re-integration, reducing empathy and acceptance of each population by the other. Unfortunately, genetic enhancement brings with it the possibility of further exacerbating these issues and driving a deeper wedge between the two populations. If war-fighters already feel separated from civilian society and unable to reintegrate, it can be hypothesized that enhanced war-fighters will face additional struggles.

How will those with enhancements fit in? Specifically, if genetic enhancements remain illegal (or unobtainable) to civilians, former service-members will be further 'othered' within broader society. This population could possess abilities that give them certain advantages in relation to manual labor, cognition, and athletics. Perhaps an argument could be made that this will ease their re-integration as they can excel and outperform their civilian counterparts. However, another (and more likely) argument is that they will only be further excluded and ostracized. Much like pharmacological performance enhancers in sports, it is plausible that these enhanced service-members will be banned from specific civilian activities where they are perceived to have an unfair advantage.

It seems obvious that any future enhancement technologies will be reasonably restricted. The arguments against the wide availability of this technology in the civilian world seem clear—we do not need, nor necessarily want, super-soldier-civilians. However, it is relevant to consider the military-specific question: what, if any, kinds of enhancement should be off-limits to the war-fighters? Ethicists have written on this topic and argued that some enhancements should be understood as ethically impermissible regardless of the military or civilian application (Mehlman and Li 2014). For these thinkers, the line is drawn when enhancement compromises human dignity. While not explicitly stated, the underlying reason for their impermissibility is directly related to our discussion as it rests on the potential harms associated with the integration of enhanced people with the unenhanced. Mehlman and Li argue that any enhancement that compromises the dignity of the war-fighter should be considered off-limits—specifically any enhancement that causes stigmatizing or disfiguring physical characteristics (Mehlman and Li 2014).

Their discussion of the impermissibility of this type of enhancement reminds us that we cannot ignore the psychological injury that could be incurred through enhancement. An argument that is particularly relevant to our discussion is that of the enhanced war-fighter. This population is already more likely to suffer psychological injuries as a result of military deployment, and the increased risks (and traumas) associated with it. These men and women would inevitably face increased deployment tempo given that their enhancements are aimed at improving skills relevant to combat. As discussed earlier, it is possible that this population will be chosen based on their ability to withstand the psychological trauma of conflict. However, this immunity from combat stress or PTSD is theoretical and not guaranteed. From a psychological perspective, it seems that many enhancements could be ethically problematic and possibly confound the current problem with psychological injuries in this group (combat veterans). There is no way to know how multiple deployments would affect these fictitious super-soldiers, however, within the current military population, multiple combat deployments are positively correlated with increased risk of psychological trauma (Shay 2010). In light of this fact, we can reasonably expect that this population will suffer increased psychological injury and increased rates of

PTSD due to the increased risk of trauma and injury that accompany prolonged time in a combat zone. This possibility only serves to complicate the military to civilian transition further, as those with PTSD already face challenges re-integrating into civilian society.

The abovementioned morally relevant issues are predicated on the assumption of permanence. However, gene-editing technology could conceivably be used to undo any military enhancements. As mentioned, funded DARPA projects specifically aim to develop reversible gene-editing technology (DARPA 2017). The possibility of temporary enhancement may alter the ethical analysis provided above, alleviating some dilemmas related to germ-line manipulation and possibly re-integration. However, temporary and reversible gene-editing also raises other ethical issues, including—who owns the enhancement, the service-member or the military? Can the service-member choose to remain enhanced upon separation? Alternatively, is the military's ownership of the technology so expansive as to over-rule the service-member's autonomy in this case?

While the bodies and lives of service-members are sometimes understood as being government property, most recognize that this population does maintain some degree of bodily autonomy. If the military has ownership over the enhancements or said enhancements are illegal to civilians, it is essential to recognize the possible ramifications of being 'normalized' through further gene–editing before separation from service. It seems that if the military owns this technology, that it could force service-members to reverse gene-editing, just as they must return other forms of enhancement discussed earlier (e.g., body armour, weapons). The possibility of a forced intervention significantly alters ethical analysis, as it would fail to meet widely accepted principles in both medical and research ethics. While some militaries can legally force treatment upon service-member patients, such as vaccination and life-saving treatment in emergencies, this ability is not unrestricted. Especially during the research stage of implementing genetic enhancements, forced experimental interventions are never ethically acceptable as informed consent is required.

Beyond issues of ownership, there are other ethical issues that are introduced by temporary or reversible gene-editing. Arguably removing or reversing any enhancement upon separation from service could also lead to unforeseen psychological injury, as war-fighters would have to make two simultaneous transitions; the first transition being from an enhanced state to a normalized one, and the second being a transition from military to civilian life. We cannot know what type of harm this could entail and must proceed cautiously. While this is speculation, it is likely that temporary enhancement does not solve the problem of re-integration and may lead to more significant problems. Adjusting to civilian life after a military career is already complicated and psychologically stressful. This truth is magnified by combat deployments and military roles that have no comparable civilian profession, transforming their transition into a change in lifestyle, culture, and profession that often entails additional training or education. Should these enhanced war-fighters also need to have their enhancements reversed, we can expect that normalization to take a toll on people who have grown accustomed to their enhancements.

10.5 Conclusion

While this article provides mostly speculative ethical analysis and does not provide many solutions to the ethical issues that have been highlighted, the discussion and identification of these issues are essential. The speculative and future-looking nature of this technology provides time to begin addressing ethical issues before they become ethical dilemmas. This type of preventive discussion has been occurring within genetic research for some time. However, this paper attempts to push the discussion further. Specifically, it is important to shift our ethical analysis and focus it beyond the scope of a service member's military career. Although we must recognize and discuss issues related to research within a military population, ethicists and researchers must acknowledge the reality that service-members will someday be civilians. Understandably, military medical ethics has traditionally focused on issues related to active-duty members, with little attention paid to either former service-members or military families. This broader forward-focus is needed. Ideally, this population will live long, happy lives after separation from service—but society at large is not doing enough to make this happen, and enhancement could make it worse. Ethical analyses of genetic-enhancement in this military context, must take a broader perspective and not reduce service-members to their role as warriors.

References

Azarow, K. 2003. Ethical use of tissue samples in genetic research. *Military Medicine* 168: 437–441.

Chang, A. 2009. Military experiment seeks to predict PTSD. *U.S. News & world report,* November 20.

Committee on Opportunities in Biotechnology for Future Army Applications, Board on Army Science and Technology, National Research Council. 2001. *Opportunities in biotechnology for future army applications*. Washington, DC: National Academy Press.

Connelly, M. 2012. Genetics of post-traumatic stress disorder: Review and recommendations for genome-wide association studies. *Current psychiarty reports*.

DARPA. 2017. *Building the safe genes toolkit,* July 19. Retrieved from Defense Advanced Research Project Agency: https://www.darpa.mil/news-events/2017-07-19

Department of Defense. 2017. DoD instruction 5154.30. *Armed Forces Medical Examiner System (AFMES) Operations,* December 21. Washington, DC.

Drew, T., and U. Mueller-Doblies. 2017. Dual use issues in research—A subject of increasing concern? *Vaccine* 35 (44): 5990–5994.

Eagan Chamberlin, S. 2012. Emasculated by trauma: A social history of post-traumatic stress disorder, stigma and masculinity. *Journal of American Culture* 35 (4): 358–365.

Evans, N., and J. Moreno. 2014. Yesterday's war; tomorrow's technology: Peer commentary on 'ethical, legal, social and policy issues in the use of genomic technologies by the US military'. *Journal of Law and the Biosciences* 2 (1): 79–84.

Huntington, S. 2008. *The soldier and the state: The theory and politics of civil-military relations*. Cambridge: The Belknap Press of Harvard University Press.

Illing, S. 2018. *Chasing Captain America: Why superhumands may not be that far away,* April 27. (Vox Media) Retrieved July 2018, from Vox: https://www.vox.com/science-and-health/2018/4/27/17263128/captain-america-avengers-infinity-war-bioengineering-technology

JASON. 2010. *The $100 genome: Implications for DOD*. Washington, DC: The MITRE Corporation.

Lázaro-Muñoz, G., and E. Juengst. 2015. Challenges for implementing a PTSD preventice genomic sequencing program in the U.S. Military. *Case West Journal of International Law* 47 (1): 87–113.

McManus, J., A. McClinton, and M. Morton. 2002. Ethical issues in conduct of research in combat and disaster operations. *American Journal of Disaster Medicine* 4: 87–93.

McManus, J., S. McClinton, A. De Lorenzo, and T. Baskin. 2005. Informed consent and ethical issues in military medical research. *Academic Emergency Medicine: Official Journal of the Society for Academic Emergency Medicine* 12: 1120–1126.

Mehlman, M., and T. Li. 2014. Ethical, legal, social, and policy issues in the use of genomic technology by the U.S. Military. *Journal of Law and the Biosciences* 1 (3): 244–280.

Miles, D. 2007. *DNA registry unlocks key to fallen servicemembers' identities,* January 25. Retrieved from U.S. Army: http://www.army.mil/article/1508/DNARegistryUnlocksKeytoFallenServicemembers039Identities/

Office of Research & Development. 2018. *Million veteran program (MVP),* July 3. Retrieved from U.S. Department of Veteran Affairs: https://www.research.va.gov/mvp/

Patrone, D., D. Resnik, and L. Chin. 2012. Biosecurity and the review and publication of dual-use research of concern. *Biosecutiry and Bioterrorism: Biodefence Strategy, Practice, and Science* 10 (3): 290–298.

Pence, C. 2014. Military genomic testing: Proportionality, expected benefits, and the connection between genotypes and phenotypes. *Journal of Law and the Biosciences* 2 (1): 85–91.

Resnik, D., D. Barner, and G. Dinse. 2011. Dual-use review policies of biomedical research journals. *Biosecurity and Bioterrorism: Biodefense Strategy, Practice, and Science* 9 (1): 49–54.

Selgelid, M. 2009. Governance of dual-use research: An ethical dilemma. *Bulletin of World Health Organization* 87 (9): 720–723.

Shay, J. 2010. *Odysseus in America: Combat trauma and the trials of homecoming*. New York: Scribner.

The National Institute of Mental Health. 2011. *Army STARRS New Soldier Study (NSS): The first days of service.* Retrieved from http://www.nimh.nih.gov/health/topics/suicide-prevention/suicide-preventionstudies/army-starrs-newsoldier-study-nss-the-first-days-of-service.shtml

VA. 2016. *PTSD: National center for PTSD,* October 3. Retrieved from U.S. Department of Veteran Affairs: https://www.ptsd.va.gov/public/PTSD-overview/basics/how-common-is-ptsd.asp

Williams, H. 2016. *Bio-hacking: Everything you need to know about DIY biology*. Retrieved from io-Based World News: https://www.biobasedworldnews.com/biohacking-everything-you-needto-know-about-diy-biology

Chapter 11
Military Medical Enhancement and Autonomous AI Systems: Requirements, Implications, Concerns

Tomislav Miletić

11.1 Introduction

"AI is the new electricity" Andrew Ng, the former Baidu Chief Scientist and Coursera co-founder, famously declared at a Stanford MSx Future Forum talk (Ng 2017). Somewhere at the same time under the same motto of "Ai is the new electricity" Finland's Ministry of Economic Affairs starts working toward an official proposal for the development of artificial intelligence (AI) in Finland. On April 10th 2018, 25 European countries signed a Declaration of cooperation on Artificial Intelligence (AI) in agreeing to work together on the most important issues raised by Artificial Intelligence, from ensuring Europe's competitiveness in the research and deployment of AI, to dealing with social, economic, ethical and legal questions (EU Declaration on Cooperation on Artificial Intelligence – JRC Science Hub Communities – European Commission 2018). It seems that the age of AI is, without a single doubt, already upon us with profound and still unforeseen effects waiting for us in the near future poised to transform the human society. Business, social services, judicial systems, education, and especially healthcare are already becoming heavily shaped by the advent and development of smart and autonomous AI systems.

Specifically, in healthcare AI systems are already supplementing, empowering and replacing specific medical skills with experts proposing how some of the medical jobs are to be replaced while others are to experience a profound change in the way they are being educated for and practiced (Brynjolfsson and Mitchell 2017; Acemoglu and Restrepo 2018). Coming AI healthcare systems equipped with prediagnosis capacities, real-time sensor monitoring, personal data assessment, and disease prediction aim to revolutionize the global healthcare system. The long-lasting

T. Miletić (✉)
Faculty of Humanities and Social Sciences, Rijeka University, Rijeka, Croatia

D. Messelken, D. Winkler (eds.), *Ethics of Medical Innovation, Experimentation, and Enhancement in Military and Humanitarian Contexts*, Military and Humanitarian Health Ethics,
https://doi.org/10.1007/978-3-030-36319-2_11

goal of medicine engrained in the maxim of the precautionary principle, (Fischer and Ghelardi 2016) summarized in that famous "better to be safe than sorry" proverb, seems close to our grasp. Yet, the development of such autonomous medical systems brings many ethical issues with which to concern ourselves. For instance, the questions of autonomy, the possibility of action explanation, the trust one can put in such autonomous systems (Hengstler et al. 2016) and their decision-making process (Bennett and Hauser 2013), are all of great importance. Many of these issues will be shared between the civilian and the military sector as the technologies used, and the crucial ethical issues such as trust or reliability (Tavani 2015), will be equally shared. Still, it is important to have in mind that this transformation will be manifested according to specific economic expectations and social goals and for this reason, the application of autonomous AI healthcare systems between the civilian and the military sector will also differ. And it will do so in the peculiar technological divergence, data privacy rules and AI system autonomy (Lawless et al. 2017) especially with regards to those AI systems working in collaboration and partnership with humans (Azevedo et al. 2017). As such, one could envision many peculiar differences pertaining to developmental paths for autonomous AI healthcare systems in the military that will not be shared by their civilian appliance.

One of such trajectories, upon which I will focus in this work, is the development of the next generation AI empowered exoskeleton smart suits that offer enhanced operational status for the service member in the mission field through AI assisted monitoring, evaluation and administration of treatments to its user. The Joint Acquisition Task Force Tactical Assault Light Operator Suit (JATF-TALOS), presented at the Special Operations Forces Industry Conference (SOFIC 2018), is the most recent example of such technological possibility. Marketed as the "next generation, technologically advanced combat operator suit" equipped with "enhanced protection, super-human performance, surgical lethality and exponential situational awareness" (SOFIC 2018, 53) the TALOS suit aims to accomplish the goal of enhanced soldiers on the mission field. And as the U.S. Army is not the only powerful military force embarking on such a developmental path[1] it can be safely presumed how exoskeleton technologies are poised to be assisted and empowered by autonomous (medical) AI systems. Consequently, the integration of AI systems into the exoskeleton suit aims to transform not only the soldier's capacity for military operations but also the work and role of the future medical officer.

To explore the possibility and point out the crucial ethical issues rising from the use of such technology I propose one such system, an AI empowered smart-suit which I call, in honor of its famous gaming namesake, the Praetor Suit.[2] The Praetor

[1] The Russian military rushes forward to develop their own third-generation Ratnik 3 suit which aims to integrate different important systems of life support, enhanced communication, protection and movement enhancement into their smart warrior suits (Ritsick 2018)

[2] The Praetor Suit is the armored suit worn by the Doom Slayer character in the popular game Doom (published in 2016). The suit is given to the player at the very beginning and is worn for the entirety of the game. Praetor Suit covers the Doom Marine's whole body, including his arms. The suit is described as being made from nearly impenetrable material, and may be responsible for what appears to be the Doom Marine's superhuman abilities (Praetor Suit 2018).

Suit is envisioned to have, projecting forward the outlined medical capacities of current exoskeleton suits, the ability to monitor the service member's physiological and psychological state, the ability to report the monitored state to medical officers supervising the suit's operations through teleoperation and, whence required, to autonomously administer medical treatments such as drugs or painkillers to the suit's user. The Praetor Suit could have many operationally enhancing capacities such as enhanced movement, durability or perception but in this investigation I will solely focus on the Praetor Suit's purported medical capacities which would, already on their own, effectively enhance the service members' operational capacity on the mission field. Consequently, with such a proposal, I wish to simultaneously and more broadly engage some of the intertwined issues of AI ethics and human enhancement by focusing specifically on the relation between AI medical autonomy and AI empowered medical enhancement inside military operational application.

11.2 Praetor Suit

The Praetor Suit has three main AI empowered capacities – monitoring, evaluation and administration of treatments. In exploring the ethical concerns arising from these capacities, I presume that each of them is distinctively delegated to a specific AI sub-system. Even if not so, the stratification of the Praetor Suit's AI system into these three sub-systems serves to clearly elucidate the ethical issues and concerns arising from the suit's application. As such, in this exploration, I will focus on the Monitor AI, the Evaluation AI, and the Administration AI. The Monitor AI would, as the name entails, monitor the service member's state psychological and physiological state, the Evaluation AI would evaluate the state received from the monitor AI and the Administration AI, if functioning autonomously, would treat the user according to the diagnosis received from the Evaluation AI. Thus, the system is tripartite with the bottom layer being the Monitor AI and the top layer the Administration AI. The system as a whole, if functioning autonomously, is causally dependable from the bottom-up for the accuracy of its decision-making processes with specific technical requirements and ethical considerations expected to be satisfied by each part of the system.

The monitor AI is the first and initial dimension of the suit's smart system which we must take into consideration. Its primary function is to continuously monitor the service member's physiological and psychological state and report that state to the Evaluation AI and/or the medical supervisors who are surveilling the suit's functioning through a teleoperated link. For the Monitor AI to function properly, one has to take into consideration the sensory range of the suit, its accuracy, and speed, as well as the probability for false readings. AI-assisted sensor technology could be intra or extra-bodily sensors, as ingested sensors (Kalantar-zadeh et al. 2017), bodily implants, heat or haptic sensors monitoring the amount and level of perspiration (Gao et al. 2016; Sim et al. 2018), body temperature, heart rate, (Warrick and Nabhan Homsi 2018) and facial expressions (Cohn et al. 2009; Lee et al. 2012) which are already outperforming human experts (Rajpurkar et al. 2018). The Monitor AI should also remember the user's psychological and physiological states and provide the personalized data set to the evaluation AI's precise medical assessment as we can expect to have many differences in the amount of psychological and physiological stress, pain and injury levels between service members utilizing the suit.

The Evaluation AI is the second and middle layer of the Praetor Suit's system and its main purpose is to, upon receiving the monitored state from the Monitor AI, create a personally tailored medical diagnosis which it then feeds forward to the Administration AI and medical operators surveilling the suit's functioning. Thus, the Evaluation AI works as a medical advisor only and is presumed to be, due to the Monitor AI precise sensor accuracy and the highly personalized data set at its disposal, highly efficient in evaluating the user's medical state. Due to the current level of AI development, and in order to showcase the ethical considerations of interest, I compare the Praetor Suit's Evaluation AI with one of the famous medical diagnostic systems – Watson AI (Watson Oncology).

The third, last, and most important of the Praetor Suit's system is the Administration AI which is the decision-making level of the system responsible for the administration of treatments to the user. Initially, there are two possible ways medical treatments could be administered through the Praetor Suit's systems dependent upon the level of the suit's autonomy. The first one is semi or non-autonomous in which case medical officers surveilling the soldier's state through the process of teleoperation remotely administer the treatment through the suit's smart systems. Here, the suit's autonomy is manifested only in the action potential that is in the capacity to receive the medical command and to transparently and reliably act upon it. The second option, on the other hand, entails full system autonomy in which the suit, due to specific circumstances or scenarios, operates autonomously in all of its parts – monitoring, evaluation and administration of treatments. Full autonomy also implies that the Administration AI, based upon the evaluation AI diagnosis, has the capacity not only to deploy highly effective treatments but also includes the possibility to mistreat the user of the suit due to system error or malfunction.[3] Since such mistreatments could theoretically result in grave consequences for the operational status and health of the user this level of the Praetor Suit's system is the ethically most demanding and requires special ethical consideration.

[3] The possibility of medical error is something which cannot be excluded from the medical profession as errors in medical diagnosis and treatment application are, unfortunately, a fact of medical life. Both humans and AI could err in the field of medical work, even though the reasons why the err may differ and differ drastically. For instance, even though the AI may never get tired or emotionally upset and may operate optimally and constantly without rest it, at least for now, has no capacities to improvise or adapt to unforeseen situations if it becomes necessary for the success of the medical operation.

11.3 Ethical Issues

11.3.1 Monitor AI

Ethical issues connected with the Monitor AI capacities can be safely divided into two categories. The first are those connected with consequences resulting from malfunctioning technology, such as sensors providing inaccurate or false readings, which can, if the suit operates in fully autonomous mode create a series of serious and unfortunate consequences for the health or life of the user. Although gravely important for a proper functioning of the entire system these issues are not specifically ethically demanding since a clear line of responsibility can be established and traced between designing and testing such technical systems and their operation on the mission field. Naturally, as with all technical equipment, rigorous testing is primarily required before the equipment's utilization in the mission field can be safely approved.

The second set of issues, on the other hand, is concerned with data privacy issues and bears greater ethical complexity. For instance, as the suit is capable of reading its user's psychological and physiological state "inside out" and the user has no capacity to hide from the suit's encompassing monitoring, one could easily call into question the intrusiveness and data privacy issues of such technology. Additionally, when we add the possibility of the system to collect and store user's personal data as learning sets for creating better prediction models, one can imagine the level and amount of ethical concerns such technology can generate. For instance, we can responsibly ask, who will use the collected user data and for which purpose exactly? Where will the personal data be stored and how will it be protected? Who owns the data? Additionally, at what point and under what terms can the user of the suit opt out of using it if she believes that the suit breaches her basic privacy rights? What is the relation between the need to use the collected data for creating better machine learning models, which ensure the suit's optimal and highly personalized functionality and the need to secure user' privacy rights? Consequently, in mirroring the fears of an algorithmic society (Balkin 2017) should we worry about an algorithmic military, where the personal, service member's, assessment for the mission field is dominated by smart algorithms rather than humans?

Finally, as was recently illustrated in the Facebook's Cambridge Analytica scandal, the personal user information collected with the goal to better understand its user and create personally tailored data-driven decisions may lead to nefarious misuse from third party members. And since, the development of such products is already including the collaboration of nonmilitary institutions and individuals, the important issue of code ownership and code inspection stands equally important if, as was explored in the famous Pro Publica report (Angwin and Mattu 2018), the algorithms showcase a glaring bias in their computations. Unfortunately, since there is no simple way to resolve these difficult issues the challenge to develop, "monitoring that is not only trustworthy, nonsubvertible, and privacy-aware, but also forensics-worthy will have to be addressed" (Neumann 2016).

11.3.2 Evaluation AI

The evaluation AI is the suit's system that, based upon provided monitored data, diagnoses the state in which the service member finds herself and advises a corresponding set of treatments. As noted, it can be compared to Watson AI, (specifically Watson Oncology) arguably the most proliferated AI medical system in the Western world. As such, in trying to illustrate the impacts of the Praetor Suit, I will rely upon the recent in-depth STAT analysis which „examined Watson for Oncology's use, marketing, and performance in hospitals across the world "(Ross and Swetlitz 2017). So, what are some of the realistic and important impacts AI empowered diagnostics creates in the medical field which could also be applied to the Praetor Suit's use?

First, similarly to Watson, the Praetor Suit could advise treatments based upon supporting evidence from official medical literature and medical practice as for instance the survival rates connected to specific treatments or the efficacy of treatments towards specific age groups or health conditions.

Second, the Praetor Suit could empower its users with robust and precise treatment information allowing them to enter into further dialogue with their medical team on the viability or efficiency of the proposed treatment (if time allows it). What is important to note is that this expert-like assistance is provided to patients in processing speed far exceeding any human capacity and without the need for additional physical presence.

Third, the suit could similarly to Watson, empower the decision-making process of less experienced doctors (engaged with the suit's medical operations) and change the medical and social relations between the less and more experienced doctors. Where up until now, the process of medical counseling and decision making was solely delegated to humans, and the experienced senior doctors held the highest authority now, with the introduction of AI assistance into the foray, the junior doctors are able to question their senior colleague's proposals with the assistance of AI analysis. For the Praetor Suit, dependent upon the system's level of medical expertise, the question to whom will the junior doctors give trust to (especially in crucial, time-limited decisions) the AI's evaluation, their own, or the senior's opinion opens up.

Fourth, in hospitals or occurrences with few or no medical experts at all, Watson serves as a constitutive part of medical decision making, as medical teams tend to rely far more on its advice than those teams which have available human experts either physically present in the hospital or available for assistance through the communication means. In a similar manner, the Praetor Suit can become the primary medical advisor in circumstances where no other human medical expert is available.

All of this shows how, through the introduction of AI-assisted medical evaluation, the nature of the *patient-medic* relation now changes into the patient-AI-medic[4]

[4] I stand inspired by Klein's patient-machine-medic relation (Klein 2015) with the noted difference I purposefuly dislike using the term "machine" when describing human-AI relations as this term,

relation in which both the receiver and the primary giver of medical assistance are profoundly impacted by the introduction of smart AI systems into the previously firmly defined human union. Similarly, if introduced into the *medic-soldier* relation, the Praetor Suit would inexorably change the way medical assessment and treatment is deployed on the field of operation and consequently change the relation of trust and reliance between the medic and the soldier. Unfortunately, the possibility for error (which is always present) cannot be excluded even from the best trained AI systems as no computational, including human comprised, system is infallible. To exclude, as much as possible, the tendency to error medical experts ought to be included not only into the design process but also the very operation of such systems within real-life occurrences as supervisors of the system's operations and final decision makers. The possibility of error (especially if it results in a diagnosis that can jeopardize the health of the user) also calls for the necessity of the PS system to explain itself, that is to offer reasons or insights into the process of why it generated such a diagnosis, especially if the error was produced by a specific AI Bias[5] resulting from incomplete, non-representative or simply poorly designed data sets as well as improper system training produced by human system trainers. Since we aim to establish a strong relation of trust between all three constitutive members of the newly established medic-AI-soldier relation, the ability of the AI to provide an explanation for its evaluations is essential, as DARPA recognizes, "if future warfighters are to understand, appropriately trust, and effectively manage an emerging generation of artificially intelligent machine partners" (Gunning 2017).

11.3.3 Administration AI

The Administration AI module is the decisive layer of the system responsible for putting into action the treatment evaluations provided by the Evaluation AI. As such, the system can work either in autonomous mode or be remotely operated through medical teleoperated surveillance.

The teleoperated surveillance entails the presence of a medical supervisor surveilling the suit's operations through a teleoperation link with the capacity to directly issue medical commands to the suit and to confirm or deny the suit's treatment evaluations. In this regard, the suit bears little or no autonomy in its decision making and this mode of operation brings no considerable ethical considerations with regards to Ai autonomy as both the surveillance of the medical diagnostic process and the final medical decision remains with the human medical expert.

---

engrained in popular Western culture, tends to produce discomforting connotations which tend to negatively impact the attitude towards the formation of human-AI relations. (Coeckelbergh 2014)

[5] AI Bias is recognized by leading institutions and experts (AI and bias – IBM Research – US 2018; Campolo et al. 2017) as one of the two biggest issues (the other problem is the "black box" or "explainable AI" issue) hindering further development and implementation of autonomous AI systems in the foreseeable future.

The autonomous mode, on the other hand, brings such considerations to the forefront especially when we have in mind that we can envision a number of circumstances and operational scenarios where it would be beneficial and even necessary for the suit to be fully autonomous in the decision for and application of medical treatments. Let us explore some of them. First, in matters of grave danger in which the system has to react instantly in order to safeguard the service members' life or treat a grave injury it should be necessary for the suit to quickly and effectively administer specific treatments to the member without waiting for human approval since the speed of reaction required for the effectiveness of such treatment is crucial. Still, it is important that the suit informs the user of the suit (and the medical supervisors) of its actions in order to transparently present its intentions and effectively build trust with its human user (Wortham et al. 2016). If the user is conscious the suit should inform her of her action for instance by auditory messages:" Susan, I am now stopping the bleeding in your right leg by pressuring the wound." or simpler visual signs. The amount and the type of presented information should primarily depend on the premise that the user should not be in cognitively burdened or overloaded by it. Naturally, this depends upon the situation in which the suit's user finds herself currently. For instance, in combat scenarios, the suit's treatment notifications should be presented to the user in the most minimal but also precise way as not to disturb the concentration required for the mission operation. Additionally, if the user is incapacitated the suit should inform her of its actions upon regaining consciousness, as for instance: "Susan, I am glad you are awake now. I had to apply pressure to your right leg in order to stop the bleeding" or by visually pointing out the applied treatment. The opposite case, in which the suit administers treatments without notifying the user the relation of trust between the suit and the user could become jeopardized as the human user could start doubting the AI's decisions and could even refuse to cooperate or communicate with the AI system. In this regard, it is paramount that the user, through constant and reliable use, builds a clear awareness of the suit's transparently presented medical intentions.

Second, the suit should autonomously act in "environment adaptive" procedures such as for instance when operating in cold environments the suit should be allowed to adjust its internal temperature autonomously. In cases of a light flare attack or when operating in desert-like environments, the suit should be able to autonomously administer visual filtering to reduce eye strain or sight damage. Such autonomous action does not require specific confirmation from the user nor from the medical supervisor as it operates on the level of adaption. That is, the suit automatically adapts its internal environmental condition towards the different (and possibly detrimental) environments in order to safeguard the user's health and enhance the body's capacity to operate in such conditions.

Third, full suit autonomy can be, as a general rule, applied to all those cases where the system as a whole is equal to or better at administering specific treatments than a human medic. To use a basic example, if the Praetor Suit is in under all possible conditions the best medical agent able to stop the external bleeding of its user than it ought to always autonomously do so. Thus, for treatments where it exemplifies the capacity to outperform human aid and does so safely, it should act autonomously. Still, it is highly important that such a level be specifically defined and rigorously tested prior to the suit's full utilization on the mission field. Consequently, such functional medical autonomy can only be achieved if medical experts together with technicians are engaged in the very design, training and the proof-testing of the system.

Fourth, in the circumstances where the suit is the only possible agent of medical assistance capable of diagnosing and administering medical treatments to its user, the suit should act autonomously in doing so but not without prior informing the suit's user of its new autonomy and awaiting his approval for the possible treatment. For instance, in cases where the communication between the medical supervision and the suit is severed or jammed due to different operational conditions, non-excluding human intervention i.e. hacking.

One might contest to this by stating that specific user approval for each applied treatment is not a necessary condition, since in this case, the suit is the only possible agent of medical assistance and its action could be treated as a medical procedure for fitness of duty. As soldiers cannot refuse participation within those standard treatments which purpose is to uphold their fitness for duty, it is also questionable if the service member has the right to refuse administration (by the Praetor Suit) of the treatments which would keep them fit for duty during mission operation,[6] if the user stands informed of the suit's range of medical operation prior to mission commencement.[7] Additionally, a soldier with no medical knowledge and a hidden psychological bias against machines could, in a specific operational scenario, distrust the suit's diagnosis and deny the suit's beneficial medical action jeopardizing not only his health but also the mission's success. In this case, it can be questioned if the soldier would have the right to deny such treatment especially when we have in mind how the requirements for the necessity of user approval can be pre-programmed within the suit dependent upon specific mission conditions and the suit's user could be fully informed of those requirements prior to mission commencement. For instance, if the mission is to be held in complete "radio silence" then it is expected that the suit will act autonomously on all medical circumstances as maintaining a network connection with medical supervision will not be possible. As such, suit users would be notified in advance for which type of medical decisions the Praetor Suit will not ask user permission and will not be startled with the suit's decision for treatment if it comes to happen.

Still, it would be paramount, to build the relation of trust, that the suit's user is knowledgeable, in advance and at least in general categories, of the possible range of treatments that can be administered to him by the Praetor Suit. Additionally, she

[6] "In other words, commanders do have the legal right to require service members to undergo certain medical procedures such as vaccinations and periodic medical examinations for fitness of duty." (McManus et al. 2005, 1124)

[7] Such pre-mission obtained consent could be attained transparently and fully if, for instance, the suit's operation would be field tested either in virtual space (through VR simulation) or real space. Such tests would not only psychologically accommodate the user to the suit's use but would also fulfill the ethical necessity of informing the user on the suit's beneficial operational capacities and the possible harmful consequences resulting from its use.

should also be knowledgeable of the scope of possible, harmful, consequences which could result from the Praetor Suit's operation of which none may, by design, result in threatening injuries. This, minimally, entails that the user of the suit, prior to consenting to its use, should be informed on the possible range of the suit's autonomous medical actions which cannot exclude the possibility of suit's system failure and the consequences resulting from it.[8] In other words, the service member has to be transparently presented with all the benefits of the Praetor Suit use not neglecting the possibility of system failure and the range of its possible outcomes. This is highly important since the last thing we wish to have is the loss of trust between the Praetor Suit and the soldier, where the relation between the Praetor Suit and the soldier might be seen as enslavement or imprisonment, the soldier being put into a living cage that he becomes subservient to. Such a relation between the Ai and the human, where the human becomes downgraded to a piece of functional equipment, like a cog in the machine, and becomes subjugated to the AI would be highly detrimental to troop morale, individual psyche[9] and mission success. Such a state could be compared to the situation where the commander coerces the soldier into harm's way and takes away his autonomy but unfortunately, "the contrast of a service member as both a vulnerable subject who must be protected from a commander's coercion and, simultaneously, a warrior who may at any time be ordered into harm's way is not well explored." (McManus et al. 2005, 1123).

The introduction of AI decision systems additionally perplexes this existing ethical conundrum as the Praetor Suit might, in specific cases, override the user's autonomy in order to safeguard its users' health.[10] For instance, the Praetor Suit might coerce its own autonomy over that of the user operating it (this possibility would also have to be transparently presented to the suit's user prior to its use), if the suit's user starts exhibiting mental states or intentions deleterious to her own or other troop member's health or life. In such cases the Praetor Suit should have the capacity to act autonomously in prohibiting or simply refusing to cooperate in such actions effectively stopping the user from doing harm to oneself or other troop members.[11]

---

[8] Although there are different types of system failures which could result in such harmful consequences, it is paramount that they are not produced from design or system administration incompetence which results in system vulnerabilities or failures. Unfortunately, there will always exist a type of rare and unpredictable high impact events, "Black Swans" (Taleb and Chandler 2007), which cannot ever be fully excluded from manifesting even with best possible system design.

[9] This is especially important if the suit could have the capacity to administer or provide the user with stimulants or other enhancement drugs such as "stimulants to attenuate the effects of sleep loss (often referred to as 'go pills') and hypnotics to improve the ability to sleep ('no-go pills')." (Liivoja 2017, 5). It is paramount that such (combat) enhancement is not done autonomously without the knowledge of the user. In this regard one might even remember the 2001's Space Odyssey HAL 9000, where HAL's secret decision for the missions' success results in an ethical disaster and loss of human life.

[10] Such a scenario also raises the question, are AIs then included into the chain of, medical, command?

[11] This could be achieved by the suit taking over or restricting movement autonomy and removing the service member from the harm's way (or from inflicting harm to oneself or others), or, if necessary, by administering simple sedatives.

Still, notwithstanding the possibility for such extreme cases, the user of the suit should never experience being coerced by the suit's AI. This should be especially noted when we have in mind that the AI can become highly personalized and efficiently trained for the fulfillment of the specific mission goal and is far more capable at environment processing, agent action prediction and sensor capacities than its human partner. Similarly to the prospective use of neurowarfare (White 2008), decision making could be, for the sake of the mission's success, relegated towards the Praetor Suit system especially in those occasions where the case is clear why the suit should bypass or override human decision making. Although such AI capacities sound promising, and perhaps even tempting to be deployed, it must not be forgotten that there always exists the possibility for the Praetor Suit to create a medical decision with adverse effects which could result in severe consequences not excluding grave injury or even death of the suit's user which would result in a sense of moral guilt and heavy distrust towards the use of the Praetor Suit. Still, if AI is properly designed for a specific issue it is capable of giving better predictions and recommendations than human experts in the same field (Agrawal et al. 2018). Unfortunately, humans are often reluctant in accepting recommendations given by AIs, a phenomenon called the "algorithm aversion" (Dietvorst et al. 2014; Taylor 2017). As such, it is highly important that automation biases are challenged through proper education and scenarios under which we should or we shouldn't allow humans to override AI evaluations are thoroughly analyzed.

Lastly, since the possibility of malfunction or system error is ever-present, even in cases where the suit outperforms the possibility of human intervention and is allowed to autonomously act, human supervision of the process should always be included. For this reason, the system's autonomy should be designed in such a way as not to burden the medic or the user with excessive information, which would only result in operational inefficiency but rather be capable to assess and critically suggest the crucial medical information in the required time interval if the situation arises (Casner et al. 2016). Although easily framed, this issue is one of the most difficult questions pertaining to effective automation design. This is due to the fact that more automation does not necessarily mean more efficiency nor fewer mistakes as we can design systems which are focused solely on multiplying automation tasks rather than to effectively manage the human potential and skill within the automation processes. Consequently, when things go awry humans, as ultimate decision makers and controllers of such systems, could become incapable to properly react once required (Neumann 2016). This means to say that the human subject, in order to effectively operate in the automation empowered environment has to be both properly trained to detect malfunctions and failures of technology and to appraise the moment in which she, if necessary, has to take over the system. Yet, this also means that she has to be aware that there are tasks and scenarios in which it is better to leave the AI to operate autonomously rather than to intervene.

Finally, since there are no and never will exist infallible machines the best fail-safe framework we can aim to develop is one of Human-AI partnership, or Human-AI symbiosis in which both the human and the AI are delegated with those tasks which they do best and in doing so are able to support, empower and check upon one another to complete their tasks correctly and efficiently.

11.4 Human-AI Symbiosis

The notion of Human-AI symbiosis is inspired by the HRSI (Human-Robot Symbiotic Interaction) framework (Rosenthal et al. 2010; Reidsma et al. 2016) where robots are designed to be autonomous in their actions (within the predefined environment) but in doing so they are transparently and constantly revealing their internal states, which includes their perceptual, physical or cognition design limitations, while communicating with humans. Symbiosis, on the other hand, denotes that both the humans and the AI are capable of supplementing and empowering each other in those, for the task required, proper capacities which are required for the achievement of their common or joint goals. The relation of symbiosis also entails the AI partner's capacity to detect and supportively respond toward the behavioral and emotional states of the human person with, at least, the minimal capacity to learn from the common interaction in order to anticipate user's future behavior. The human user, on the other hand, is advised by the AI, assisted in her cognitive or physical tasks, and can even receive psychological or emotional support. As such, both partners in the symbiotic relation influence and learn from one another to reap the joint benefit as they learn to cooperate on their common tasks by a coordinated and distributed division of work in which internal states and intentions of both partners are transparent to each other in order to build trust and accountability. Accountability also entails that both partners, especially the AI, have the capacity for error correction and action explanation. This design principle is especially important when we have in mind that symbiotic AIs are not built to be human/like, but rather they are built to compliment, adapt and enhance that what humans are and what they do in imaginative and novel ways. Finally, the symbiotic relation makes both AIs and humans more effective than when both of the partners function solely on their own, as autonomous AIs or autonomous humans. And this means that AIs built for symbiotic agency need to share a basic design tenet of adaption to, rather than dominance over, human counterparts. In other words, the human partner in the symbiotic relation must not be forced to change the fundamental way in which it acts or thinks by having to forcefully adapt to the system's automation demands. On the opposite, the design goal ought to foster the human to utilize the fullness of its capacities in the most effective way rather than to become squished within a non-optimized system.

Admitting, in military applications, a specific type of AI automation, which encourages and helps improve military behavior through AI assisted partnership, could be accepted but never so at the cost of the psychological or psychological health of the service member. Additionally, the sense of the soldier's specific type of personal autonomy within the military hierarchy and the mission parameters should be preserved. As prior noted, military AI agents are poised to become real partners to their human counterparts and developers should aim to build agents which can support, empower and form relations of trust and reliability with human members. This is especially important for healthcare automation as the next generation of collaborative AI partners in the medical camp should entail AI agents capable of complimenting to and adding valuable assistance to the medic-patient relation. The opposite case, where the medic hides behind the AI, similarly to hiding behind a computer screen, rather than to directly engage his patient through a rational and trusted dialogue should always be evaded. The patient, in his relation to the medic, must not be denigrated by an AI proxy, no matter the degree of its medical expertise. Finally, the AI medical partner has to build bridges of partnership and empowering relation between the patient and the medic rather than to become a wall of division between them and a force of subjugation for both of them. For this reason de-skilling of medical experts cannot be accepted as a conclusion of the AI automation process but rather further medical specialization and education of human medics in line with AI empowering technologies ought to be sought for. Through proper education and training the human agent learns to work with the AI partner by having clear expectations of its capacities but also retaining the awareness of her own decision-making authority.

To illustrate this with a narrative example, I will recall the "Docking" scene from the movie Interstellar. The scene depicts three active agents, the first is the mission commander and main pilot of the human (Cooper) and the two AI robotic agents (Tars and Case). The fourth human agent, Dr. Brand as a non-pilot, observes the situation from her strapped seat. At the beginning of the scene, just after the initial blast caused by Dr. Mann which sends the ship "Endurance" dropping down to the planet's stratosphere, Cooper silently evaluates the gravity of the situation and without communicating his intention to the rest of his teammates in the cockpit, starts the thrusters and sends the shuttle towards the Endurance to initiate the docking procedure. Case, the robotic AI partner, recognizes Cooper's intention and immediately advises him on the futility of such action (how there is no point in using fuel to attempt the docking) but Cooper cuts him off with: "Just analyze the Endurance spin". Case obediently follows the command but when notified by Cooper how it should "Get ready to match it on the retro-thrusters" it tries to give its last objection on the futility of such action by stating: "It's not possible!" on which Cooper famously answers "No. It's necessary". What this first part reveals is twofold. First, it shows that the first and final decision for the mission's goal stands upon the human (Cooper) who takes upon himself the full responsibility for the success or failure of the mission. Case, the AI partner, accepts the mission goal set by the human group leader but shares his assessment of the success of such an endeavor as impossible, according to his computation. This demonstrates how the AI partner has to have the capacity to evaluate the proposed action plan according to its sensory and computational capacities and has to be able to challenge it, if found lacking. Still, and very much important, after Cooper as the leader of the team confirms the necessity of such action Case does not continue with further objections but gives his full cooperation and support for the mission's successful conclusion. What this second part demonstrates is that, after the initial evaluation of the situation, if the team leader firmly establishes (under the chain of command) the mission's goal, the AI partner should be relentless in pursuing it by using the fullness of its capacities. In other words, upon accepting the mission goal the AI partner does not falter, does not waver and does not tire until the mission is successfully concluded. This also, importantly, entails that the AI partner isn't only giving his all through the use of its, above human level, computational or physical capacities but also through sharing its motivational and emotional support for other human team members if the need arises. All of this was beautifully exemplified in the scene's most stressful moment which shows how Cooper after aligning the shuttle towards the "Endurance" and calling for his AI partner's final confirmation: "Case, you ready?" (on which it responds: "Ready!") becomes momentarily gripped by uncertainty and freezes in place for a few moments jeopardizing the success of the mission and the lives of the shuttle's crew. But, luckily for Cooper, he wasn't alone in carrying the missions' burden as Case, his AI partner immediately recognizes the gravity of the situation and supports Cooper's initial decision by stating:" Cooper? This is no time for caution." On this Cooper, reinvigorated by Case's motivational support acknowledges his own frailty and responds valiantly: "If I black out, take the stick", and then to the other AI partner: "Tars, get ready to engage the docking mechanism." After these final instructions the team goes forward to accomplish the mission's goal and a few moments later successfully manages to dock the shuttle with the Endurance.

This conclusion of events elucidates a clear example of the human-Ai symbiotic relation where the AI partner, in the most crucial moment for the mission's success, is capable of supporting its human partner with the full range of its capacities. The human partner, on the other hand, is able to acknowledge his own frailty and confirm the role of his AI partner as his successor in order to secure the mission's successful end, if he ought to fail in doing so. Both of the partners are functioning as team members knowledgeable of each other's advantages and lacks and are fully committed to the mission's goal by taking on those tasks which are best suited for them due to their specific cognitive or physical capacities. The AI partner, additionally, due to its inability to experience fear or doubt in pursuing the mission's success is also able to motivationally and emotionally support its human partner if the need arises.

11.5 Military Medic or Military Enhancer?

As previously explored, the human-AI symbiosis leads to the birth of the medic-machine-soldier relation where the usual medic-soldier relation is now co-joined by the Praetor Suit's AI system as an intermediary agency of medical assistance and empowerment. This introduction opens up numerous AI automation issues of which perhaps the most important one is the ability to, through the nature of its appliance, change the important aspects of military medical work and with it, perhaps, the role of the military medic. To portray this possible change we must first remember how the use of the Praetor Suit will require specifically educated and trained medical officers working in tandem, as medical partners, with the suit's AI and its user. This entails that, if military medical AI assistance continues within the trajectory of the Praetor Suit example, the coming stratification and specialization of military medical work will also have to include those medics who will be able to work with both the suit's AI and its user as joint but distinctive partners in a symbiotic relation.

To illustrate the possibility of this novel medical role I will use a popular gaming concept, that of a healer class. In many computer games, especially massively multiplayer online games (MMO/MMORPG), there exists a stable diversification of character classes from which the players can choose upon starting to play a new game. The oldest three of these are the so-called "holy trinity" classes, the staple classes for every player cooperative mission centered game. The first of these is the tank whose purpose is to stand in the harm's way and redirect (soak) damage upon herself while the damage dealer, as it is usually termed – the DPS[12] or damage-per-second class, inflicts as much damage as possible in the shortest amount of time to the enemy. While these two classes are basically securing the enemy defeat, the third class, the so-called medic or healer class, is securing that both the tank and the DPS are healed from their injuries or malign status effects (for instance paralyzation or stun effect) which the enemy may inflict upon them as they endure the battle. To ensure that his team members stay alive and healthy through the duration of the entire fight the medic has to keep a close eye on his team and constantly heal their injuries through the capacities provided by her class. The peculiarity of such a task means that the player playing the medic class has to watch over the health status of his companions constantly and with great diligence in order to dispense healings when necessary. As combat situations in such games (especially in difficult missions) are highly complex scenarios the medic of the team has to be an experienced and highly skilled player who is not only highly knowledgeable of the capacities of his own class but also of the lacks and advantages of the other two classes – the tank and the DPS. Perhaps even more, the medic needs to be knowledgeable of the hindering and/or damage capacities of the enemy since it is her role to heal injuries, remove impediments and ameliorate negative states from the other players which are caused by the enemy or the environment in which the mission is occurring. All of this requires constant attention, focus and quick reaction by the player as she has to keep a constant eye on the situation's current status and the possible ways it can be resolved in the future.[13]

[12] In gaming worlds, the „DPS" abbreviation is a colloquial term used to design a wide variety of player classes, playing styles or character builds aimed to fulfill a single specific goal – to deal as much damage as possible to the enemy. As such, those players who build their characters to become damage dealers usually forego all other character traits, for instance strong defense, in order to ensure their characters maximum offensive power.

[13] It is often the case how players have to pass through the same mission more than once as combat usually occurs in stages with different enemy behavior or environment rules occurring for each separate stage. For this reason, experienced medic players are highly sought for especially by inexperienced, or first-time, players venturing into the same mission.

Similarly, the future work of the military medic might include a medical presence which is not directly and physically occupying the battlefield but will rather, through the connection with and in cooperation with the Praetor Suit, support and heal the soldiers on the field. Like the player who controls the character class in the game and watches over the team, the future medic might surveil the Praetor Suit's functioning, and if required intervene within its autonomy, to keep the user of the suit constantly out of harm's way and ensure her operational fitness for the mission's success. It is also important to have in mind that such support will not only encompass medical treatments but will also through the synergy provided by the suit's AI capacities produce a sum benefit of an effective combat or operational enhancement of the party member's mission prowess.

And this brings in yet another similarity with the healer or medic class which among the repertoire of its abilities usually counts not only the capacity to provide healing or mitigate/remove adverse status effects but also the capacity to empower, or enhance the party members with different enhancements or "buffs" as they are called in gaming terms. These can be seen as permanent or temporary boosts to party member's abilities, as the ability to attack with greater precision or faster with greater force or to upgrade their defenses for evasion or damage resistance etc. In similar regard, the Praetor Suit would become like the medic class, while the medical operator's role surveilling the Praetor Suit's system would be akin to the role of the human player controlling the gaming avatar. As the human player presses the button for heal and trusts that the healing process provided by the medical class will function as intended she also monitors over the process by (perhaps even subconsciously) being aware how, even though rare, glitches or bugs may always occur and disrupt the healing process provided by the class. Similarly, the medical operator surveilling the suit's medical operations trusts the suit's AI in its autonomous medical operations but also works in tandem with it for those operations where a joint action and attention is required. In all of this, she is being aware of and prepared for the possibility of system malfunction. As such the medical operator becomes something similar to that age-old portrait of the guardian angel who in this case watches over the suit's autonomous actions and remotely administers or instructs the suit to apply specific heals or buffs to its user when necessary.

Finally, this leads us to the main question, will the role of the medical operator surveilling the suit become perceived by the enemy side as one which has changed from purely non-combative towards the *enhancer* and indirectly combative? And would it result in removing the scope and level of protection military medical personnel holds in military operations?[14] Possibly so, since the suit's system is designed to enhance the soldier's operative capacities by safeguarding his optimal psychological and physiological health and through the provision of fast, safe and effective

[14] In this regard, I agree with Liivoja's (2017) lucid analysis of the reasons why medical personnel engaged in the biomedical enhancement of soldiers would suffer a "loss of special protection that such personnel and units enjoy" (Liivoja 2017, 27). Still, I take that the prospect of AI empowered enhancement has the potential to generate more fundamental changes to the role and purpose of the military medic than the ones exemplified in cases of biomedical enhancement.

treatment enhance the user's capacities for a prolonged period of time. And since, as noted, these systems should not be left without medical human supervision then the role of the medic in charge of the suit's medical systems inadvertently changes from a purely medical role to that of an "enhancer", or in gaming terms a "buffer", one that has capacities to enhance or buff the operational or combat abilities of his team members. Unfortunately, as experienced players are aware, the first class or role that gets targeted by the enemy team in order to ensure victory is the healer/support class since the prospects of winning are greatly increased if the enemy healer is removed from the play first, since the opposite entails that the healer constantly heals the injuries of his team members or buffs, that is enhances, their abilities.

If the portrayed scenario becomes a reality, will the enemy team first aim to target the medic supervising the suits operational capacities in order to remove their influence from the combat? And does this mean that such development will inadvertently push us towards delegating more power towards AI autonomy or will the future create a novel medical role hybrid in nature and its application? It remains to be seen. Still, it stands certainly that the introduction of AI partnership into military operations will surely change the landscape of military medical cooperation for which we ought to be properly prepared.

11.6 Conclusion

The use of automated AI systems in the military medical field has the potential to drastically improve medical assistance on the mission field. Smart AI monitoring, detection, and evaluation together with precise and effective administration of treatments holds the promise of an effective operational enhancement in variant environments and circumstances. To explore this possibility, I have portrayed the use of the Praetor Suit, an AI-empowered exoskeleton suit with the capacity to monitor, diagnose and administer medical treatments to its user. The design, implementation, and use of such a system open a number of important ethical issues, such as questions of data privacy, autonomy, and machine trust challenge us to create new relational models for human-AI collaboration and partnership. *Human-AI symbiosis* is one such promising model capable of sustaining and empowering Human-AI relations in order to effectively address the important ethical issues given birth through the introduction of AI medical assistants into the mission field, of which some have the power to change not only the future work of military medics but also their role and operational purpose in the mission field.

References

Acemoglu, Daron, and Pascual Restrepo. 2018. *Artificial intelligence, automation and work.* Cambridge, MA: National Bureau of Economic Research. Ssrn. https://doi.org/10.2139/ssrn.3098384.

AI and bias – IBM Research – US. 2018. Research.ibm.com

Agrawal, Ajay, Joshua S. Gans, and Avi Goldfarb. 2018. Exploring the Impact of Artificial Intelligence: Prediction Versus Judgment.. *Rotman School of Management,* https://doi.org/10.2139/ssrn.3177467.

Angwin, Julia, and Surya Mattu. 2018. *Machine Bias — ProPublica.* ProPublica.

Azevedo, Carlos R.B., Klaus Raizer, and Ricardo Souza. 2017. *A vision for human-machine mutual understanding, trust establishment, and collaboration.* 2017 IEEE conference on cognitive and computational aspects of situation management, CogSIMA 2017, 9–11. https://doi.org/10.1109/COGSIMA.2017.7929606.

Balkin, Jack M. 2017. Free speech in the algorithmic society: Big data, private governance, and new school speech regulation. *SSRN Electronic Journal.* Elsevier BV. https://doi.org/10.2139/ssrn.3038939.

Bennett, C.C., and K. Hauser. 2013. Artificial intelligence framework for simulating clinical decision-making: A Markov decision process approach. *Artificial Intelligence in Medicine* 57: 9–19. https://doi.org/10.1016/j.artmed.2012.12.003.

Brynjolfsson, Erik, and Tom Mitchell. 2017. What can machine learning do? Workforce implications: Profound change is coming, but roles for humans remain. *Science* 358: 1530–1534. https://doi.org/10.1126/science.aap8062.

Casner, Stephen M., Edwin L. Hutchins, and Don Norman. 2016. The challenges of partially automated driving. *Communications of the ACM* 59: 70–77. https://doi.org/10.1145/2830565.

Coeckelbergh, Mark. 2014. The moral standing of machines: Towards a relational and non-Cartesian moral hermeneutics. *Philosophy and Technology* 27: 61–77. https://doi.org/10.1007/s13347-013-0133-8.

Cohn, Jeffrey F., Tomas Simon Kruez, Iain Matthews, Ying Yang, Minh Hoai Nguyen, Margara Tejera Padilla, Feng Zhou, and Fernando De La Torre. 2009. *Detecting depression from facial actions and vocal prosody.* Proceedings – 2009 3rd international conference on affective computing and intelligent interaction and workshops, ACII 2009. https://doi.org/10.1109/ACII.2009.5349358.

Campolo, Alex, Madelyn Sanfilippo, Meredith Whittaker, and Kate Crawford. 2017. AI Now 2017 Report. https://ainowinstitute.org/AI_Now_2017_Report.pdf.

Dietvorst, Berkeley J., Joseph P. Simmons, and Cade Massey. 2014. Algorithm aversion: People erroneously avoid algorithms after seeing them err. *SSRN Electronic Journal.* Elsevier BV. https://doi.org/10.2139/ssrn.2466040.

EU Declaration on Cooperation on Artificial Intelligence – JRC Science Hub Communities – European Commission. 2018. JRC Science Hub Communities.

Fischer, Alastair J., and Gemma Ghelardi. 2016. The precautionary principle, evidence-based medicine, and decision theory in public health evaluation. *Frontiers in Public Health* 4: 1–7. https://doi.org/10.3389/fpubh.2016.00107.

Gao, Wei, Sam Emaminejad, Hnin Yin Yin Nyein, Samyuktha Challa, Kevin Chen, Austin Peck, Hossain M. Fahad, Hiroki Ota, Hiroshi Shiraki, Daisuke Kiriya, Der-Hsien Lien, George A. Brooks, Ronald W. Davis, and Ali Javey. 2016. Fully integrated wearable sensor arrays for multiplexed in situ perspiration analysis. *Nature* 529 (7587): 509–514. https://doi.org/10.1038/nature16521.

Gunning, David. 2017. *Explainable artificial intelligence (xai).* Defense Advanced Research Projects Agency (DARPA).

Hengstler, Monika, Ellen Enkel, and Selina Duelli. 2016. Technological forecasting & social change applied arti fi cial intelligence and trust — The case of autonomous vehicles and medical assistance devices. *Technological Forecasting and Social Change* 105: 105–120. https://doi.org/10.1016/j.techfore.2015.12.014.

Lawless, W.F., Ranjeev Mittu, Stephen Russell, and Donald Sofge. 2017. *Autonomy and artificial intelligence: A threat or savior?* Cham: Springer. https://doi.org/10.1007/978-3-319-59719-5.

Lee, Chien-Cheng, Cheng-Yuan Shih, Wen-Ping Lai, and Po-Chiang Lin. 2012. An improved boosting algorithm and its application to facial emotion recognition. *Journal of Ambient Intelligence and Humanized Computing* 3: 11–17. https://doi.org/10.1007/s12652-011-0085-8.

Liivoja, Rain. 2017. Biomedical enhancement of warfighters and the legal protection of military medical personnel in armed conflict. *Medical Law Review* 0: 1–28. https://doi.org/10.1093/medlaw/fwx046.

Kalantar-zadeh, Kourosh, Nam Ha, Jian Zhen Ou, and Kyle J. Berean. 2017. Ingestible Sensors. *ACS Sensors* 2 (4): 468–483. https://doi.org/10.1021/acssensors.7b00045.

Klein, Eran. 2015. Models of the Patient-Machine-Clinician Relationship in Closed-Loop Machine Neuromodulation. In *Machine Medical Ethics. Intelligent Systems, Control and Automation: Science and Engineering*, ed. S. van Rysewyk and M. Pontier, vol. 74. Cham: Springer. https://doi.org/10.1007/978-3-319-08108-3_17.

McManus, John, Sumeru G. Mehta, Annette R. McClinton, Robert A. De Lorenzo, and Toney W. Baskin. 2005. Informed consent and ethical issues in military medical research. *Academic Emergency Medicine* 12: 1120–1126. https://doi.org/10.1197/j.aem.2005.05.037.

Neumann, Peter G. 2016. Risks of automation. *Communications of the ACM* 59: 26–30. Association for Computing Machinery (ACM). https://doi.org/10.1145/2988445.

Ng, Andrew. 2017. *Andrew Ng: Artificial intelligence is the new electricity*. YouTube.

Praetor Suit. 2018. Doom Wiki.

Rajpurkar, Pranav, Jeremy Irvin, Robyn L. Ball, Kaylie Zhu, Brandon Yang, Hershel Mehta, and Tony Duan et al. 2018. Deep Learning For Chest Radiograph Diagnosis: A Retrospective Comparison Of The Chexnext Algorithm To Practicing Radiologists. *PLOS Medicine* 15 (11): e1002686. doi:https://doi.org/10.1371/journal.pmed.1002686.

Reidsma, D., V. Charisi, D.P. Davison, F.M. Wijnen, J. van der Meij, V. Evers, et al. 2016. The EASEL project: Towards educational human-robot symbiotic interaction. In *Proceedings of the 5th international conference on living machines*, Lecture notes in computer science; Vol. 9793, ed. N.F. Lepora, A. Mura, M. Mangan, P.F.M.J. Verschure, M. Desmulliez, and T.J. Prescott, 297–306. Cham: Springer International Publishing. https://doi.org/10.1007/978-3-319-42417-0_27.

Ritsick, Colin. 2018. Ratnik 3 – Russian Combat Suit | Future infantry exoskeleton combat system. Military Machine.

Rosenthal, Stephanie, J. Biswas, and M. Veloso. 2010. *An effective personal mobile robot agent through symbiotic human-robot interaction*. Proceedings of the 9th international conference on autonomous agents and multiagent systems (AAMAS 2010), 915–922.

Ross, Casey, and Ike Swetlitz. 2017. IBM pitched Watson as a revolution in cancer care. It's nowhere close. *STAT*.

Sim, Jai Kyoung, Sunghyun Yoon, and Young-Ho Cho. 2018. Wearable sweat rate sensors for human thermal comfort monitoring. *Scientific Reports*, 8. Springer Nature. https://doi.org/10.1038/s41598-018-19239-8.

Taleb, Nassim, and David Chandler. 2007. *The black swan*. Rearsby: W.F. Howes.

Tavani, Herman T. 2015. Levels of trust in the context of machine ethics. *Philosophy and Technology* 28: 75–90. https://doi.org/10.1007/s13347-014-0165-8.

Taylor, Earl L. 2017. Making sense of "algorithm aversion". *Research World* 2017: 57. Wiley. https://doi.org/10.1002/rwm3.20528.

The United States Special Operations Command. 2018. SOFIC 2018 conference program & exhibits guide 2018. In 2018 *United States special operations forces industry conference (SOFIC) and exhibition*. https://www.sofic.org/-media/sites/sofic/documents/sofic_2018_final-low-res3.ashx.

Warrick, Philip A, and Masun Nabhan Homsi. 2018. Ensembling convolutional and long short-term memory networks for electrocardiogram arrhythmia detection. *Physiological Measurement*. IOP Publishing. https://doi.org/10.1088/1361-6579/aad386.

White, Stephen. 2008. Brave new world: Neurowarfare and the limits of international humanitarian law. *Cornell International Law Journal* 41 (1): 177–210.

Wortham, Robert H., Andreas Theodorou, and Joanna J. Bryson. 2016. *What does the robot think? Transparency as a fundamental design requirement for intelligent systems*. IJCAI-2016 ethics for artificial intelligence workshop.

Chapter 12
Experimental Usage of AI Brain-Computer Interfaces: Computerized Errors, Side-Effects, and Alteration of Personality

Ian Stevens and Frédéric Gilbert

12.1 Introduction

12.1.1 Military Medicine Moral Obligation

Moral obligations within the context of military medicine come to fruition in various ways. They often crystalize as a necessary response to the consequences of military-related actions and decisions. For example, the Army's decision to use specific chemical deterrents during armed conflict has led to military personnel developing chronic symptoms later in life; which obligate the Army to then alleviate. This can be seen in U.S. veterans who develop Parkinson's disease from an exposure to Agent Orange during military while serving in Vietnam. Interestingly, these soldiers do not have to prove a connection between their disease and service to be eligible to receive Veterans Affairs health care and disability compensation (U.S. Department of Veterans Affairs). This example is a clear illustration of how military decisions and actions requires a moral obligation from military medicine to develop treatment solutions in response to Parkinson's symptoms (U.S. Department of Veterans Affairs). Therefore, since serving as military personnel during armed conflict entails risks of suffering from various pathologies later in life, the Army ought to innovate treatment addressing potential neurological and psychiatric pathologies. Other

I. Stevens
Northern Arizona University, Flagstaff, AZ, USA
e-mail: ims28@nau.edu

F. Gilbert (✉)
University of Tasmania, Hobart, Australia

University of Washington, Seattle, USA
e-mail: fgilbert@utas.edu.au

D. Messelken, D. Winkler (eds.), *Ethics of Medical Innovation, Experimentation, and Enhancement in Military and Humanitarian Contexts*, Military and Humanitarian Health Ethics,
https://doi.org/10.1007/978-3-030-36319-2_12

examples of innovative treatments that military medicine is obligated to develop, include, but are not limited to, neurological treatments for paralysis, spinal cord injuries, limb loss, and psychiatric treatments for psychological trauma induced during warfare, such as treatment for Post-Traumatic Stress disorder, mood disorder, cognitive deficit and memory impairment (Talan 2014).

There has been a surge by the US Defence Advanced Research Projects Agency (DARPA) in funding experimental trials testing in human novel medical brain-computer interfaces (BCIs) operated by Artificial Intelligence (AI) to treat a range of the neurological and psychiatric conditions listed above (Reardon 2017; Talan 2014). Among them, novel AI BCI devices mesh the neural and digital worlds to provide innovative treatments for patients with paralysis and mental disorders. In brief, AI BCI consists of any invasive or non-invasive system (i.e. a computer chip or headset) that can read neural signals from an individual and translate them into commands for a computerized device (Krucoff et al. 2016).· Most noticeably, studies funded by DARPA have shown that BCIs can connect human thought to the movement of a digital cursor, paralysis limb or prosthetic limb (Downey et al. 2017; Bouton et al. 2016). This novel ability to translate brain states into computerized movement is accomplished through the interpretation of neural signals from the sensorimotor cortex of a research subject often using an electroencephalogram (EEG) (Fabiani et al. 2004). Like learning to play the piano late in life, research subjects must learn how to operate the computerized signal. Simultaneously BCI system then reads and learns from the brain signals of the subject (Kostov and M. Polak 2000). As a result, a unique dynamic emerges from the ongoing interaction between the subject and BCI that eventually leads to a therapeutic response. Aside from enabling patients to move prosthetic or paralyzed limbs, there is a substantial research effort to develop BCIs to treat other types of neurological conditions, such as movement disorders, Parkinson's disease and treatment-resistant epilepsy (Gilbert et al. 2019a; Holtzheimer and Mayberg 2011). In recent years, DARPA has developed devices showing promise in treating psychiatric disorders such as treatment-resistant depression and Obsessive-Compulsive Disorder (OCD) (Ezzyat et al. 2018; Reardon 2017; Mayberg et al. 2005). From sensory restoration to cognitive restoration, hopes and expectations surrounding clinical applications of BCI are substantial (Rajesh 2013).

12.1.2 Military Medicine and Experimental Usage of AI BCI

The military efforts to test experimental research in the human brain could potentially lead to treatment interventions and create an urgent need to examine the ethical and regulatory implications of such nascent AI BCI devices. There are at least four main reasons why special protections for patient's need to be considered during the investigational phase of these military-funded experimental trials of AI BCI devices.

A first ethical question is whether patients enrolling in these invasive experimental trials possess the appropriate cognitive competence to consent to experimental trial. This challenge may be exacerbated due to the nature of their disease. For instance, can feelings of despair motivate a depressive patient to consent to implantation? Can former military personnel afflicted by PTSD consent to be implanted while having a therapeutic misconception (believing they are enrolling for treatment while the study is strictly for research) (Leykin et al. 2011)?

The second group of ethical concerns surrounds how to value informed consent and a patient's autonomy or ability to make choices over time in these treatments and research (Glannon 2008). How is informed consent maintained under circumstances where the former employer of the patient is financially supporting the trial? Could military personnel be under the undue influence to consent by their employer? How can military personnel be totally free to consent when the agency financing the trial is also their own employer?

The third grouping of ethical issues relates to the interventions of the implantation of AI BCIs creating a skewed risk and benefit assessment with the risk of harm having a magnitude greater than any potential foreseeable benefit. This is because the devices currently belong to the category of non-therapeutic risky research where safety and effectiveness have not been yet fully demonstrated, at least not in a human, and sometimes not even in a preclinical study (Viaña et al. 2017a, b).

The fourth group of ethical concerns revolves around the ability for AI BCI devices to uniquely respond to each patient's neural signals. This unique trait has provided insightful medical innovations, however, even if AI BCI technologies offer greater control at the level of neural circuits, the extent to which this new personalization and resulting grasp on neuronal functions affects the patient's psychological level is still uncharted territory. These AI BCI devices contrast the conventional relationship between the brain and computer that has included the brain being aided by the computer to execute its potentials that are currently hindered by pathological impairments. These new devices present a greater degree of the control over the human that needs further ethical examination. This degree of control over a patient's brain function and how they alter the phenomenological (i.e. self and personality) responses of these patients demands ethical examination now. This manuscript focuses on exploring this fourth group of ethical issues raised above.

In this chapter, we will be examining the potential safety issues, as well as exploring the side effects of AI BCIs, more specifically novel closed-loop deep brain stimulation implants. Namely, we will describe that potential safety issues may have adverse effects on a patient's psychological life (sense of self and/or personality). However, the resulting conclusions will also touch on the second and third group of ethical concerns. They will examine how a patient's psychological risks with AI BCIs will demonstrate an altered definition of safety. Similarly, the chapter investigates practical procedures needed to use these devices ethically and maximize a patient's autonomy. Interacting with all these ethical concerns will hopefully give both a deep philosophical examination of the field and a comprehensive argument.

12.2 From Open to Closed Loop

Though AI BCI devices are still considered experimental, they provide a new generation of therapeutic responses to disrupt neurologic symptoms. Current BCI technology, such as deep brain stimulation (DBS) treatments have emerged from past developments as it was US Food & Drug Administration approved for the treatment of Parkinson's disease back in 2002; and currently approved under the humanitarian exemption for OCD (Fins et al. 2011). Similarly, DBS treatment procedures for seizures can be traced back to the 1970's where it was, and still is, an innovative way to manage treatment-resistant epilepsy (Irving S. Cooper et al. 1976). DBS counteracts dysfunctional neural networks by the direct stimulation of nervous tissue. Some varying number electrodes are surgically placed on brain regions to conduct a pulse created by a generator implanted below the clavicle. These brain regions are the target of oscillating electrical stimulation (Hz) with the amount depending on the kind of device and physiological disorder present. DBS can then be split into two kinds of devices; opened-loop stimulation (oDBS) and closed-loop stimulation (cDBS) based off how neural stimulation is initiated.

While subtle, the difference in electrical stimulation initiation between oDBS and cDBS devices is a major difference. oDBS provides patients with a stimulation on a set interval of time. For example, a study published in 2010 had over a hundred-subject sample to test the effect of anterior nucleus stimulation for treatment-resistant epilepsy. This double-blind study showed the reduction of epileptic episodes when its oDBS program consisted of a 1-min stimulation "ON" and a 5-min "OFF". The "ON" stimulation had a 5 V potential over 90 μs pulses at 145 pulses per second (Fisher et al. 2010). cDBS uses similar parameters for stimulation, however, this stimulation only results when a particular "diseased" neural state is recognized by an implanted neurostimulator (ECoG). The ECoG examines the changes in the EEG activity to infer when a stimulation from the electrodes is necessary. This is similar to the transformation of brain states into movement commands with the aid BCIs as described above. cDBS maintains the same voltages, pulses, and pulses per second as the oDBS but the parameters of sensitivity and frequency ultimately are dependent on the device. Said differently, at this point in time, cDBS devices do not dictate the voltage and pulse since these variables are set by the researchers. Specifically, only the frequency of stimulation is determined by a cDBS device. Most importantly, each cDBS device delivers a personalized stimulation frequency on the subjects and their unique diseased neural states (Thomas and Jobst 2015). This allows each patient to receive an efficient amount of stimulation for effective seizure prevention. However, this personalization quality of cDBS is only possible with the implantation of an AI-controlled system that can facilitate a recognizable ECoG reading to stimulation event. This is done independently from either direct researcher or subject manipulation. In patients with cDBS, the computer acts autonomously to the actions of each brain and is therefore understood as an AI-tailored treatment. While cDBS treatments show promise in the reduction of epileptic seizures over the original oDBS (or even for Parkinson's) (Arlotti et al. 2016), the implementation of both are not without significant psychological side effects (Gilbert et al. 2019a; 2018a, b; Gilbert 2015a, b). In light of this, and because of seeing an agenda being pushed for the use of cDBS technologies to target psychiatric conditions (Widge et al. 2018; Ezzyat et al. 2018; Reardon 2017), many safety and efficacy concerns need to be addressed, particularly with respect to cDBS effects on personhood.

12.3 The Concerns for Safety

Safety is a common, while not exclusive, measurement of the degree of medical beneficence. Bioethically speaking, this *prima facie* principle needs concrete reasons for drifting beyond its limitations (i.e. in circumstances where the benefits outweigh the high risks of treatment). Towards one end of the risk-benefit spectrum, implanting brain devices into patients carries severe risks. The intracranial surgery may result in edema and intracerebral hemorrhages (Glannon and Ineichen 2016; Christen et al. 2012; Müller and Christen 2011). In the clinical examination of a cDBS device known as NeuroPace, involving a patient poll of 191 participants, reported implant site pain, dysesthesias, and headaches in the first year after treatment with a reduction in epileptic event frequency of 44%. While the most-common side-effect reported after the second year of implantation was site infection (Morrell 2011). Other reports articulate the acute effects of neural stimulation on the brain include the disruption of neural synchronicity, while longer-term effects show the increase glial cell expression of mRNA, used in the production of cellular proteins. These differences in gene expression have only been described in those neurons associated with neural stimulation leaving the opportunity for further longer-term neurobiological effect research studies to be conducted (Sohal and Sun 2011; Thomas and Jobst 2015). In addition, out of a sample of 160 participants being treated for a movement disorder, there was a 4.3% suicide rate and, similarly, patients being treated for Parkinson's Disease have been found to have a ~15-fold increase in suicide attempts. Therefore, it has been recommended patients with a history of major depression and successive oDBS surgeries should be excluded as candidates for the procedure (Burkhard et al. 2004). These drawbacks show a low benefit to high risk of DBS treatments on the physiological level, however, there are other emerging theoretical safety risks with cDBS that have yet to be explored.

How treatments are theoretically understood as safe is important to understand how cDBS treatments might pose ethical concerns. The notion of treatment safety in medical contexts can be understood as a "freedom from accidental injury"(Institute of Medicine Committee on Quality of Health Care in America 2000; Leape et al. 2002). This broad definition incorporates the fact that AI-tailored cDBS patients are warranted a treatment without intended harmful effects. The accidental nature of injury can come from known side-effects, human error, or, of more concern here, *computational error*. The injuries from these sources can reach into the neural, mental, and physiological realms. However, on this definition, if there are severe unknown side-effects, the treatment would not be considered safe. Along with this, safe medical practices are understood as a process that reduces the deleterious effects of a disease. Thus, to qualify as a safe treatment, the risks must be known, and the effects ought to be positive (Shojania et al. 2002). Importantly, the implicit commonality within both criteria is that both the risks (i.e. surgical site infection) and positives (i.e. reduced number of seizures) of the treatment are both known to be the same for each patient. To focus on the risks, this is like the toxic dose of pharmacological substances, where all parties are generally understood as having the same resulting deleterious effects to an amount of substance.[1] If identical twins were to consume a certain amount of lead, the same safety standard (i.e. toxic dose) would be present for both. The same could be said for one of these twins and a total stranger. Therefore safety, beyond a guide to avoid harm, is understood medically as universal. The cDBS devices, however, present situations where there is not inherently a known universal safety standard.

12.4 Rita and Roberta: A Twin Analogy to Highlight cDBS Safety Issues for Personhood

A twin analogy can be consulted to see how cDBS can cause issues with a universal safety standard; as being implanted in the brain create a multitude of complex body-technology-environment relations (Pateraki 2018). Imagine, one day a pair of military personnel, incidentally identical twin sisters, Rita and Roberta, enroll in a cDBS clinical trial for treatment-resistant depression. After serving the same amount of time and in the same armed conflict areas, each twin developed chronic depressive symptoms. Each twin has their depressive events occurring in the same brain locations, which suggest a strong genetic predisposition. Accordingly, cDBS electrodes were implanted in the same brain areas. These electrodes were correctly trained to recognize Rita and Roberta's brain states and deliver the same voltage at a certainly prescribed pulse. When implanted with these devices each device resolves ~50% of each twin's depressive symptoms with no surgical complications over the course of a year. However, when asked at any point in that year whether each twin has the same frequency ($F_{Rita} = F_{Roberta}$) of stimulation, the answer is ambiguous. Unlike oDBS (where the frequency is held constant), there is no compulsion in the cDBS system to present a certain stimulation frequency. While $F_{Rita} = F_{Roberta}$ could be true, they could also be unequal. Rita could experience five stimulations per day, while Roberta could be experiencing 20 stimulations a day, yet they both result in a 50% reduction of depressive events. This discrepancy could be larger, smaller or caused by simple differences in diseased neural patterns. Regardless of these facts, there is no guide to immediately quantify the frequency of stimulation. This statement may be anticlimactic since it has already been

[1] Accounting for body mass, gender, age, etc.

described as the primary innovation that allowed cDBS to surpass oDBS. What this also implies is that between Rita and Roberta, beyond reading the device or neural states in real time, there is no way to know what frequency is being delivered for the same treatment relief. Thus, if one twin presented a side-effect while receiving less or more stimulation than the other twin, then it would seem their personal tolerance for stimulation is different. If then each twin's tolerance of stimulation is personal then, unlike lead, there is likely no universal toxic level of stimulation between them. Said differently, this shows that there is likely no one frequency were Rita *and* Roberta will present side-effects.

Imagine a clinical setting where qualitative interviews are conducted with Roberta, Rita, and their respective partner. Roberta's partner reports, since the implantation of cDBS, that Roberta has suddenly started gambling at the local casino their family inheritance, for no apparent reason. Roberta confesses spending time at the local casino but rationalizes the rush of gambling has helped her experience less severe depressive symptoms. On her side, Rita's partner didn't notice any behavioral changes. As stated above, even if each patients' symptoms resolve by 50%, their simulation frequency could highly vary. Rita could be experiencing five stimulations per day, while Roberta could be experiencing 20 stimulations a day. While this threshold of 20 stimulations is Roberta's toxic level, hypothetically Rita's could be 30 stimulations. Here, reaching a toxic level of stimulation may translate in behavioral changes for Roberta that were generally unpredictable. Thus, concerns can then be raised about if there even is even a "toxic level" of stimulation with cDBS. This point could undermine the notion of a universal medical safety standard described earlier.

In the traditional clinical setting, each drug treatment will have a toxic dose. When a psychiatrist prescribes a pharmacological treatment to a patient, they will have the prior knowledge of what amount will likely be successful and the maximum dosage a patient should take. Once the patient is on this regime, results can be monitored, and the dose changed if the disorder is not adequately being treated, or the side effects become overbearing. Lastly, at any given moment, in theory, the psychiatrist could describe the amount of drug in the patient's system since they have the previous knowledge of dosage. Now, compared to a patient on anti-depressants, Roberta as a cDBS patient will have a few key differences. Like the previous paragraphs have demonstrated, there is little ability to know the frequency of a cDBS, unlike the patient on pharmacological interventions. Secondly, the cDBS has been trained to recognize a certain neural state to intervene on, but it and the researchers have no knowledge of a maximum stimulation or even an effective stimulation. The cDBS simply reacts to the diseased neural state and however much the brain can handle the stimulation, will guide the treatment regime. That is since there is no internal regulatory guide for the cDBS, the therapy is only successful based on how much each person's brain can handle. Then, since each brain could theoretically handle different amounts of stimulation there looks to be no way to label these devices as having a universal toxic level of frequency. Beyond the differences of sex, height, and weight that the pharmacological regimes use as a guide, there is little if any level of intervention deemed excessive and unsafe in cDBS regimes.

Thus, this ability of cDBS to vary between patients undermines the notion of medical standards leaving neurotechnological, neurosurgical, and neuroethical researchers without a guide to fully interpret beneficence measurements. An inability to account for future side effects undermines the first criteria of safety which is freedom from accidental harms. If this is true, then how should beneficence now be understood for neurotechnological treatments involving cDBS devices now that it is the founding notion of safety is undermined? One solution may be to extend the notions of patient autonomy beyond the threshold of a present moment to allow a consent process that encompasses the discovery of personality side-effects. However, before this idea can be explored further, the empirical understandings of putative personality changes ought to be defined.

12.5 Reaching the "Toxic Level" of Stimulation

Even if Rita and Roberta symptoms are gone or diminish by 50%, there are still other side effects to be concerned with. The adverse effects of brain stimulation may include hypomania, gambling, and hypersexuality (Glannon and Ineichen 2016; Christen et al. 2012; Müller and Christen 2011). Most pressing, our example suggests the stimulation side-effects may result in putative personality changes. There have been a plethora of ethical articles examining the putative effects of DBS on a patient's personality. This can differ in both how a DBS putatively changes a patient's personality initially with the interface and after the treatment has been successful. What has been described as the 'Burden of Normality' (how to understand one's self now that they have been successfully treated) has proved to be a hurdle in the coherence of a patients self-narrative (Gilbert 2012). In the examination of DBS, phenomenological interviews with patients afflicted with Parkinson's Disease have shown a correlation between those patients who previously were estranged by their disease and those who then reported being alienated by their devices. From this same study, patients were found to have either restorative or deteriorative self-estrangement effects with the implantation of a DBS. Deteriorative self-estrangement was recorded to include lack of emotional impulse control, however, this was found to be the minority of the two discovered estrangement types. Restorative self-estrangement showed to pose the greatest threat to a patient's sense of self through practically returning patients to their previous selves. This could be because while some preoperative characteristics may have returned, new postoperative differences have emerged. It should be noted that these findings contrasted with common arguments for DBS stimulation crediting its power to restore autonomy (Gilbert et al. 2019a; Gilbert et al. 2017). These findings have been discovered and analyzed in similar AI-tailored BCI treatment technologies.

Though not DBS, similar putative changes to personality have been seen in patients receiving BCI monitors for epilepsy. These devices are the first of their kind to be formally referred to as artificially intelligent due to their predictive and advisory roles for patients. The device also must intelligently learn about the pathological states of their unique patient. However, AI advisory technologies should not be thought of as synonymous with the autonomous cDBS devices described above. The AI advisory BCI devices monitor a patient's mental states to provide them with an accurate knowledge of when an epileptic event will occur and not intervene in these states. Within the study one subject talked about feeling invincible, while another described the harm this device did to their sense of self (Gilbert et al. 2019a; 2017). The four other patients, of this six participate study, supported the improved level of control this gave their lives. "[The BCI] was me, it became me, […] with this device I found myself". Together with this device, cDBS devices are the first of their kind to present a relationship with the brain were in varying degrees the computer is responsible, as a physician, for the treatment. The difficult part about this relationship could be in the contestation of a device or disorder into a patient's personality.

A common string among these studies has been the putative impact on the sense of the self (i.e. restorative or deteriorative self-estrangement) independent of whether direct stimulation is part of the treatment or not. This could point to some intrinsic aspects of having the brain interact with a machine that inherently calls for further self-examination. Two aspects of this AI-tailored research that should be drawn is firstly the magnitude of physiological and phenomenological harms that could, and sometimes do occur in DBS treatments. These potential deleterious phenomenological changes to subjects show future risks that are part of these devices beyond what the initial studies on NeuroPace may publish (Osorio 2014). Secondly, that the putative personality changes beyond simple "physiological" risks, while possibly statistically predictable, present unknown changes in a patient's personality and therefore unknown changes in how they might perceive themselves.

These putative changes in personality are what should be understood as largely unpredictable harms that may affect the patient and their autonomy. For the second point has, implicitly, been derived from the idea that there could be a possible risk when the computer has autonomous AI-tailored systems that deliver a personalized treatment for diseased neural states. This along with the knowledge that everyone's brains have no known "toxic threshold" in which to provide a guiding rule. The clear lack of empirical samples to support these overlapping concerns has not been without effort. In summary then, since these devices provide a frequency and there is no knowledge of if this frequency could either be none side-effect producing or side effect producing then it seems reasonable to conclude that putative personality changes are largely unpredictable but severely possible. However, while these points are grounded in what these devices do, new ethical questions must then be asked. These questions can address concerns around what cDBS devices do to a patient's autonomy, as others have done (Glannon and Ineichen 2016). On first approximation, cDBS appears to increase and augment patient autonomy; however, in which respect this increase of autonomy may lead to harm remain to be explored (Gilbert 2015a, b, c).

The more pressing questions of this chapter emerge when the clinical ramifications of these devices are asked ethically. For example, how can a patient consent to this treatment if they don't have tangible knowledge of its risks (i.e. putative personality changes) in which to make a competent decision? Then, after having the device for some time, what happens if Roberta is accepting of her new gambling habits considering her reduced symptoms and doesn't want the device to be turned off? Which decisions should prevail, the pre-operative or post-operative one? These questions might normally be rationally answered with the notion of beneficence, but a lack of this priniciple creates a conundrum in the risk-benefit analysis. As the previous section described since we are not considering cDBS devices beneficial per se yet because of their unpredictable putative personality changes, then there is also no way to use beneficence. For in expressing autonomy a patient must look out for future harms (i.e beneficence). Yet, if they cannot utilize this principle then their autonomy seems limited. Answering this new ambiguity might allow the ethical dilemmas that surround DBS to be examined in a redefined neurotechnological form of Principlism (Beauchamp and Childress 2013).

12.6 Ethical Implications of Roberta's Case and the Importance of Autonomy

Roberta's case can provide a pragmatic guide to the theoretical conundrum present in assessing the risks of cDBS treatments. Since patients, researchers, and clinicians cannot reasonably know what the resulting symptoms of a cDBS treatment will be, and if informed consent must be upheld, then they should wait until after the surgery has been conducted for the patient to express their autonomy. That is, until the putative personality side effects present themselves, there is no reasonable way to access the "safeness" of cDBS, and thus there is a need to wait until the unpredictable personality changes are known. Roberta must then be confident with her decision to have a surgery performed since its physiological complications are well known, but she should not be required to at that moment accept the consequences which a cDBS treatment could have on her personality. This can be demonstrated if Roberta's initial enrollment conditions were modified. Instead of a Roberta enrolling in an invasive surgical procedure, researchers had another pharmacological treatment that would result in the same unknown health outcomes to her personality and have the same 50% reduction in epileptic seizures. However, this pill will have the horrific side effect of irreversibly destroying Roberta's sense of taste. Roberta then should have the ability to consent to two things. Firstly, she must consent to the pill's irreversible physiological side effects. Secondly, she must consent to whatever the effects of the pill are to her personality. Since she cannot consent to the latter, she must suspend her consent to the effects of the pill until *after* they have presented themselves. Knowing how to understand what aspects of this procedure can be consented to at what time, the final concern surrounding Roberta's autonomy can be answered.

Roberta can choose, while in a stable mind, what to do with the cDBS device after the symptoms have presented themselves. Regardless of whether Roberta accepts or dislikes the side-effects of her cDBS when the symptoms develop she will then express her autonomous wishes. It is not until this point that an accurate risk assessment of the cDBS can be conducted. Unless these side-effects somehow undermine Roberta's ability to be autonomous, she will have the ability to continue or discontinue treatment as she sees fit. For at that point, there will no longer be an ambiguous risk analysis on the possibilities of these personalized cDBS devices. However, this is under the assumption that philosophically, and ethically speaking, the post-operative Roberta is not being undermined in her decision by this device.

This recommendation of understanding the need for a two-stage longitudinal informed consent process must note previous neuroethical debates around informed consent in a patient expressing putative personality changes. Since Roberta is now gambling, the concern is that this device is undermining other aspects of her cognition and altering Roberta into an entirely new person with separate ethical privileges and preferences. These debates cite ethical conundrums such as if the post-operative Roberta wants to accept the side-effects, but the pre-operative Roberta was described her desire for the device to be turned off in such a case, and they are technically now two different people, who should be listened to? While these concerns are pressing, the point of this chapter is that because the pre-operative Roberta could not accurately access the risks and benefits of her post-operative life, she already has limited autonomy. That is, she has a limited sense of autonomy to consent prior to the procedure and should be allowed to longitudinally, over the course of her care, express her full autonomy knowing the putative changes in her personality. Therefore, post-operative Roberta should be consulted now knowing the putative effects of the treatment. We understand this form of autonomy as her "longitudinal autonomy" and note that it has been contrastingly inspired by other papers focusing on how post-operative autonomy is undermined by philosophical definitions of the self (Witt et al. 2013; Stich and Warfield 2008; Schechtman 2007; Müller et al. 2017; Witt 2017). However, instead of focusing on the philosophical debate about a true self, this arguments stance is more in line with a practical and pragmatic use of the sense of self through time (i.e longitudinally) to maintain a patient's autonomy in their care (Müller et al. 2017).

12.7 Conclusion

Military medicine has responsibilities to develop an innovative treatment to respond to symptoms caused by serving in an armed conflict, however, it must be noted that this responsibility extends to the public and humanitarian sectors. Innovative treatment involves enrolling military personnel within experimental trials. These trials carry severe and unknown risks of harms to patients. AI BCI is currently being tested in humans and target a range of neurological and psychiatric symptoms; including major depressive disorder, treatment-resistant epilepsy, and OCD, and is being positively portrayed in media without much ethical considerations (Gilbert et al 2019b). Our chapter has shown that AI BCI may induce computerized errors which may be difficult to identify as such given there are technical challenges to establish toxicity level of stimulation. It has also shown that these cDBS treatments have a degree of independence in dictating a patient's treatment with each patient, in theory, receiving a different amount of stimulation from their device in response to a diseased brain state. This was demonstrated through the Rita and Roberta analogy to express that since conventional notions of toxic doses no longer apply, prior notions of safely become undermined. With a patient lacking a risk analysis their ability to then consent to these surgical procedures was shown to be limited if not possibly undermined. These issues are aligning with existing ones observed with other neurotechnologies and neurointerventions (Burwell et al 2017; Pugh et al. 2018; Gilbert et al. 2014, 2018a, b, 2019c; Vranic and Gilbert 2014; Gilbert and Tubig 2018; Gilbert 2015c, 2018). Thus, to account for lack of pre-operative autonomy, a longitudinal autonomy has been suggested. The longitudinal autonomy was also described to show how a patient's relationship to these devices works, like both a drug and physician, by having a semi-independent role in their brains.

While both the oDBS devices and cDBS have caused concern for potentially altered sense of the self and physiological harms, the cDBS has undermined the ability of patients to make informed choices about what to expect. The AI system is like a physician in controlling the brain, semi-independent in its role of trying to provide relief, but without a guide on the harms it might be inflicting. Therefore, they are also like a drug in following what they were designed to do. This shows that patients are in a relationship with a semi-independent entity and accommodating for this device into a trustworthy and reliable brain-computer relationship will require a level of patience. Since trusting these devices with our brains will transpire by understanding how to make these devices more reliable. Thus, we recommend the need to prioritize a patient's longitudinal autonomy when they interact and develop with these devices (Gilbert et al. 2018a, b). This is because, while these devices originated in an oDBS device structure, recent research has shown the success of cDBS in treating these ailments and therefore their importance in future research. For although these devices may be intelligent enough to "know" how to perform their job, they cannot account for the consequences of their actions. Therefore it is important that we spend time with these devices, develop them scientifically, and pause to understand their ethical implications so that one day we can trust them with our minds.

Acknowledgement Frederic Gilbert is supported by a grant from the Australian Research Council (DECRA award Project Number DE150101390).

References

Arlotti, Mattia, Lorenzo Rossi, Manuela Rosa, Sara Marceglia, and Alberto Priori. 2016. An external portable device for adaptive deep brain stimulation (aDBS) clinical research in advanced Parkinson's disease. *Medical Engineering & Physics* 38 (5): 498–505. https://doi.org/10.1016/j.medengphy.2016.02.007.

Beauchamp, Tom L., and James F. Childress. 2013. *Principles of biomedical ethics*. Oxford/New York: Oxford University Press.

Bouton, Chad E., Ammar Shaikhouni, Nicholas V. Annetta, Marcia A. Bockbrader, David A. Friedenberg, Dylan M. Nielson, Gaurav Sharma, et al. 2016. Restoring cortical control of functional movement in a human with quadriplegia. *Nature* 533 (7602): 247–250. https://doi.org/10.1038/nature17435.

Burkhard, P.R., F.J.G. Vingerhoets, A. Berney, J. Bogousslavsky, J.-G. Villemure, and J. Ghika. 2004. Suicide after successful deep brain stimulation for movement disorders. *Neurology* 63 (11): 2170–2172.

Burwell, Sarah, Matthew Sample, and Eric Racine. 2017. Ethical aspects of brain computer interfaces: a scoping review. *BMC Medical Ethics* 18 (1).

Christen, Markus, Merlin Bittlinger, Henrik Walter, Peter Brugger, and Sabine Müller. 2012. Dealing with side effects of deep brain stimulation: Lessons learned from stimulating the STN. *AJOB Neuroscience* 3 (1): 37–43. https://doi.org/10.1080/21507740.2011.635627.

Cooper, Irving S. 1976. Chronic Cerebellar Stimulation in Epilepsy. *Archives of Neurology* 33 (8): 559.

Downey, John E., Lucas Brane, Robert A. Gaunt, Elizabeth C. Tyler-Kabara, Michael L. Boninger, and Jennifer L. Collinger. 2017. Motor cortical activity changes during neuroprosthetic-controlled object interaction. *Scientific Reports* 7 (1).

Ezzyat, Youssef, Paul A. Wanda, Deborah F. Levy, Allison Kadel, Ada Aka, Isaac Pedisich, Michael R. Sperling, et al. 2018. Closed-loop stimulation of temporal cortex rescues functional networks and improves memory. *Nature Communications* 9 (1): 365. https://doi.org/10.1038/s41467-017-02753-0.

Fabiani, G.E., D.J. McFarland, J.R. Wolpaw, and G. Pfurtscheller. 2004. Conversion of EEG Activity Into Cursor Movement by a Brain–Computer Interface (BCI). *IEEE Transactions on Neural Systems and Rehabilitation Engineering* 12 (3): 331–338.

Fins, Joseph J., Helen S. Mayberg, Bart Nuttin, Cynthia S. Kubu, Thorsten Galert, Volker Sturm, Katja Stoppenbrink, Reinhard Merkel, and Thomas E. Schlaepfer. 2011. Misuse of the FDA's humanitarian device exemption in deep brain stimulation for obsessive-compulsive disorder. *Health Affairs (Project Hope)* 30 (2): 302–311. https://doi.org/10.1377/hlthaff.2010.0157.

Fisher, Robert, Vicenta Salanova, Thomas Witt, Robert Worth, Thomas Henry, Robert Gross, Kalarickal Oommen, et al. 2010. Electrical stimulation of the anterior nucleus of thalamus for treatment of refractory epilepsy. *Epilepsia* 51 (5): 899–908. https://doi.org/10.1111/j.1528-1167.2010.02536.x.

Gilbert, Frederic. 2012. The burden of normality: From 'chronically ill' to 'symptom free'. New ethical challenges for deep brain stimulation postoperative treatment. *Journal of Medical Ethics* 38 (7): 408–412. https://doi.org/10.1136/medethics-2011-100044.

———. 2015a. A threat to autonomy? The intrusion of predictive brain implants. *AJOB Neuroscience* 6 (4): 4–11. https://doi.org/10.1080/21507740.2015.1076087.

———. 2015b. Are predictive brain implants an indispensable feature of autonomy? *Bioethica Forum* 8: 121–127.

———. 2015c. Self-estrangement & deep brain stimulation: Ethical issues related to forced explantation. *Neuroethics* 8 (2): 107–114. https://doi.org/10.1080/2326263X.2019.1655837.

———. 2018. Deep brain stimulation: Inducing self-estrangement. *Neuroethics* 11: 157–165. https://doi.org/10.1007/s12152-017-9334-7.

Gilbert, Frederic, and Paul Tubig. 2018. Cognitive enhancement with brain implants: The burden of abnormality. *Journal of Cognitive Enhancement* 2 (4): 364–368. https://doi.org/10.1007/s41465-018-0105-0.

Gilbert, Frederic, Alexandre Harris, and Robert Kapsa. 2014. Controlling brain cells with light: Ethical considerations for optogenetics trials. *American Journal of Bioethics Neuroscience* 5 (3): 3–11. https://doi.org/10.1080/21507740.2014.911213.

Gilbert, Frederic. Brown, Dasgupta, Martens, Klein, Goering, An Instrument to Capture the Phenomenology of Implantable Brain Device Use. 2019c. *Neuroethics*. https://doi.org/10.1007/s12152-019-09422-7.

Gilbert, Frederic, Eliza Goddard, John Noel M. Viaña, Adrian Carter, and Malcolm Horne. 2017. I miss being me: Phenomenological effects of deep brain stimulation. *AJOB Neuroscience* 8 (2): 96–109. https://doi.org/10.1080/21507740.2017.1320319.
Gilbert, Frederic, Terence O'brien, and Mark Cook. 2018a. The effects of closed-loop brain implants on autonomy and deliberation: What are the risks of being kept in the loop? *Cambridge Quarterly of Healthcare Ethics* 27 (2): 316–325. https://doi.org/10.1017/S0963180117000640.
Gilbert, Frederic, John Noel M. Viaña, and Christian Ineichen. 2018b. Deflating the "DBS caused personality changes" bubble. *Neuroethics*. https://doi.org/10.1007/s12152-018-9373-8.
Gilbert, F., Mark Cook, Terrence O'Brien, and Judy Illes. 2019a. Embodiment and estrangement: Results from a first-in-human 'Intelligent BCI' trial. *Science and Engineering Ethics*, November: 1–14. https://doi.org/10.1007/s11948-017-0001-5.
Gilbert, Frederic. C. Pham, Jnm Viaña, and W. Gillam. 2019b. Increasing brain-computer interface media depictions: pressing ethical concerns. *Brain-Computer Interfaces* 6 (3): 49–70.
Glannon, Walter. 2008. Deep-brain stimulation for depression. *HEC Forum* 20 (4): 325–335. https://doi.org/10.1007/s10730-008-9084-3.
Glannon, Walter, and Christian Ineichen. 2016. Philosophical aspects of closed-loop neuroscience. In *Closed loop neuroscience*, 259–270. Amsterdam: Academic.
Holtzheimer, Paul E., and Helen S. Mayberg. 2011. Deep Brain Stimulation for Psychiatric Disorders. *Annual Review of Neuroscience* 34 (1): 289–307.
Institute of Medicine (US) Committee on Quality of Health Care in America. 2000. In *To err is human: Building a safer health system*, ed. Linda T. Kohn, Janet M. Corrigan, and Molla S. Donaldson. Washington, DC: National Academy Press. www.ncbi.nlm.nih.gov/books/NBK225182/.
Kostov, A., and M. Polak. 2000. Parallel man-machine training in development of EEG-based cursor control. *IEEE Transactions on Rehabilitation Engineering* 8 (2): 203–205. https://doi.org/10.1109/86.847816.
Krucoff, Max O., Shervin Rahimpour, Marc W. Slutzky, V. Reggie Edgerton, and Dennis A. Turner. 2016. Enhancing nervous system recovery through neurobiologics, neural interface training, and neurorehabilitation. *Frontiers in Neuroscience* 10: 584. https://doi.org/10.3389/fnins.2016.00584.
Leape, Lucian L., Donald M. Berwick, and David W. Bates. 2002. What practices will most improve safety? Evidence-based medicine meets patient safety. *JAMA* 288 (4): 501–507.
Leykin, Yan, Paul P. Christopher, Paul E. Holtzheimer, Paul S. Appelbaum, Helen S. Mayberg, Sarah H. Lisanby, and Laura B. Dunn. 2011. Participants' perceptions of deep brain stimulation research for treatment-resistant depression: Risks, benefits, and therapeutic misconception. *AJOB Primary Research* 2 (4): 33–41. https://doi.org/10.1080/21507716.2011.627579.
Mayberg, Helen S., Andres M. Lozano, Valerie Voon, Heather E. McNeely, David Seminowicz, Clement Hamani, Jason M. Schwalb, and Sidney H. Kennedy. 2005. Deep Brain Stimulation for Treatment-Resistant Depression. *Neuron* 45 (5): 651–660.
Morrell, Martha J., and RNS System in Epilepsy Study Group. 2011. Responsive cortical stimulation for the treatment of medically intractable partial epilepsy. *Neurology* 77 (13): 1295–1304. https://doi.org/10.1212/WNL.0b013e3182302056.
Müller, Sabine, and Markus Christen. 2011. Deep brain stimulation in Parkinsonian patients—Ethical evaluation of cognitive, affective, and behavioral sequelae. *AJOB Neuroscience* 2 (1): 3–13. https://doi.org/10.1080/21507740.2010.533151.
Müller, Sabine, Merlin Bittlinger, and Henrik Walter. 2017. Threats to neurosurgical patients posed by the personal identity debate. *Neuroethics* 10 (2): 299–310. https://doi.org/10.1007/s12152-017-9304-0.
Osorio, Ivan. 2014. The neuroPace trial: Missing knowledge and insights. *Epilepsia* 55 (9): 1469–1470. https://doi.org/10.1111/epi.12701.
Pateraki, Marilena. 2018. The multiple temporalities of deep brain stimulation (DBS) in Greece. *Medicine, Health Care, and Philosophy* 22 (3): 353–362. https://doi.org/10.1007/s11019-018-9861-y.

Pugh, Jonathan, Laurie Pycroft, Hannah Maslen, Tipu Aziz, and Julian Savulescu. Evidence-Based Neuroethics, Deep Brain Stimulation and Personality - Deflating, but not Bursting, the Bubble. *Neuroethics*. https://doi.org/10.1007/s12152-018-9392-5.

Rao, Rajesh P.N. 2013. *Brain-Computer Interfacing: An Introduction*. 1st ed. New York: Cambridge University Press.

Reardon, Sara. 2017. AI-controlled brain implants for mood disorders tested in people. *Nature* 551 (7682): 549–550. https://doi.org/10.1038/nature.2017.23031.

Schechtman, Marya. 2007. *The constitution of selves*. Ithaca: Cornell University Press.

Shojania, Kaveh G., Bradford W. Duncan, Kathryn M. McDonald, and Robert M. Wachter. 2002. Safe but sound: Patient safety meets evidence-based medicine. *JAMA* 288 (4): 508–513. https://doi.org/10.1001/jama.288.4.508.

Sohal, Vikaas S., and Felice T. Sun. 2011. Responsive neurostimulation suppresses synchronized cortical rhythms in patients with epilepsy. *Neurosurgery Clinics of North America* 22 (4): 481–488, vi. https://doi.org/10.1016/j.nec.2011.07.007.

Stich, Stephen P., and Ted A. Warfield. 2008. *The Blackwell guide to philosophy of mind*. Hoboken: Wiley.

Talan, Jamie. 2014. DARPA. Neurology Today 14 (20):8–10

Thomas, George P., and Barbara C. Jobst. 2015. Critical review of the responsive neurostimulator system for epilepsy. *Medical Devices* 8: 405–411. https://doi.org/10.2147/MDER.S62853.

US Department of Veterans Affairs, Veterans Health. *Parkinson's disease and agent orange – Public health*. General Information. www.publichealth.va.gov/exposures/agentorange/conditions/parkinsonsdisease.asp. Accessed 12 Aug 2018.

Viaña, John Noel M., Merlin Bittlinger, and Frederic Gilbert. 2017a. Ethical considerations for deep brain stimulation trials in patients with early-onset Alzheimer's disease. *Journal of Alzheimer's Disease: JAD* 58 (2): 289–301. https://doi.org/10.3233/JAD-161073.

Viaña, John Noel M., James C. Vickers, Mark J. Cook, and Frederic Gilbert. 2017b. Currents of memory: Recent progress, translational challenges, and ethical considerations in fornix deep brain stimulation trials for Alzheimer's disease. *Neurobiology of Aging* 56 (August): 202–210. https://doi.org/10.1016/j.neurobiolaging.2017.03.001.

Vranic, Andrej, and Gilbert Frederic, 2014. Prognostic implication of preoperative behavior changes in patients with primary high-grade Meningiomas. *The Scientific World Journal* 2014: 398295, 5 pages https://doi.org/10.1155/2014/398295.

Widge, Alik S., Donald A. Jr Malone, and Darin D. Dougherty. 2018. Closing the loop on deep brain stimulation for treatment-resistant depression. *Frontiers in Neuroscience* 12: 175. https://doi.org/10.3389/fnins.2018.00175.

Witt, Karsten. 2017. Identity change and informed consent. *Journal of Medical Ethics* 223 (4): 254–254. https://doi.org/10.1136/medethics-2016-103684.

Witt, Karsten, Jens Kuhn, Lars Timmermann, Mateusz Zurowski, and Christiane Woopen. 2013. Deep brain stimulation and the search for identity. *Neuroethics* 6 (3): 499–511. https://doi.org/10.1007/s12152-011-9100-1.

Wolkenstein, Andreas, Ralf Jox, and Orsolya Friedrich. 2018. Brain–Computer Interfaces: Lessons to Be Learned from the Ethics of Algorithms. *Cambridge Quarterly of Healthcare Ethics* 27 (4): 635–646.

Chapter 13
Memory Modification as Treatment for PTSD: Neuroscientific Reality and Ethical Concerns

Rain Liivoja and Marijn C. W. Kroes

13.1 Introduction

Exposure to traumatic events can trigger mental disorders, such as posttraumatic stress disorder (PTSD). People with PTSD are haunted by intrusive traumatic memories that evoke severe fear responses, causing great suffering to affected individuals. Members of the armed forces are at an increased risk of experiencing traumatic events and therefore developing PTSD. In the US, overall lifetime prevalence of PTSD in the general population has been found to be about 8% (Kessler et al. 2005; Keane et al. 2006), whereas the prevalence of PTSD among US military personnel returning from Iraq has been estimated to be 23% (Fulton et al. 2015).

While some current treatments of PTSD have proven effective, many patients do not benefit from them or experience a return of symptoms even after initially

A slightly different version of this paper appeared as Marijn C. W. Kroes and Rain Liivoja. 2019. Eradicating war memories: Neuroscientific reality and ethical concerns. *International Review of the Red Cross* 101: 69–95. https://doi.org/10.1017/S1816383118000437. Sections of this article have been reproduced here by kind permission of Cambridge University Press. We are grateful to Chris Jenks and James Wolfe for comments on an earlier version of this paper. While writing the paper, both authors were supported by Branco Weiss Fellowships. Marijn Kroes was also supported by an H2020 Marie Skłodowska-Curie Fellowship.

R. Liivoja (✉)
School of Law, University of Queensland, Brisbane, Australia

Erik Castrén Institute of International Law and Human Rights, University of Helsinki, Helsinki, Finland
e-mail: r.liivoja@uq.edu.au

M. C. W. Kroes
Radboud University Nijmegen Medical Center, Nijmegen, The Netherlands
e-mail: marijn.kroes@donders.ru.nl

D. Messelken, D. Winkler (eds.), *Ethics of Medical Innovation, Experimentation, and Enhancement in Military and Humanitarian Contexts*, Military and Humanitarian Health Ethics,
https://doi.org/10.1007/978-3-030-36319-2_13

successful treatment (Vervliet et al. 2013; Hendriks et al. 2018). This highlights a need to develop more effective and persistent treatments. Memory modification techniques (MMTs) hold great potential to prevent or treat PTSD as they could be used to target traumatic memories. However, given the intimate connection between memories and personal identity, and social significance of some memories, MMTs also have ethical, legal and social implications (ELSI).

Bioethicists have debated these implications extensively. Much of that debate, however, has taken place in broad terms, without being entirely clear about what concerns relate to all uses of MMTs, including therapeutic uses, and what concerns only apply to abuses or misuses.

Also, the bioethical discussion has sometimes lost track of what is scientifically possible or probable. Yet, to develop a defensible policy, one should consider the actual effects (both intended effects and side effects) of MMTs. We do not deny the value of speculating about scientific developments, which may help identify problems worthy of further contemplation (e.g. Roache 2008). Also, we agree with the idea that "ethical reflection should precede technological [and scientific] progress and possible future applications" (Cabrera and Elger 2016, 96). But judgments about the propriety or otherwise of biomedical interventions should be passed on the merits of those interventions, rather than based on what other (more potent, more dangerous, etc.) interventions might be developed in the future. After all, to a sufficiently conservative observer, every advance in science and technology looks like the thin edge of some wedge. Moreover, an overly speculative approach can result in scientists discarding ethical concerns as unrealistic and abandoning the debate.

We begin our discussion by providing an overview of how traumatic memories contribute to PTSD, current treatment methods, their limitations, and the state-of-the-art of MMTs. Then, drawing on these neuroscience insights, we discuss some ELSI of utilizing MMTs to treat PTSD in military populations. We focus on three major sets of issues: safety and social justice concerns, concerns about threat to authenticity and identity, and possible legal and moral duties to retain certain memories.

13.2 Posttraumatic Stress Disorder

PTSD is a mental disorder that can develop after exposure to severely distressing events, such as death, serious injury or sexual violence, and is characterised by clusters of symptoms persisting over a significant period of time (American Psychiatric Association 2013). People with PTSD persistently re-experience the traumatic events via intrusive thoughts, nightmares, and "flashbacks". They suffer negative thoughts and feelings, avoid trauma-related reminders, and experience hyper-arousal symptoms such as irritability, difficulties concentrating or sleeping. In response to trauma-related stimuli they often experience dissociative symptoms such as depersonalization or derealization. Patients often report feeling as if the traumatic event is happening in "real time" instead of the past, which evokes a sense of current threat. Beyond psychological symptoms, people with PTSD often experience interpersonal, psychosocial, and physical health problems (Keane et al. 2006).

There has been a shift, particularly in the United States, to "drop the 'D' from PTSD" – to refrain from referring to the condition as a disorder (Itkowitz 2015). The destigmatisation pursued by this move must be commended but it can lead to confusion. Many people develop psychological symptoms as a normal reaction to traumatic experience, which usually dissipate over time. This normal experience of posttraumatic stress (PTS) can be distinguished from PTSD (Jia 2017). The latter involves more severe and persistent symptoms, which are so debilitating as to require treatment.

The defining onset of PTSD is a traumatic experience that results in the formation of a traumatic memory. The severity and the perceived threat of the traumatic experience predict PTSD severity (Brewin et al. 2000; Ozer et al. 2003), and are also factors known to strengthen memory formation (see LaBar and Cabeza 2006). Moreover, PTSD may develop because the traumatic experience shatters our learned assumptions and beliefs about the safety of our world (Janoff-Bulman 1992; Horowitz 2014).

While a number of psychological theories on PTSD have been put forward, memory plays a critical role in all of them (Brewin and Holmes 2003). Indeed, PTSD might even be considered a memory disorder (Foa et al. 1989). PTSD is characterized by intrusive memories. The content of intrusive memories often includes trivial stimuli or situations that preceded the traumatic event (Ehlers et al. 2002). For example, a war veteran may have intrusive memories of rustling leaves that were seen or heard prior to the emergence of enemy soldiers from the jungle. Such memories may serve as warning signals that are later interpreted as signals of impeding danger and thus evoke a sense of current threat. As such, intrusive memories are learned predictors of danger that come to evoke defensive responses, avoidance behaviours, and involuntary retrieval of thoughts, feelings, and memories of the traumatic event.

People with PTSD involuntarily re-experience intrusive memories as if happening in real-time, but often have difficulty purposefully recollecting the trauma memory (Foa et al. 1989; Ehlers and Clark 2000; Brewin and Holmes 2003). Recall is often not chronological but jumps back-and-forth in time between events and, unlike in ordinary memory retrieval, people often get stuck or hung-up on particular details and feelings. As such, traumatic memories in PTSD may be processed differently from ordinary emotional memories and may be qualitatively different and possibly stored differently in the brain (Brewin et al. 2010).

Many who experience psychological trauma will initially develop PTSD-like symptoms but most will learn to overcome these symptoms over time (Bisson et al. 2015). Only a portion of people who experience trauma will not learn to control traumatic symptoms and develop PTSD. Thus, disturbances in learning to control emotional responses and memory for situations of safety also contribute to PTSD.

13.3 Current Treatments and Their Limitations

Treatments for PTSD continue to be developed based on advancing psychological and neuroscientific insights. Antidepressants, specifically selective serotonin reuptake inhibitors (SSRIs), are the most commonly prescribed pharmacotherapies for PTSD. However, SSRIs are only moderately effective for treating PTSD and less effective than psychotherapy (van Etten and Taylor 1998). The primary psychological intervention for PTSD is exposure treatment, in which patients are guided to vividly imagine and describe the traumatic experience, and to re-evaluate and reinterpret stimuli, their meaning, and responses (Lang 1977), all with the aim of reducing emotional responses and increasing the feeling of control.

Most modern psychotherapies have integrated exposure treatment with other behavioural and cognitive approaches (see Brewin and Holmes 2003). Psychotherapy is effective in reducing PTSD symptoms and reaching remission (van Etten and Taylor 1998). However, the majority of patients experience some return of symptoms even after initially successful treatment (Vervliet et al. 2013). This may be due to exposure treatment being based on the principles of extinction learning in Pavlovian threat conditioning, which forms a separate safety memory that comes to inhibit the expression of a threat memory but does not modify the threat memory itself (see Sect. 13.4.3). This indicates that although psychotherapy for PTSD aims to restructure memory, it probably also does not change the trauma memory itself, leaving the risk for the return of symptoms.

13.4 Neuroscience of Memory Modification

Modern neuroscience is discovering techniques to permanently modify the original threat memory itself, which has great potential to develop novel treatments for psychological trauma.

13.4.1 Concept and Taxonomy of Memory

Memory can be defined as an internal representation of an experience captured in a physiological change in the brain, enabling the expression of the earlier experience in thought or behaviour (Dudai 2007). This definition contains two components: the expression of memory in thought or behaviour, and its neural underpinning. The latter component is called an engram or memory trace (Semon 1921).

Different neural systems support different behaviours and thoughts, while having the capacity for memory (Henke 2010). As a result, psychologists distinguish between different types of memories (see Tulving 1972; Squire 1992). Memory types that primarily contribute to PTSD are conditioned memories and episodic memories (Squire 2004). These different types of memories contribute to distinct symptoms in PTSD.

Aversive conditioned memories can be formed via Pavlovian threat conditioning: pairing a stimulus (such as a sound) with an aversive outcome (such as pain) can come to evoke defensive responses (for example, changes in heart rate) indicating the formation of a memory association between the stimulus and the outcome (LeDoux 2000). Such conditioned threat memories may contribute to the hyper-arousal and re-experiencing symptoms evoked by warning signals in PTSD (Ehlers and Clark 2000).

Episodic memory relates to particular experiences that include associations between who, what, where, when, and why (Squire 2004) – for instance, recalling "in front of our minds' eye" a particularly distressing war experience. Episodic memories play a role in the re-experiencing of autobiographical events of the traumatic experience in PTSD. For example, flashbacks involve the reliving of the traumatic episode as if happening in real time. Furthermore, episodic memory of the traumatic experience is often fragmented in PTSD and in extreme cases people may have no episodic memory of the traumatic event at all (amnesia).

13.4.2 Memory Formation and Consolidation

How are these memories formed? Experiences create patterns of neural activation in the brain via our senses. The formation of a memory of an experience involves the strengthening of connections between brain cells activated by an experience and requires neurotransmitter signalling, gene transcription, and protein synthesis. Drugs or other interventions administered right before or after learning – the moment of acquisition of information of an experience – can impair memory (McGaugh 2000). However, the same interventions administered hours after learning no longer have an effect on memory. This has led to a standard view on memory that suggests that memories are initially labile (meaning they are sensitive to modification by interventions) but stabilize over time during a period of consolidation after which memories are stable and can no longer be modified (McGaugh 2000). Neurotransmitters and hormones that are released during emotional experiences, such as noradrenaline and cortisol, can strengthen consolidation and result in an emotional memory enhancement (McGaugh 2000).

This implies that, immediately before and after a traumatic experience there may be a brief window of opportunity to prevent a traumatic memory from becoming permanently stored or to minimize its emotional enhancement. This has been experimentally attempted in clinical practice. The administration of a beta-blocker to people admitted to an Emergency Department after a traumatic experience reduced threat responses to trauma reminders and PTSD symptoms 1 month later (Pitman et al. 2002). Thus, beta-blockers may impair the consolidation of trauma memory and prevent the development of PTSD. However, the usefulness of this approach is limited, as the treatment needs to take place right after the traumatic experience to be effective whereas most people with PTSD do not seek treatment till months after the trauma.

13.4.3 Extinction Learning

Patients often come into a therapist's office long after a trauma memory has formed and been consolidated. The often-observed return of PTSD symptoms even after initially successful psychotherapy can be explained because treatment (particularly exposure treatment) is based on principles of extinction learning of Pavlovian conditioning (Vervliet et al. 2013).

During extinction training, a threatening stimulus is repeatedly presented without an aversive outcome so that over time the person will stop displaying threat-related defensive responses. However, extinction learning does not modify the original threat memory. It rather forms a novel safety memory that inhibits the expression of threat responses, which can give way to the return of threat responses with the passage of time, changes in context, or increases in arousal (Myers and Davis 2002; Bouton 2004; Quirk and Mueller 2008).

From an evolutionary perspective it makes sense not to overwrite a conditioned threat memory as the threat memory is adaptive and protects us from danger. However, the unfortunate result is that psychotherapy most likely also does not alter the original trauma memory but forms a novel safety memory (Vervliet et al. 2013), which leaves the risk of the return of symptoms even after initially successful treatment.

13.4.4 Flexibility of Memories

Memories, particularly episodic memories, can be flexible (Kroes and Fernández 2012). Most of what we initially remember we forget within 24 h. What we still remember after 24 h we forget at a much slower rate (Hirst et al. 2009). For distressing experiences such as the Challenger Space Shuttle explosion or the September 11 attacks, we are often very sure about the accuracy of our episodic memories whereas in fact we accurately remember only around 30% (Neisser and Harsch 1992; Hirst et al. 2009). Furthermore, with a bit of suggestion, it is possible to make people remember things that never happened, like being lost in a mall as a child (Loftus and Pickrell 1995).

We thus forget most of what we initially remember, and our memories can be highly inaccurate or even completely false. This flexibility of episodic memory is adaptive as it helps us to survive. Our environments continuously change and forgetting may allow us dispose of out-dated and unimportant information and keep our memory fresh. At the same time, updating of episodic memories by new experiences and integration of memory from different experiences help to better describe regularities of our environment. What is still unclear is whether the memory flexibility described here results from a modification of the original memory or confusion between different memories at the time of retrieval. Regardless, when discussing the ELSI of MMTs, it is critical to realize that memories are not a faithful reflection of the past but serve to support adaptive responses and decision-making in the future.

13.4.5 Memory Reconsolidation

The classical view on memory holds that memories are initially labile but stabilize over time during a period of consolidation after which they remain essentially unchanged (McGaugh 2000). This suggests that once memory is consolidated, it becomes resistant to MMTs.

Contemporary neuroscience has challenged this view on memory and suggests that it is possible for MMTs to modify consolidated memories (Nader and Hardt 2009; Kroes et al. 2016). In a seminal study, Karim Nader et al. (2000) used rats to show that a brief reminder can reactivate a consolidated threat conditioned memory and temporarily return the memory to a labile state requiring re-stabilization processes (such as protein synthesis), referred to as reconsolidation. It was found that disrupting reconsolidation processes by blocking protein synthesis can result in the loss of the conditioned threat responses and prevent their return.

To disrupt reconsolidation, Nader et al. (2000) injected a toxic protein-synthesis inhibitor into the brain of rats, which is clearly not safe for use in humans. Subsequent laboratory experiments showed that the administration of beta-blockers, such as propranolol could also disrupt reactivated memories and prevent the return of threat conditioned responses in rodents and humans (Dębiec and LeDoux 2004; Kindt et al. 2009).

The disruption of consolidated memory by interventions such as beta-blockers only occurs under specific circumstances, namely, when memory is reactivated by a brief reminder and when the intervention is administered within a short time period after the reminder – that is, during the reconsolidation window. Furthermore, behavioural interventions may also affect reconsolidation: reactivating a conditioned threat memory to return the memory to a labile state and then administering extinction training during the reconsolidation window (reactivation-extinction paradigm) can also prevent the return of conditioned threat responses (Monfils et al. 2009; Schiller et al. 2010). Here the idea is that when memory is returned to a labile state, extinction can overwrite or update the original threat memory and thus negate the formation of a separate safety memory. Hence, the reactivation-extinction paradigm suggests that existing exposure treatments could be optimized by a minor change in procedures. Collectively, these laboratory experiments suggest that pharmacological and behavioural interventions can disturb reconsolidation of reactivated threat-conditioned memories (e.g., memory for tone-shock associations) resulting in the loss of the expected reaction to a learned threat.

Interestingly, initial laboratory studies in humans indicated that reconsolidation interventions specifically disrupted threat conditioned defensive responses (e.g. sweating or startle responses) but left intact participants' ability to explicitly recall the threatening experience to mind (Kindt et al. 2009; Schiller et al. 2010). This led to the suggestion that reconsolidation interventions only affect the emotional component of memory but preserve episodic memory (Dębiec and Altemus 2006; Soeter and Kindt 2010; Elsey and Kindt 2016). However, subsequent studies indicated that reconsolidation interventions with beta-blockers and behavioural manipulations can diminish the emotional enhancement of episodic memory and that electrical brain stimulation can even fully eradicate specific episodic memories in humans (see Kroes et al. 2016). In one study, participants who received electroconvulsive treatment (ECT) for unipolar depression learned two slide-show stories a week prior to treatment (Kroes et al. 2014). Right before ECT the participants were briefly reminded of one of the two stories to reactivate memory for this specific story. One day after the reminder and ECT, participants could explicitly remember the story that they were not reminded of but could no longer remember the story of which they had been reminded. Thus, reconsolidation interventions can modify specific reactivated conditioned memories as well as episodic memories. That said, reconsolidation interventions do not cause general memory impairments – meaning they do not impair all memories that we have (Nader and Hardt 2009; Kroes et al. 2016). Only the specific memory that is reactivated and returned to a labile state can be modified.

From a clinical perspective, there are several limitations to reconsolidation interventions. First, evidence for reconsolidation has been obtained across many experimental paradigms and species, but not all studies have yielded positive results (Nader and Hardt 2009; Kroes et al. 2016). Second, older memories, especially episodic memories, appear less sensitive to reconsolidation interventions (Kroes et al. 2016). Third, much is still unknown about the conditions under which memories do and do not return to a labile state and can be modified. Fourth, if memory is reactivated but no intervention is administered or if interventions fail, reconsolidation strengthens memory. Hence, the opportunity to translate reconsolidation-intervention techniques to treat patients who often have had traumatic memories for many years may be limited and, if interventions fail, reconsolidation may inadvertently strengthen trauma memories (Kroes et al. 2016). Much research is still needed to translate this laboratory research to effective clinical applications.

Reactivating a memory can provide an opportunity for interventions to steer the reconsolidation of a specific maladaptive memory in a particular direction and potentially treat trauma-, stressor- and anxiety-related disorders, including PTSD. Importantly, reconsolidation interventions would require only minor changes to existing psychotherapeutic behavioural procedures (for example adding a brief memory reactivation prior to standard exposure treatment) or involve a precise combination of psychotherapy and administration of a single dose of medication.

Based on laboratory experiments, several clinical trials have investigated the potential to use MMTs to enhance exposure treatment or impair reconsolidation of trauma memory to treat patients.

The advantage of reconsolidation interventions is that they theoretically allow for the modification of specific memories at any time after learning and that the intervention can be applied after controlled memory reactivation. Initial clinical trials found that targeting reconsolidation of trauma memories with beta-blockers in PTSD patients can subsequently reduce threat responses to trauma reminders and reduce PTSD symptoms (Brunet et al. 2008, 2011, 2014). However, there are limitations to these studies (Kroes et al. 2016) and subsequent studies failed to replicate (Wood et al. 2015). The effectiveness of a beta-blocker reconsolidation intervention to treat PTSD thus appears limited. Note that PTSD patients are haunted by intrusive emotional *episodic* memories (Brewin et al. 2010) that they have often had for many years, which may be particularly difficult to modify as explained above. Targeting reconsolidation to treat specific phobias, which mainly involve conditioned responses, may be more effective (Soeter and Kindt 2015).

In sum, clinical trials have shown that MMTs may prevent the formation of traumatic memories, can enhance exposure treatment, and may disrupt reconsolidation of consolidated memories. However, the efficacy of clinical translations has so far been limited, potentially because it is unclear how to optimally target trauma memories or because not all types of memory may be equally sensitive to MMTs (Kroes et al. 2016).

13.5 Ethical, Legal and Social Issues

The possible use of MMTs raises important ethical, legal and social questions. Unfortunately, these questions do not lend themselves to comprehensive and universal answers.[1]

For one, any analysis would hinge upon whether an MMT is used (merely) to rid a person of an unpleasant and undesirable but adaptive memory or whether it seeks to address a serious malady such as PTSD in which trauma memory is maladaptive to normal functioning and survival. Regrettably, much of the bioethical debate concerning MMTs has taken place in broad terms, leaving room for speculation as to whether the concerns raised apply to the use of MMTs therapeutically (however that might be delimited) (President's Council on Bioethics 2003 being a case in point). In this chapter, we have specifically restricted our focus to the use of MMTs in the prevention or treatment of PTSD in members of armed forces. In doing so, we do not deny that MMTs could be misused or abused, but we want to avoid overgeneralising problems associated with the potential abuse to the intervention as such (cf. Levy 2007, 131). After all, that the recreational use of some medications (for example, amphetamines) can potentially be dangerous and a source of serious social ills surely cannot mean that treating recognized maladies (such as narcolepsy) with that medication becomes objectionable. The problem of abuse would need to be man-

[1] Others have more broadly noted the need for a contextualised case-by-case assessment (see, e.g., President's Council on Bioethics 2003, 208; Levy 2007, 131; Parens 2010, 106).

aged with appropriate regulation and the professional ethics of medical practitioners, as is the case with many other medical interventions.

Furthermore, the juncture of intervention has considerable normative significance. As we noted earlier, MMTs could be used prophylactically to prevent symptoms of PTSD from ever developing, or as treatment when symptoms of PTSD have already manifested. These two options present somewhat different dilemmas. Much of the bioethical debate so far has focused on the prophylactic use of MMTs (but see Elsey and Kindt 2016). We want to draw attention to the differences and the similarities of the diverse approaches from an ethical, legal and social perspective.

We begin with two foundational questions, namely whether MMTs are safe and effective, and whether equitable access to them can be ensured. We then turn to a set of issues that we think are at the core of the debate around MMTs – though we doubt whether those issues are strictly speaking ethical, as they appear to be more broadly societal. The first of these is the notion that by modifying memories we jeopardize identities and fail to live an authentic life. The second is that MMTs interfere in normal psychological coping processes and deny benefits of learning to deal with trauma. The third and final issue is that, for different reasons, we might be duty bound to retain certain memories – that is, society may require us to retain certain memories.

13.5.1 Safety, Effectiveness and Equitable Access

MMTs inevitably raise questions that pertain to all new medications or therapeutic devices. First, is the intervention safe – is it relatively free of serious adverse effects? Second, is the intervention effective – does it achieve its intended purpose in clinical practice? Taken together, these questions are about whether the benefits of the intervention in addressing a malady (here, PTSD) outweigh the known risks.

We addressed the effectiveness of different interventions in the previous section. To recapitulate, while more research is necessary, there is cause for cautious optimism that certain memory-modifying interventions may indeed bring relief from symptoms of PTSD.

As for safety, the question becomes more contextual. Different interventions, each with a different side effect profile, can be used to interfere with the consolidation and reconsolidation of memories. Even propranolol, generally regarded a relatively benign medication (Hall and Carter 2007, 23–4), has some side effects. Indeed, because it is used to treat certain cardiovascular conditions, it must necessarily have cardiovascular side effects, such as bradycardia, when used to produce neurocognitive effects.

Safety is a particularly serious concern when it comes to prophylaxis. An overgenerous prophylactic use would mean that many people would be exposed to the side effects of the intervention without gaining any benefits. Unfortunately, it is difficult to predict if and when PTSD might develop: not all traumatic events are so serious as to trigger PTSD and not all people experiencing the same event develop PTSD (President's Council on Bioethics 2003, 226, 228; Bell 2007, 29). Thus, it is not clear who should receive prophylaxis.

The magnitude of this problem depends on whether we are concerned with pre-exposure prophylaxis (in anticipation of a traumatic event) or post-exposure prophylaxis (after a traumatic event but before symptoms of PTSD have developed). Pre-exposure prophylaxis is clearly more challenging due to an added variable: not knowing if and when a traumatic event might take place. Thus, effective prophylaxis would require the use of long-acting interventions or keeping the person on a particular medication for days on end. Adverse effects of many medications are dose-dependent, and so prolonged administration resulting in a higher cumulative dose can be more likely to produce adverse effects (see, e.g., Moret et al. 2009).

Aside from the consequences of long-term medication use, the medications that are likely candidates for MMTs may have an immediate operational impact. That is, medications may impact on the ability of a soldier to perform tactically important tasks at a predictably high level. We can use beta-blockers as an illustration.

First, beta-blockers have physical side effects because they reduce heart rate. Beta-blockers would therefore likely improve the accuracy of a sniper, as shots could be fired between heart-beats. Yet, through the same mechanism, beta-blockers may reduce the volume of oxygen that the cardiovascular system can deliver to muscles. Thus, beta-blockers would inhibit exercise performance and impair the ability of soldiers to meet the physical demands of combat (Donovan 2010, 70).

Secondly, beta-blockers can affect behaviour and cognition (Aston-Jones and Cohen 2005; Bouret and Sara 2005). They interfere with the actions of stress hormones and as such can reduce arousal and feelings of anxiety. Stress hormones and arousal, however, "are central to the fight-or-flight response, and they trigger the heightened awareness necessary for soldiers to survive in combat situations" (Donovan 2010, 70). Thus, beta-blockers might alter the agility of a soldier to a degree that it would place them in greater danger in threatening circumstances (Henry et al. 2007, 16; Aoki 2008, 356).

Finally, beta-blockers also affect decision-making (Rogers et al. 2004; Doya 2008; Sokol-Hessner et al. 2015). This raises the question as to whether they affect the way people resolve morally significant problems (Craigie 2007, 31; Levy 2007, 187–195). There is evidence that propranolol leads to more deontological and less utilitarian decisions (at least in certain circumstances), and that it decreases response times and increases decisiveness (Terbeck et al. 2013, 325). More impulsive and less consequentialist decision-making can pose problems for compliance with the law of armed conflict. In particular, it might influence decision-making in circumstances where the law requires a fine consequentialist calculation, such as with the principle of proportionality, which requires balancing the anticipated military effect to be gained from an attack against the incidental harm caused to civilians and civilian objects.[2] Prospectively, medications might cause people to make different decisions that they would otherwise. Retrospectively, this might affect the degree of moral responsibility that could be assigned to them afterwards (Wolfendale 2008, 30).

The situation with post-exposure prophylaxis is slightly different. In effect, one variable – whether or not a traumatic event will occur – has been removed. Thus, a short-term intervention immediately after a distressing event may be sufficient. Also, the effects on physical performance and decision-making can be discounted as long as the person does not need to engage in physically strenuous or morally taxing activities while undergoing post-exposure prophylaxis.

That does not, of course, obviate the problem of not knowing whom to treat, but there are other measures that can be taken to reduce that uncertainty. In other contexts, for example as concerns infectious diseases, decisions about post-exposure prophylaxis are frequently made by means of a probabilistic risk assessment and on previously adopted guidelines. This is the case, for instance, in the event of a suspected exposure to the human immunodeficiency virus (HIV) (e.g. Benn et al. 2011) or the rabies virus (e.g. Brown 2017).

The risk factors for developing PTSD are not as well understood as those for developing an HIV infection or rabies. However, already in one of the earliest clinical trials of propranolol as post-exposure prophylaxis for PTSD – the administration of medication to people in a hospital emergency department after a traumatic event – was based on psychological and physiological risk factors (Pitman et al. 2002). As our understanding of risk factors improves, more reliable guidelines for PTSD prophylaxis might be developed.

In any event, post-exposure prophylaxis of PTSD and the treatment of PTSD once symptoms have emerged would require the administration of medication for a limited number of times. This would help mitigate some of the concerns about safety. Thus, as a general matter, the use of medications for memory modification is likely to be safer than the prolonged use of antidepressants, anxiolytics, antipsychotics and hypnotics currently used in the management of PTSD symptoms.

All biomedical interventions also raise the question of equitable access. Would all those who could benefit from the intervention be able to gain access to it? Again, the fairly limited number of times a medication would need to be administered as an MMT would likely mean that it is cheaper than prolonged symptomatic treatment with psychoactive medications. This could make it a more equitable treatment option (Bell 2008, 4).

In conclusion, there are reasons to be cautious about PTSD prophylaxis, especially pre-exposure prophylaxis. The long-term effects of such prophylaxis are not

[2] See Protocol Additional (I) to the Geneva Conventions of 12 August 1949, and Relating to the Protection of Victims of International Armed Conflicts, 1125 UNTS 3, 8 June 1977 (entered into force 7 December 1978), Art. 51(5)(b). The article prohibits the launching of "an attack which may be expected to cause incidental loss of civilian life, injury to civilians, damage to civilian objects, or a combination thereof, which would be excessive in relation to the concrete and direct military advantage anticipated."

necessarily well understood, and the immediate side effects could be problematic in the military context. Therefore, the benefits of such prophylaxis might not necessarily outweigh the risks. If post-exposure prophylaxis and, even more so, treatment at a stage where PTSD symptoms first appear, is no less effective than pre-exposure prophylaxis, the former seems to be the more ethically defensible option.

13.5.2 Identity and Authenticity

Concerns around memory modification go well beyond these relatively technical matters, which, moreover, are not unique to the interventions that might be used as MMTs. Perhaps the most prominent of the broader issues is the worry that by permitting our memories to be modified, "we might succeed in erasing real suffering at the risk of falsifying our perception of the world and undermining our true identity" (President's Council on Bioethics 2003, 227). Much of this concern seems to be premised on the idea advanced by John Locke (1690 bk. II, ch. 27) that our memories are what define us as persons – what give us identities that persist in time. The problem that arises here relates to two interconnected philosophical notions, authenticity and narrative identity, which we cannot fully unpack here (but see, in particular, Erler 2011; Vukov 2017). The basic idea that we largely are what we remember about ourselves and the world around us makes intuitive sense. Most people would probably agree that by erasing all our memories we would commit a kind of cognitive suicide. Even by modifying some of our memories we would transform ourselves.

An argument frequently advanced to mitigate these concerns is that MMTs only affect conditioned defensive responses to threats ("emotional responses") but leave episodic memories completely intact (Dębiec and Altemus 2006; Soeter and Kindt 2010; Elsey and Kindt 2016). The premise here is that mainly episodic memory contributes to our sense of self and if MMTs affect only conditioned responses, our identities would remain essentially unharmed. This argument is largely based on studies that have investigated beta-blockers. However, as discussed above, beta-blockers can also reduce the emotional enhancement of episodic memory. Other MMTs, including different medications or brain stimulation, can eradicate specific episodic memories altogether. MMTs can thus also impact episodic memories that contribute to our sense of identity. What is more, a problem in the argumentation is the assumption that emotional responses contribute less to our identity than episodic memories. This Cartesian view of separation between reason and emotion is false. Emotion and cognition are necessarily intertwined to the degree that one cannot exist without the other (Damasio 1994; Phelps et al. 2014). Changing learned emotional responses would also alter reasoning and our personal identity.

Regardless of the effects of MMTs on both conditioned responses and episodic memory, we submit that MMTs do not necessarily impinge on identity and authenticity to such a degree that we should shun the treatment. As discussed above, by nature memories are flexible: we forget most of what we learn, memories can be highly inaccurate, and may even be entirely false. Yet none of this has been a source of major philosophical concern. The inability to account for each moment of one's waking hours with complete accuracy and full emotional vigour neither undermines our identity nor hampers functioning normally in daily life (see, along the same lines, Kolber 2006, 1604; Bell 2008, 3; Donovan 2010, 68). Quite the contrary, the flexibility of memory is adaptive and aids optimal decision-making in the future. Memory flexibility thus also constitutes a major way in which we build our autobiography and, by extension, our identity, which is fluid over time.

Even if concerns about identity might lead us to conclude that we ought not to have unfettered access to MMTs, this does not mean that they should not be used to treat PTSD. Indeed, the symptoms of PTSD can become so overwhelming as to fully consume a person's life: daily existence becomes haunted by memories of the past, resulting in major changes in personality and withdrawal from society to avoid stimuli that might trigger episodes of anxiety. Moreover, PTSD and suicidal behaviour are strongly correlated.[3] Thus, PTSD poses a risk not only to personal identity, but also to life. In PTSD, trauma memory is clearly maladaptive. MMTs may allow people with PTSD to regain adaptive responses and return to normal life and facilitate the maintenance of identity rather than undermine it (see, e.g., Wasserman 2004, 14; Kolber 2006, 1604; Bell 2008, 4; Donovan 2010, 72).

13.5.3 Normal Recovery and Traumatic Growth

Another common concern about memory modification is that it would interfere with normal recovery from trauma, or "working things through" (Schacter 2001, 183; President's Council on Bioethics 2003, 226; Holmes et al. 2010). Moreover, going through such a process has certain adaptive consequences, which have been conceptualized as post-traumatic growth (PTG). This manifests in different ways, including "an increased appreciation for life in general, more meaningful interpersonal relationships, an increased sense of personal strength, changed priorities, and a richer existential and spiritual life" (Tedeschi and Calhoun 2004, 1). MMTs would seem to deny traumatized persons the benefits of experiencing PTG (Warnick 2007, 37), which is said to be far more common in the wake of traumatic events than PTSD (Parens 2010, 102).

For persons who suffer from PTSD, however, traumatic memories and the associated emotions are so powerful as to make it impossible to work things through (Henry et al. 2007, 16). Their "experiences are simply tragic and terrifying, offering virtually no opportunity for redemption or transformation" and "even if it is better to weave traumatic events into positive, life-affirming narratives, many people are never able to do so" (Kolber 2006, 1599, 1600). Also, an individual who is afflicted to the point of functional loss or self-harm may simply be incapable of experiencing PTG (Donovan 2010, 70).

[3]This is true even after controlling for physical illness and other mental disorders (see Sareen et al. 2005, 2007).

Furthermore, MMTs and PTG need not be mutually exclusive. In fact, MMTs may lay the groundwork for recovery and PTG. It is perfectly possible that MMTs "might make it easier for trauma survivors to face and incorporate traumatic recollections, and in that sense could facilitate long-term adaptation" (Schacter 2001, 183), "may enable such people to make life transformations that they would be *incapable* of making in the absence of the medications" (Kolber 2006, 1600) and "may aid in induction of PTG as well as relieve PTSD" (Donovan 2010, 70).

The concern about circumventing natural processes has more merit when MMTs are used prophylactically. The question does arise whether we should be prepared to "replace this near-universal feature of human life [i.e. PTG] with a mass preventative pharmacotherapy that benefits a small minority of the population" (Warnick 2007, 37). But again, it is not clear whether MMTs would necessarily replace PTG; in persons at risk of PTSD, prophylactic MMTs may well contribute to ensuring that natural processes (including PTG) take place. To use an analogy, if a person fractures a bone, we do not allow nature to simply take its course. We may need to realign the fracture and set a cast in order for optimal healing to take place. Likewise, MMTs may return patients to a natural path to recovery (Holmes et al. 2010).

13.5.4 A Duty to Remember?

Another major concern about MMTs is the risk of altering memories that we might be under a duty to preserve for the common good. Arguably, collective memories of atrocities, and of the carnage of war more generally, depend upon individuals retaining undiluted recollections of these events (President's Council on Bioethics 2003, 231). Thus, modifying our memories of such events poses a risk not only to our personal identity, "but also prevents the sharing of these narratives, which could potentially help others in society change and evolve" (Aoki 2008, 357).

Lieutenant-General Roméo Dallaire, the commander of the United Nations Assistance Mission for Rwanda during the genocide, is sometimes used as an example (Henig 2004; Wasserman 2004, 12; Aoki 2008, 356–7). Dallaire had been put in an impossible situation – the wholly inadequate forces that had been placed under his command were unable to stop the slaughter of hundreds of thousands of Tutsis and moderate Hutus. As Dallaire himself put it in a poignant book about the genocide, he and his troops were "reduced to the role of accountants keeping track of how many were being killed" (2004, 374).

Through the book and many public appearances, Dallaire became a powerful advocate for humanitarian intervention. But he also suffered, and continues to suffer, from PTSD; indeed, his anguish has led him to self-harm (see Dallaire and Humphreys 2016). One commentator speculates, though, that had Dallaire "taken memory-dampening agents, [he] may not have been able to achieve the same level of influence on society" (Aoki 2008, 357). Another author suggests that Dallaire may have succeeded so well in telling the world about the plight because he is "the most powerful and untainted witness" to the genocide (Wasserman 2004, 12). On an alternative (much more troubling) view, some of Dallaire's effectiveness as an advocate may have derived from his own suffering. Thus, Dallaire's suffering might have been in some ways symbolic and served as a reminder to the international community how it had failed the Rwandans. This point could be formulated more broadly, suggesting that having struggling veterans in our midst serves to remind the society of the horrors of war.

While we sympathize with the idea that society should not be disconnected from the conflicts that are fought on its behalf, treating service members as instruments in obtaining that goal fundamentally dehumanizes them. We agree with Arthur Caplan, who thinks that "[t]he notion that we need to have suffering martyrs among us is cruel and exploitative" (quoted in Miller 2004, 36). There is also undoubtedly "some hypocrisy in the contention that soldiers ought to bear painful trauma for what others have commanded them to do" (Bublitz and Dresler 2015, 1300).

From a legal perspective, the problem with memory modification is that it may limit society's access to memory as evidence, for example eyewitness testimony (such concerns are summarised in, e.g., Kolber 2006, 1579–82). While this point is well taken, it should not be overemphasized. For one, the value of eyewitness testimony is probably overstated in the first instance. Individuals' recollections of events are less reliable than one might think. It is all too easy to think of memory as some sort of a documentary film that can be replayed in court as necessary. The ability of humans to remember has evolved not so as to forensically document the past, but so as to prepare us for the future. Thus, memories are reinterpreted and reconfigured as new experiences become integrated into an autobiography. For this reason, exclusive reliance on eyewitness testimonies in judicial proceedings carries significant risks of injustice.

In any event, even recognising that society sometimes has a reasonable expectation about accessing someone's memories, that right cannot be absolute. The interests of the society in obtaining the memory and the individual's interest not to suffer from traumatic memories need to be balanced. What is more, for post-trauma MMTs to work, first the details of a traumatic memory would have to be identified by a therapist prior to treatment. As such there would be a detailed archive of memory prior to modification.

Furthermore, MMTs are unlikely to completely eradicate a memory. Realistically, Dallaire would no longer suffer (as much) but still remember what happened and that he suffered, so as to be able to appreciate the importance of the memory. This likely would still leave him as a strong spokesperson for humanitarian interventions. Also, persons with PTSD often have difficulty recalling particular events and articulating their experiences as a coherent narrative. Thus, PTSD treatment might not undermine but enhance people's ability to meet the duty to remember.

13.5.5 A Duty to Suffer?

A slightly different need to preserve memories arguably arises with respect to people who have committed objectionable acts and feel pangs of guilt afterwards. Lady MacBeth has become a recurrent character in bioethical discussions on MMTs

(President's Council on Bioethics 2003, 206–7, 212, 232; Wasserman 2004, 14–15; Parens 2010; Erler 2011; Bublitz and Dresler 2015, 1299; Vukov 2017, 243). There appears to be broad agreement that people should not have access to MMTs to "relieve anguish that is proportionate to their own actions" (Parens 2010, 106). This seems uncontroversial inasmuch as such interventions are not meant to be available to anyone who simply wants to dampen undesirable or troubling memories, but to people with maladaptive memories such as in the case of PTSD.

Some of the commentary on this point might be interpreted as doubting the appropriateness of providing MMTs to persons who have developed PTSD as a result of their own wrongdoing.[4] This line of thinking may be confusing PTSD with some especially sharp form of guilt or remorse. PTSD is a serious and potentially debilitating mental health condition, not merely a feeling or a state of mind. Leaving it untreated is problematic both from a prudential and ethical perspective. As for the former, a strong association exists between PTSD symptoms and the risk of re-offending (Ardino et al. 2013). Thus, however attractive PTSD symptoms may seem to some as a form of retribution, perpetuating the condition seems wholly counter-productive from the perspective of rehabilitating offenders and reintegrating them into society. From an ethical perspective, a hallmark of a civilized society is that it provides adequate health care to those who it has convicted of wrongdoing.[5] Conversely, the idea that a medical practitioner would deny treatment to patients not because of futility or the shortage of resources but simply because of legal or ethical misgivings about their prior conduct flies in the face of medical ethics. There is broad support for the "principle of equivalence of care", which requires prisoners to be provided health care equivalent to that provided to the general public.[6]

Indeed, it would not be acceptable to modify the standard of care so as to increase or maintain suffering that has been caused by the antisocial conduct of the person. For example, it would be inappropriate for a medical practitioner to remove a bullet without anaesthesia simply because the person was shot in a firefight with police. In fact, refusal to provide anaesthesia to a person on the basis of their criminal history would almost certainly breach the prohibition of torture or cruel, inhuman or degrading treatment or punishment.[7]

Even more dubiously, PTSD treatment has been questioned in the context of warfare. For example, one commentator, Paul McHugh, has asked – rhetorically, we presume – "If soldiers did something that ended up with children getting killed, do you want to give them beta-blockers so that they can do it again?" (quoted in Giles 2005). The question itself is problematic. No one would deny that the death of children – indeed, anyone – in conflict is horrific and regrettable. Yet even the death of children does not necessarily amount to a wrongdoing on the part of the individual

[4] For a careful examination of this issue, see Kreitmair (2016).

[5] On prison health care ethics generally, see Wolff (2012), and Lehtmets and Pont (2014)

[6] For critical discussions of this concept, see Niveau (2007), and Jotterand and Wangmo (2014).

[7] Amon and Lohman (2011) provide a useful discussion of the associated human rights issues in a different context.

soldier. For example, under the law of armed conflict, children taking a direct part in hostilities can be lawfully targeted (see, e.g., Provost 2016). Soldiers who find themselves in the awful position where their only viable course of action is to use lethal force against a child soldier would, no doubt, be seriously scarred and potentially at risk of PTSD.

In any event, the two problems identified with respect to criminals arise with even more vigour when it comes to soldiers. For one, veterans with PTSD are statistically more likely to engage in antisocial behaviour than veterans who do not have PTSD (Booth-Kewley et al. 2010). Thus, again, leaving PTSD untreated could be highly counterproductive both in terms of soldiers continuing military service and re-entering civilian society. Furthermore, one expects armed forces to provide every medical assistance available to physically wounded soldiers in an attempt to restore them to health and to allow them to fight another day. With this in mind, to deny PTSD treatment to a soldier because the treatment might permit them to return to combat is simply preposterous. A serious ethical problem would arise, however, if some form of MMT was applied prior to conflict with a view to generally morally desensitise soldiers. Yet this would no longer be a problem about prevention or treatment of PTSD, which is the focus of this chapter.

13.6 Conclusion

MMTs may provide an opportunity to relieve severe suffering from mental disease in military populations, restore people's identity and authenticity, return them to a path of natural recovery and personal growth, and improve their memories to societies' benefit. The risks of using MMTs are limited since they target specific maladaptive memories and the modification of specific memories does not jeopardize personal identity, opportunity for personal growth, or social demand for memory preservation.

Based on the evidence available, we categorically reject any broad claim that "the costs to individuals and to society in using … memory-dampening agents would significantly outweigh their potential benefits" (Aoki 2008, 356). In order to reach defensible ethical conclusions, MMTs would need to be assessed in a context-specific manner, and in light of their primary effects and likely side effects. Knee-jerk reactions to MMTs on the basis of their possible abuse are counterproductive.

We recognize that MMTs may have other ethical implications for potential misuse, which require their own ethical discussion in the future. However, the likelihood of MMTs being misused is low within the near future and does not outweigh potential benefits for patients, while legal regulations of professional ethics for medical practitioners are already in place. Assuming the safety and efficacy of a particular intervention, we see nothing strikingly unethical about treating soldiers who have developed PTSD with MMTs. If PTSD is construed as a health condition, it should be treated with the most effective means available, which might at some point in time be MMTs.

Moreover, we agree with those who have suggested that investigating the viability of MMTs as a treatment in the military population is not only ethically permissible but required (see, e.g. Donovan 2010, 70, 72). A society that in the interest of its own security is prepared to place individuals in harm's way must be prepared to succour those individuals when they sustain physical or psychological injuries. If MMTs prove to be a safe and effective means of treating PTSD, their use must be considered. That said, where there is a real risk to societally significant memories, there clearly arises a need to balance the interests of a person to be free from suffering and the society's (narrowly construed) right to access the memories that are the cause of the suffering.

References

American Psychiatric Association. 2013. *Diagnostic and statistical manual of mental disorders: DSM-5*. Washington, DC: American Psychiatric Publishing.

Amon, Joseph, and Diederik Lohman. 2011. Denial of pain treatment and the prohibition of torture, cruel, inhuman or degrading treatment or punishment. *Interights Bulletin* 16: 172–184.

Aoki, Cynthia R.A. 2008. Rewriting my autobiography: The legal and ethical implications of memory-dampening agents. *Bulletin of Science, Technology & Society* 28: 349–359. https://doi.org/10.1177/0270467608320223.

Ardino, Vittoria, Luca Milani, and Paola di Blasio. 2013. PTSD and re-offending risk: The mediating role of worry and a negative perception of other people's support. *European Journal of Psychotraumatology* 4: 21382. https://doi.org/10.3402/ejpt.v4i0.21382.

Aston-Jones, Gary, and Jonathan D. Cohen. 2005. An integrative theory of locus Coeruleus-norepinephrine function: Adaptive gain and optimal performance. *Annual Review of Neuroscience* 28: 403–450. https://doi.org/10.1146/annurev.neuro.28.061604.135709.

Bell, Jennifer A. 2007. Preventing post-traumatic stress disorder or pathologizing bad memories? *American Journal of Bioethics* 7: 29–30. https://doi.org/10.1080/15265160701518540.

Bell, J. 2008. Propranolol, post-traumatic stress disorder and narrative identity. *Journal of Medical Ethics* 34: e23–e23. https://doi.org/10.1136/jme.2008.024752.

Benn, P., M. Fisher, and R. Kulasegaram. 2011. UK guideline for the use of post-exposure prophylaxis for HIV following sexual exposure. *International Journal of STD & AIDS* 22: 695–708. https://doi.org/10.1258/ijsa.2011.171011.

Bisson, Jonathan I., Sarah Cosgrove, Catrin Lewis, and Neil P. Roberts. 2015. Post traumatic stress disorder. *British Medical Journal* 351: h6161. https://doi.org/10.1136/bmj.h6161.

Booth-Kewley, Stephanie, Gerald E. Larson, Robyn M. Highfill-McRoy, Cedric F. Garland, and Thomas A. Gaskin. 2010. Factors associated with antisocial behavior in combat veterans. *Aggressive Behavior* 36: 330–337. https://doi.org/10.1002/ab.20355.

Bouret, Sebastien, and Susan J. Sara. 2005. Network reset: A simplified overarching theory of locus Coeruleus noradrenaline function. *Trends in Neurosciences* 28: 574–582. https://doi.org/10.1016/j.tins.2005.09.002.

Bouton, M.E. 2004. Context and behavioral processes in extinction. *Learning & Memory* 11: 485–494. https://doi.org/10.1101/lm.78804.

Brewin, Chris R., and Emily A. Holmes. 2003. Psychological theories of posttraumatic stress disorder. *Clinical Psychology Review* 23: 339–376. https://doi.org/10.1016/S0272-7358(03)00033-3.

Brewin, Chris R., Bernice Andrews, and John D. Valentine. 2000. Meta-analysis of risk factors for posttraumatic stress disorder in trauma-exposed adults. *Journal of Consulting and Clinical Psychology* 68: 748–766. https://doi.org/10.1037/0022-006X.68.5.748.

Brewin, Chris R., James D. Gregory, Michelle Lipton, and Neil Burgess. 2010. Intrusive images in psychological disorders: Characteristics, neural mechanisms, and treatment implications. *Psychological Review* 117: 210–232. https://doi.org/10.1037/a0018113.

Brown, Kevin. 2017. *PHE guidelines on rabies post-exposure treatment*. London: Public Health England.

Brunet, Alain, Scott P. Orr, Jacques Tremblay, Kate Robertson, Karim Nader, and Roger K. Pitman. 2008. Effect of post-retrieval propranolol on psychophysiologic responding during subsequent script-driven traumatic imagery in post-traumatic stress disorder. *Journal of Psychiatric Research* 42: 503–506. https://doi.org/10.1016/j.jpsychires.2007.05.006.

Brunet, Alain, Joaquin Poundja, Jacques Tremblay, Éric Bui, Émilie Thomas, Scott P. Orr, Abdelmadjid Azzoug, Philippe Birmes, and Roger K. Pitman. 2011. Trauma reactivation under the influence of propranolol decreases posttraumatic stress symptoms and disorder: 3 open-label trials. *Journal of Clinical Psychopharmacology* 31: 547–550. https://doi.org/10.1097/JCP.0b013e318222f360.

Brunet, Alain, Émilie Thomas, Daniel Saumier, Andrea R. Ashbaugh, Abdelmadjid Azzoug, Roger K. Pitman, Scott P. Orr, and Jacques Tremblay. 2014. Trauma reactivation plus propranolol is associated with durably low physiological responding during subsequent script-driven traumatic imagery. *Canadian Journal of Psychiatry* 59: 228–232. https://doi.org/10.1177/070674371405900408.

Bublitz, Christoph, and Martin Dresler. 2015. A duty to remember, a right to forget? Memory manipulations and the law. In *Handbook of neuroethics*, ed. Jens Clausen and Neil Levy, 1279–1307. Dordrecht: Springer. https://doi.org/10.1007/978-94-007-4707-4_167.

Cabrera, Laura Y., and Bernice S. Elger. 2016. Memory interventions in the criminal justice system: Some practical ethical considerations. *Journal of Bioethical Inquiry* 13: 95–103. https://doi.org/10.1007/s11673-015-9680-2.

Craigie, Jillian. 2007. Propranolol, cognitive biases, and practical decision-making. *American Journal of Bioethics* 7: 31–32. https://doi.org/10.1080/15265160701518565.

Dallaire, Roméo. 2004. *Shake hands with the devil: The failure of humanity in Rwanda*. New York: Carroll & Graf.

Dallaire, Roméo, and Jessica Dee Humphreys. 2016. *Waiting for first light: My ongoing battle with PTSD*. Toronto: Random House.

Damasio, Antonio R. 1994. *Descartes' error: Emotion, reason and the human brain*. New York: Avon.

Dębiec, Jacek, and Margaret Altemus. 2006. Toward a new treatment for traumatic memories. *Cerebrum* 2006: 2–11.

Dębiec, Jacek, and Joseph E. LeDoux. 2004. Disruption of reconsolidation but not consolidation of auditory fear conditioning by noradrenergic blockade in the amygdala. *Neuroscience* 129: 267–272. https://doi.org/10.1016/j.neuroscience.2004.08.018.

Donovan, Elise. 2010. Propranolol use in the prevention and treatment of posttraumatic stress disorder in military veterans: Forgetting therapy revisited. *Perspectives in Biology and Medicine* 53: 61–74. https://doi.org/10.1353/pbm.0.0140.

Doya, Kenji. 2008. Modulators of decision making. *Nature Neuroscience* 11: 410–416. https://doi.org/10.1038/nn2077.

Dudai, Yadin. 2007. Memory concepts. In *Science of memory: Concepts*, ed. Henry L. Roediger, Yadin Dudai, and Susan M. Fitzpatrick. Oxford: Oxford University Press.

Ehlers, Anke, and David M. Clark. 2000. A cognitive model of posttraumatic stress disorder. *Behaviour Research and Therapy* 38: 319–345. https://doi.org/10.1016/S0005-7967(99)00123-0.

Ehlers, Anke, Ann Hackmann, Regina Steil, Sue Clohessy, Kerstin Wenninger, and Heike Winter. 2002. The nature of intrusive memories after trauma: The warning signal hypothesis. *Behaviour Research and Therapy* 40: 995–1002. https://doi.org/10.1016/S0005-7967(01)00077-8.

Elsey, James, and Merel Kindt. 2016. Manipulating human memory through reconsolidation: Ethical implications of a new therapeutic approach. *AJOB Neuroscience* 7: 225–236. https://doi.org/10.1080/21507740.2016.1218377.

Erler, Alexandre. 2011. Does memory modification threaten our authenticity? *Neuroethics* 4: 235–249. https://doi.org/10.1007/s12152-010-9090-4.

Foa, Edna B., Gail Steketee, and Barbara Olasov Rothbaum. 1989. Behavioral/cognitive conceptualizations of post-traumatic stress disorder. *Behavior Therapy* 20: 155–176. https://doi.org/10.1016/S0005-7894(89)80067-X.

Fulton, Jessica J., Patrick S. Calhoun, H. Ryan Wagner, Amie R. Schry, Lauren P. Hair, Nicole Feeling, Eric Elbogen, and Jean C. Beckham. 2015. The prevalence of posttraumatic stress disorder in operation enduring freedom/operation Iraqi freedom (OEF/OIF) veterans: A meta-analysis. *Journal of Anxiety Disorders* 31: 98–107. https://doi.org/10.1016/j.janxdis.2015.02.003.

Giles, Jim. 2005. Beta-blockers tackle memories of horror. *Nature* 436: 448–449. https://doi.org/10.1038/436448a.

Hall, Wayne, and Adrian Carter. 2007. Debunking alarmist objections to the pharmacological prevention of PTSD. *American Journal of Bioethics* 7: 23–25. https://doi.org/10.1080/15265160701551244.

Hendriks, Lotte, Rianne A. de Kleine, Theo G. Broekman, Gert-Jan Hendriks, and Agnes van Minnen. 2018. Intensive prolonged exposure therapy for chronic PTSD patients following multiple trauma and multiple treatment attempts. *European Journal of Psychotraumatology* 9: 1425574. https://doi.org/10.1080/20008198.2018.1425574.

Henig, Robin Marantz. 2004. The quest to forget. *New York Times Magazine*, April 4. https://www.nytimes.com/2004/04/04/magazine/the-quest-to-forget.html.

Henke, Katharina. 2010. A model for memory systems based on processing modes rather than consciousness. *Nature Reviews Neuroscience* 11: 523–532. https://doi.org/10.1038/nrn2850.

Henry, Michael, Jennifer R. Fishman, and Stuart J. Youngner. 2007. Propranolol and the prevention of post-traumatic stress disorder: Is it wrong to erase the "sting" of bad memories? *AJOB Neuroscience* 7: 12–20. https://doi.org/10.1080/15265160701518474.

Hirst, William, Elizabeth A. Phelps, Randy L. Buckner, Andrew E. Budson, Alexandru Cuc, John D.E. Gabrieli, Marcia K. Johnson, et al. 2009. Long-term memory for the terrorist attack of September 11: Flashbulb memories, event memories, and the factors that influence their retention. *Journal of Experimental Psychology: General* 138: 161–176. https://doi.org/10.1037/a0015527.

Holmes, Emily A., Anders Sandberg, and Lalitha Iyadurai. 2010. Erasing trauma memories. *British Journal of Psychiatry* 197: 414–415. https://doi.org/10.1192/bjp.197.5.414b.

Horowitz, Mardi J. 2014. *Stress response syndromes: PTSD, grief, adjustment, and dissociative disorders*. 5th ed. Lanham: Jason Aronson.

Itkowitz, Colby. 2015. Dropping the 'D' in PTSD is becoming the norm in Washington. *Washington Post*, June 30. https://www.washingtonpost.com/news/powerpost/wp/2015/06/30/dropping-the-d-in-ptsd-is-becoming-the-norm/.

Janoff-Bulman, Ronnie. 1992. *Shattered assumptions: Towards a new psychology of trauma*. New York: Free Press.

Jia, Elizabeth. 2017. What is post traumatic stress vs. PTSD? *WUSA9*, July 3. https://www.wusa9.com/article/news/national/military-news/what-is-post-traumatic-stress-vs-ptsd/65-453889759.

Jotterand, Fabrice, and Tenzin Wangmo. 2014. The principle of equivalence reconsidered: Assessing the relevance of the principle of equivalence in prison medicine. *American Journal of Bioethics* 14: 4–12. https://doi.org/10.1080/15265161.2014.919365.

Keane, Terence M., Amy D. Marshall, and Casey T. Taft. 2006. Posttraumatic stress disorder: Etiology, epidemiology, and treatment outcome. *Annual Review of Clinical Psychology* 2: 161–197. https://doi.org/10.1146/annurev.clinpsy.2.022305.095305.

Kessler, Ronald C., Patricia Berglund, Olga Demler, Robert Jin, Kathleen R. Merikangas, and Ellen E. Walters. 2005. Lifetime prevalence and age-of-onset distributions of DSM-IV disorders in the National Comorbidity Survey Replication. *Archives of General Psychiatry* 62: 593. https://doi.org/10.1001/archpsyc.62.6.593.

Kindt, Merel, Marieke Soeter, and Bram Vervliet. 2009. Beyond extinction: Erasing human fear responses and preventing the return of fear. *Nature Neuroscience* 12: 256–258. https://doi.org/10.1038/nn.2271.

Kolber, Adam J. 2006. Therapeutic forgetting: The legal and ethical implications of memory dampening. *Vanderbilt Law Review* 59: 1561–1626.

Kreitmair, Karola. 2016. Memory manipulation in the context of punishment and atonement. *AJOB Neuroscience* 7: 238–240. https://doi.org/10.1080/21507740.2016.1251993.

Kroes, Marijn C.W., and Guillén Fernández. 2012. Dynamic neural systems enable adaptive, flexible memories. *Neuroscience & Biobehavioral Reviews* 36: 1646–1666. https://doi.org/10.1016/j.neubiorev.2012.02.014.

Kroes, Marijn C.W., Indira Tendolkar, Guido A. van Wingen, Jeroen A. van Waarde, Bryan A. Strange, and Guillén Fernández. 2014. An electroconvulsive therapy procedure impairs reconsolidation of episodic memories in humans. *Nature Neuroscience* 17: 204–206. https://doi.org/10.1038/nn.3609.

Kroes, Marijn C.W., Daniela Schiller, Joseph E. LeDoux, and Elizabeth A. Phelps. 2016. Translational approaches targeting reconsolidation. In *Translational neuropsychopharmacology*, ed. Trevor W. Robbins and Barbara J. Sahakian, 197–230. Cham: Springer. https://doi.org/10.1007/7854_2015_5008.

LaBar, Kevin S., and Roberto Cabeza. 2006. Cognitive neuroscience of emotional memory. *Nature Reviews Neuroscience* 7: 54–64. https://doi.org/10.1038/nrn1825.

Lang, Peter J. 1977. Imagery in therapy: An information processing analysis of fear. *Behavior Therapy* 8: 862–886. https://doi.org/10.1016/S0005-7894(77)80157-3.

LeDoux, Joseph E. 2000. Emotion circuits in the brain. *Annual Review of Neuroscience* 23: 155–184. https://doi.org/10.1146/annurev.neuro.23.1.155.

Lehtmets, Andres, and Jörg Pont. 2014. *Prison health care and medical ethics: A manual for health-care workers and other prison staff with responsibility for prisoners' well-being*. Strasbourg: Council of Europe.

Levy, Neil. 2007. *Neuroethics: Challenges for the 21st century*. Cambridge: Cambridge University Press.

Locke, John. 1690. *An essay concerning humane understanding*. London: A. and J. Churchill; and Samuel Manship.

Loftus, Elizabeth F., and Jacqueline E. Pickrell. 1995. The formation of false memories. *Psychiatric Annals* 25: 720–725. https://doi.org/10.3928/0048-5713-19951201-07.

McGaugh, J.L. 2000. Memory: A century of consolidation. *Science* 287: 248–251. https://doi.org/10.1126/science.287.5451.248.

Miller, Greg. 2004. Learning to forget. *Science* 304: 34–36. https://doi.org/10.1126/science.304.5667.34.

Monfils, Marie-H., Kiriana K. Cowansage, Eric Klann, and Joseph E. LeDoux. 2009. Extinction-reconsolidation boundaries: Key to persistent attenuation of fear memories. *Science* 324: 951–955. https://doi.org/10.1126/science.1167975.

Moret, C., M. Isaac, and M. Briley. 2009. Problems associated with long-term treatment with selective serotonin reuptake inhibitors. *Journal of Psychopharmacology* 23: 967–974. https://doi.org/10.1177/0269881108093582.

Myers, Karyn M., and Michael Davis. 2002. Behavioral and neural analysis of extinction. *Neuron* 36: 567–584. https://doi.org/10.1016/S0896-6273(02)01064-4.

Nader, Karim, and Oliver Hardt. 2009. A single standard for memory: The case for reconsolidation. *Nature Reviews Neuroscience* 10: 224–234. https://doi.org/10.1038/nrn2590.

Nader, Karim, Glenn E. Schafe, and Joseph E. Le Doux. 2000. Fear memories require protein synthesis in the amygdala for reconsolidation after retrieval. *Nature* 406: 722–726. https://doi.org/10.1038/35021052.

Neisser, Ulric, and Nicole Harsch. 1992. Phantom flashbulbs: False recollections of hearing the news about challenger. In *Affect and accuracy in recall*, ed. Eugene Winograd and Ulric Neisser, 9–31. Cambridge: Cambridge University Press. https://doi.org/10.1017/CBO9780511664069.003.

Niveau, Gérard. 2007. Relevance and limits of the principle of "equivalence of care" in prison medicine. *Journal of Medical Ethics* 33: 610–613. https://doi.org/10.1136/jme.2006.018077.

Ozer, Emily J., Suzanne R. Best, Tami L. Lipsey, and Daniel S. Weiss. 2003. Predictors of posttraumatic stress disorder and symptoms in adults: A meta-analysis. *Psychological Bulletin* 129: 52–73. https://doi.org/10.1037/0033-2909.129.1.52.

Parens, Erik. 2010. The ethics of memory blunting and the narcissism of small differences. *Neuroethics* 3: 99–107. https://doi.org/10.1007/s12152-010-9070-8.

Phelps, Elizabeth A., Karolina M. Lempert, and Peter Sokol-Hessner. 2014. Emotion and decision making: Multiple modulatory neural circuits. *Annual Review of Neuroscience* 37: 263–287. https://doi.org/10.1146/annurev-neuro-071013-014119.

Pitman, Roger K., Kathy M. Sanders, Randall M. Zusman, Anna R. Healy, Farah Cheema, Natasha B. Lasko, Larry Cahill, and Scott P. Orr. 2002. Pilot study of secondary prevention of posttraumatic stress disorder with propranolol. *Biological Psychiatry* 51: 189–192. https://doi.org/10.1016/S0006-3223(01)01279-3.

President's Council on Bioethics. 2003. *Beyond therapy: Biotechnology and the pursuit of happiness*. Washington, DC: The President's Council on Bioethics.

Provost, René. 2016. Targeting child soldiers. *EJIL:Talk!*, January 12. https://www.ejiltalk.org/targeting-child-soldiers/.

Quirk, Gregory J., and Devin Mueller. 2008. Neural mechanisms of extinction learning and retrieval. *Neuropsychopharmacology* 33: 56–72. https://doi.org/10.1038/sj.npp.1301555.

Roache, Rebecca. 2008. Ethics, speculation, and values. *NanoEthics* 2: 317–327. https://doi.org/10.1007/s11569-008-0050-y.

Rogers, R.D., M. Lancaster, J. Wakeley, and Z. Bhagwagar. 2004. Effects of beta-adrenoceptor blockade on components of human decision-making. *Psychopharmacology* 172: 157–164. https://doi.org/10.1007/s00213-003-1641-5.

Sareen, Jitender, Tanya Houlahan, Brian J. Cox, and Gordon J.G. Asmundson. 2005. Anxiety disorders associated with suicidal ideation and suicide attempts in the National Comorbidity Survey. *Journal of Nervous and Mental Disease* 193: 450–454. https://doi.org/10.1097/01.nmd.0000168263.89652.6b.

Sareen, Jitender, Brian J. Cox, Murray B. Stein, Tracie O. Afifi, Claire Fleet, and Gordon J.G. Asmundson. 2007. Physical and mental comorbidity, disability, and suicidal behavior associated with posttraumatic stress disorder in a large community sample. *Psychosomatic Medicine* 69: 242–248. https://doi.org/10.1097/PSY.0b013e31803146d8.

Schacter, Daniel L. 2001. *The seven sins of memory: How the mind forgets and remembers*. Boston: Houghton Mifflin.

Schiller, Daniela, Marie-H. Monfils, Candace M. Raio, David C. Johnson, Joseph E. LeDoux, and Elizabeth A. Phelps. 2010. Preventing the return of fear in humans using reconsolidation update mechanisms. *Nature* 463: 49–53. https://doi.org/10.1038/nature08637.

Semon, Richard. 1921. *The mneme*. London: Allen & Unwin.

Soeter, Marieke, and Merel Kindt. 2010. Dissociating response systems: Erasing fear from memory. *Neurobiology of Learning and Memory* 94: 30–41. https://doi.org/10.1016/j.nlm.2010.03.004.

———. 2015. An abrupt transformation of phobic behavior after a post-retrieval amnesic agent. *Biological Psychiatry* 78: 880–886. https://doi.org/10.1016/j.biopsych.2015.04.006.

Sokol-Hessner, Peter, Sandra F. Lackovic, Russell H. Tobe, Colin F. Camerer, Bennett L. Leventhal, and Elizabeth A. Phelps. 2015. Determinants of propranolol's selective effect on loss aversion. *Psychological Science* 26: 1123–1130. https://doi.org/10.1177/0956797615582026.

Squire, Larry R. 1992. Memory and the Hippocampus: A synthesis from findings with rats, monkeys, and humans. *Psychological Review* 99: 195–231. https://doi.org/10.1037/0033-295X.99.2.195.

———. 2004. Memory systems of the brain: A brief history and current perspective. *Neurobiology of Learning and Memory* 82: 171–177. https://doi.org/10.1016/j.nlm.2004.06.005.

Tedeschi, Richard G., and Lawrence G. Calhoun. 2004. Posttraumatic growth: Conceptual foundations and empirical evidence. *Psychological Inquiry* 15: 1–18.

Terbeck, Sylvia, Kahane Guy, McTavish Sarah, Savulescu Julian, Levy Neil, Hewstone Miles, and Philip J. Cowen. 2013. Beta adrenergic blockade reduces utilitarian judgement. *Biological Psychology* 92: 323–328. https://doi.org/10.1016/j.biopsycho.2012.09.005.

Tulving, Endel. 1972. Episodic and semantic memory. In *Organization of memory*, ed. Endel Tulving and Wayne Donaldson, 381–402. New York: Academic.

van Etten, Michelle L., and Steven Taylor. 1998. Comparative efficacy of treatments for post-traumatic stress disorder: A meta-analysis. *Clinical Psychology & Psychotherapy* 5: 126–144. https://doi.org/10.1002/(SICI)1099-0879(199809)5:3<126::AID-CPP153>3.0.CO;2-H.

Vervliet, Bram, Michelle G. Craske, and Dirk Hermans. 2013. Fear extinction and relapse: State of the art. *Annual Review of Clinical Psychology* 9: 215–248. https://doi.org/10.1146/annurev-clinpsy-050212-185542.

Vukov, Joseph. 2017. Enduring questions and the ethics of memory blunting. *Journal of the American Philosophical Association* 3: 227–246. https://doi.org/10.1017/apa.2017.23.

Warnick, Jason E. 2007. Propranolol and its potential inhibition of positive post-traumatic growth. *American Journal of Bioethics* 7: 37–38. https://doi.org/10.1080/15265160701518615.

Wasserman, David. 2004. Making memory lose its sting. *Philosophy & Public Policy Quarterly* 24: 12–18.

Wolfendale, Jessica. 2008. Performance-enhancing technologies and moral responsibility in the military. *American Journal of Bioethics* 8: 28–38. https://doi.org/10.1080/15265160802014969.

Wolff, Hans, Alejandra Casillas, Jean-Pierre Rieder, and Laurent Gétaz. 2012. Health care in custody: Ethical fundamentals. *Bioethica Forum* 5: 7.

Wood, Nellie E., Maria L. Rosasco, Alina M. Suris, Justin D. Spring, Marie-France Marin, Natasha B. Lasko, Jared M. Goetz, Avital M. Fischer, Scott P. Orr, and Roger K. Pitman. 2015. Pharmacological blockade of memory reconsolidation in posttraumatic stress disorder: Three negative psychophysiological studies. *Psychiatry Research* 225: 31–39. https://doi.org/10.1016/j.psychres.2014.09.005.

Chapter 14
"A Difficult Weapon to Confiscate" – Ethical Implications of Military Human Enhancement as Reflected in the Science Fiction Genre, Taking Star Trek as an Example

Frederik Vongehr

14.1 Introduction

With its exceptional prevalence and a broad reading public, the science fiction genre takes up current scientific topics and extrapolates future developments in science and their applications in a narrative way. But advanced, possible future technology is only seemingly at the fore. In truth, it comes second because the focus of future technology and science is on human existence and the changes that will potentially be triggered in human society, and this is also where its influence will be felt. Consequently, the manifold genre provides a sound board for discourses on society, for example, on human enhancement, a topic that has recently become increasingly important. Star Trek represents a special example of science fiction, as it has expanded its universe as a series for more than 50 years, enjoying great popularity and always having been an example of the genre considered to be particularly critical of society.

Amid the constant consideration being given to ways of improving military force performance, aspects concerned purely with weapons technology are increasingly being joined by others that concern the human component, the soldier, and go

This paper has been published originally in German language. Cf. Frederik Vongehr: "A difficult weapon to confiscate". Ethische Implikationen von Military Human Enhancement im Spiegel des Science Fiction-Genres am Beispiel von Star Trek". It is a secondary publication kindly permitted by the editor. In: Florian Steger (ed.) Jahrbuch Literatur und Medizin. Heidelberg 2017 (Jahrbuch Literatur und Medizin; 9), pp. 89–125. The translation was kindly supported in part by the Bundessprachenamt (Federal Office of Languages).

F. Vongehr (✉)
Central Medical Service, University of Marburg, Marburg, Germany
e-mail: frederikvongehr@bundeswehr.org

D. Messelken, D. Winkler (eds.), *Ethics of Medical Innovation, Experimentation, and Enhancement in Military and Humanitarian Contexts*, Military and Humanitarian Health Ethics,
https://doi.org/10.1007/978-3-030-36319-2_14

beyond the purely physiological preservation of health. The application of human enhancement[1] (HE) in the military sector is neither fiction nor a topic that is only found in a possible distant future or in future development analyses, but is a historical fact and is found today in the armed forces of a variety of nations.

Though maybe not quite so much in the public mind as robotics, artificial intelligence (AI) (Planungsamt der Bundeswehr 2013a), cyber warfare and cyber security, HE is nevertheless one of the topics that figure prominently in present-day civil and military future development studies and can thus be considered as an upcoming challenge. Today's knowledge on biology, nervous systems, computers, robotics and materials allows the body to be altered and redesigned to a certain degree (Lin 2013).

HE is an emerging technology that is known as a disruptive technology. This means that structures that have hitherto been decisive for strategic advantages or disadvantages must be devalued or reconsidered due to these innovations (Reschke et al. 2009). Military human enhancement (MHE) is consequently a so-called game-changer – in the same way as the introduction of gunpowder, the automobile, the computer or the atomic bomb once were (Allhoff et al. 2009, 6). This results in challenges for the future.

HE covers all medical, medical engineering or biotechnological measures that improve the performance, appearance or capabilities of humans aside from what is necessary to achieve, maintain or restore health. These measures go beyond the restoration or maintenance of a normal state and so exceed therapeutic or preventive purposes (Juengst 2009; Lin et al. 2013, 17 f; Koch 2015, 44f). A distinction is made between intracorporeal approaches like biotechnology and pharmacology and extracorporeal approaches like exoskeletons, augmented reality and combat suits (Daum and Grove 2015, 301; Bartz and Jäschner 2015; Ley 2010).

HE is not, however, a very concrete collective term for technologies that are in some cases very different. MHE is the term for military applications and includes targeted development. Some of them have already been used, for example, in WWII. Others are being developed by allies and potential adversaries. As its possible application is not limited to the military sector, developments in the civil sector also have to be taken into account. Consequences and ensuring discussions are equally important for the military and civil society (Planungsamt der Bundeswehr 2013b; Haupt and Friedrich 2016).[2]

Though the technologies summarized under the term MHE are numerous and diverse, and in some cases blend into each other, it is possible to classify them according to the way in which they bring about enhancement; there is pharmacological, genetic, technically non-invasive and technically invasive enhancement

[1] In addition to the term "human enhancement", there is the synonymic term "personal augmentation". This is used, for example, by the US military to avoid negative reactions in studies (Michaud-Shields 2014). Another term is "human performance optimization" (Lin 2010).

[2] Planungsamt der Bundeswehr (Bundeswehr Office for Defence Planning): Future Topic. Human Enhancement. Eine neue Herausforderung für Streitkräfte? Berlin 2013, p. 10. The term "doping" has established itself in the area of competitive sports. For details on this, see Oliver Haupt, Christoph Friedrich: Zur Geschichte der Dopingmittel. In: Pharmakon 4 (2016), pp. 8–16.

(Planungsamt der Bundeswehr 2013b).[3] On the other hand, a distinction can be made on the basis of the capabilities that have to be altered. For instance, the effects can concern psychological, cognitive, sensory and physical capabilities (Lin et al. 2013, 21–27). These two perspectives are not necessarily complementary, but often overlap. A pharmaceutical can influence physical endurance and cognitive performance at the same time. And physical capabilities can be manipulated by non-invasive measures, like muscle-enhancing exoskeletons, as well as by genetic engineering or pharmaceuticals.

Some authors also distinguish between human performance optimization (HPO), human-systems integration (HSI) and human performance enhancement (HPE). HPO comprises intense physiological and psychological training measures without grave interference in a soldier's anatomy or physiology. HSI includes improved equipment, for instance, in the form of environmental suits or portable computers, that supports the soldiers in the performance of their tasks. HPE includes more radical measures like genetic engineering and metabolism changes or intra-cortical computer interfaces (Friedl 2009).

From the medical perspective, a relevant issue in enhancement application is whether it is irreversible or only brings about a temporary state (Koch 2015, 47). In pharmacologic enhancement, pharmaco-kinetic parameters define the state, while components of brain-computer interfaces interacting with nerve endings of the brain often require a great deal of work to be done to implant them and risks are posed when they are removed (Merkel et al. 2007, 117–160; Hoffmann 2011, 645–651).

The requirement for quick, effective and decisive action in the military is fueling MHE implementation (Allhoff et al. 2009, 31f.). It is known that major investments are being made in research projects, for example, at the US Defense Advanced Research Projects Agency (DARPA) (Dvorsky 2014). One result could be a shift in warfare: Armies without advanced enhancement and so prone to easy defeat by an enhanced adversary might alternatively choose to increase their arsenal of nuclear or biochemical weapons (Allhoff et al. 2009, 35).

The changes to the body finally render it necessary to look at this topic from the military medicine perspective. This also applies in situations where armed forces reject MHE for their soldiers, for they could face an adversary who has a different view on the matter. Perhaps his forces will behave differently with regard to physical exhaustion, vulnerability and mental behavior. Furthermore, the application of MHE in allied multinational operations creates another point of commonality as both adversary and allied forces can get into each side's effective

[3] Then there are human performance degradation and biomonitoring. Biomonitoring is the term used to denote feedback mechanisms for administrating drugs and monitoring of the vital parameters at the same time. (Planungsamt der Bundeswehr 2013b, 7f). Biomonitoring is often a topic in science fiction, for example, in *Star Trek*. The Enterprise's onboard computer was able to monitor the vital parameters of all the crew members anytime. See Star Trek [The Original Series] [Television]: "Mudd's Women" (1966). USA (1966–1969), Gene Roddenberry (Creator). Cf. also Star Trek: The Next Generation [Television]: "Remember Me" (1990), USA (1987–1994), Gene Roddenberry (Creator). The hospital beds were furthermore able to independently administer intravenous shots of emergency drugs. For details on this, see Sternbach and Okuda 1991.

radius as casualties – or in the first case also as prisoners of war (Planungsamt der Bundeswehr 2013b; Michaud-Shields 2014, 30f.).

Although MHE is not currently a central topic in the German armed forces, it is useful and necessary to take a look at the topic. The rapid advances in innovation in technology and medicine make it likely to be applied in future conflicts. Subordinates will justifiably debate with their superiors on why soldiers from other nations are being granted enhancements while they are not. It is therefore advisable to involve all military personnel in the MHE debate at an early stage (Planungsamt der Bundeswehr 2013b).

Military developments in HE do not necessarily mean a mono-directional change for civil society. It may be the other way around, with innovations spreading widely in civil society before they find their way into the military. The latter would then ultimately have to bow to change (Michaud-Shields 2014, 32). Military medical researchers may also provide input for the civilian sector. For example, the effort put into creating prosthetics for disabled veterans also have an effect on the treatment of patients who have lost limbs in work-related accidents (Tucker 2014).

In the wake of changes in human biology, there may be a change in the interpretation of existing laws of warfare and human ethics itself (Koch 2015, 49). The subject must therefore be looked into before advances in technology hit society unprepared, without there having been any debate and assessment (Michaud-Shields 2014, 32). As some time always passes between the development of new technologies and a debate on them in society, this should be done early on (Lin et al. 2013, 21f.).

In the following, this study primarily examines such aspects reflected in science fiction that are particularly relevant from the medical perspective. They are pharmacological, genetic and technical-invasive enhancement and 3D bioprinting. Some technologies are still primarily in their early days, while there are already historical and recent examples of pharmacological enhancement.

14.2 *Star Trek* as an Object of Medical Ethics Study

Science fiction in general and *Star Trek* in particular are a rewarding object of study from the medical ethics perspective (von Engelhardt 2002). Science fiction can both provide inspiration for new developments and point out ethical implications in a generally comprehensible way (Pomidor and Pomidor 2006). *Star Trek* episodes are even used as educational material in medical ethics classes. One aspect that is seen as an advantage is that there is no bias caused by the viewer's prejudices, as the scenarios are fictional. Viewers perceive religious attitudes or social groups in science fiction differently to how they do in real life. Still, the episodes are tangible and make the topics presented easier to discuss than anything abstract (Hughes and Lantos 2001).

During the five decades between 1966 and 2016 alone, 726 TV episodes of *Star Trek* and 13 movies have been created, making *Star Trek,* with its global prevalence, one of the most popular TV products ever. The action initially revolves around the crew of the starship *Enterprise*,[4] which far from Earth encounters not only cosmic phenomena, but alien forms of society and again and again has to face social and ethical problems despite or precisely because of its advanced technology. Production suffered from a certain degree of underfunding at the beginning, but the producers still succeeded in conveying a credible and coherent image of the possible future. This solid canonical fictional sphere prompted viewers to think up other adventures – so-called fan fiction –, which come in a remarkable form in *Star Trek* and testify to a special response among the audience. The crew comes from a utopian federation in which hunger, poverty or discrimination do not exist. In addition to an alien, it comprises a black woman as an officer and even a member from Russia – both quite a provocation for a conservative US audience in the 1960s. The creator Gene Roddenberry (1921–1991) skillfully used the adventures to provocatively address social issues such as racism, the Cold War and the Vietnam War, gender discrimination or dangers of technical advancement (Vongehr 2016a, b; Perkowitz 2016; Rogotzki et al. 2009).[5] The series influenced the treatment of minorities in a variety of social areas. In the 1990s, an African American and a woman even assumed command, but this was less striking in the social context at that time than the constellation 30 years previously.

Besides the *Enterprise* herself, space stations portrayed as conflict-laden ergotopes or sociotopes constitute the scene of the action, this providing a contrast to the idealized conditions aboard the *Starfleet* spaceships. After Roddenberry's death, the producers ventured into this more progressive representation of interpersonal or interspecies tensions in a move that meant a departure from the originally utopian orientation of the canon. This also completed a transformation from the stereotype plot according to which the spaceship is used as a vehicle for superior ethical and social norms and an instrument for spreading them. The intercultural tensions are now inherent to the crew. There is a comparable development in *Star Trek: Voyager*, where a terrorist group has to be integrated into the crew, and there are tensions between the crew and individuals which are picked up during the starship's long flight home.

Medicine and pharmacy are not reduced to a supporting role in *Star Trek*, whole episodes often being dedicated to medical topics. This is not only reflected in episode titles like "Life Support", "Ethics" or "Hippocratic Oath". The spaceship's surgeon is a central character in several episodes, tasked with dealing with delicate problems while maintaining the observance of ethical principles. Due to his attitude and action, the surgeon frequently comes into conflict with his immediate superiors and the military leadership. They for their part are bound by civil rules and social values, for example, the Prime Directive, which prohibits the protagonists from meddling in any way in matters of alien societies. The suspense created allows the

[4]In later adaptations, it involves the space station *Deep Space Nine* and spaceships *Defiant* and *Voyager*.

[5]Transcriptions of any episode are available at www.chakotay.net. Further extensive sources, also offering episode analyses with background information, are provided by www.ex-astris-scientiae.org and www.memory-alpha.wikia.com

audience to reflect about their own values. They are contextualized in different ways, as various problems and various surgeons are portrayed during the long period in which the *Star Trek* canon is screened. Leonard McCoy and Beverly Crusher both maintain a close friendship with the commander and are regarded as competent and empathic advisers. In Julian Bashir, however, the crew has a surgeon who is initially insecure and inexperienced, but academically very successful augments the crew and who only gains emotional confidence in the course of *Deep Space Nine*. By contrast, in *Voyager* and *Enterprise*, both surgeons are obvious outsiders, as they are a nameless computer program or, in the case of the civilian surgeon, Doctor Phlox, an alien. The distinctive capabilities of the holographic surgeon and his becoming a sentient being – with all too human weaknesses – are addressed in much detail. In the course of the various series, a differentiating description of medical science and the relationship of the (military) physician and patient is built up.

At second glance, however, the prominent appearance of medicine and pharmacy is not very surprising. While futuristic technology is usually a key aspect in science fiction, the genre often focuses on how we humans handle this new technology and how it can affect our society and values. In this regard, there are close parallels between this genre and medicine. The confrontation with technological innovations and their significance for the health and well-being of humans is indispensable for this discipline as well. Some of today's emerging MHE technologies were already introduced in science fiction long before their realization seemed even technically possible (Reschke et al. 2009).

14.3 Therapy Versus Human Enhancement

There is a definition vagueness in the relationship between therapy and human enhancement (HE). The traditional definition differentiates therapeutic purposes from non-therapeutic purposes (Allhoff et al. 2009, 8f.) though the boundaries are blurred (Koch 2015, 41f.). This definition refers to the term illness, to the professional understanding of medical action or to measures taken to ensure the "species-typical functioning" of an individual, but it has no claim to universality (Juengst 2009). The differentiation only becomes evident during the application, or in the original state of the individual concerned – and thus is in the eye of the beholder (Allhoff et al. 2009, 11–13).

The invasions in physiological processes or in the human anatomy connected with HE or a therapy partially only differ due to the diverging intentions. An artificial access to the body – for example, through a neural system interface – allows both "enhancing" and therapeutic applications to be conducted. When applied to a healthy individual, systems that are originally therapeutic may result in an enhancement (Schöne-Seifert 2007, 99–102), while, vice versa, approaches that originally serve enhancement can be repurposed for therapeutic ends. For example, a technology designed to support people with speech impairments has been developed into the military's silent speech interface that is in the soundless transmission of speech

(Planungsamt der Bundeswehr 2013b, 6). Thought on HE must consequently include innovations that at first served solely therapeutic purposes. On the other hand, research in the enhancement sector can help to gain scientific knowledge for therapeutic applications.

Fictive technologies from the *Star Trek* canon also trigger similar thought: The episode "The Menagerie" screened in 1966 shows one of the first examples of the application of a brain-computer interface. The original commander of the Enterprise, an invalid completely immobilized, is sitting in a wheelchair controlled by thoughts: "His wheelchair is constructed to respond to his brain waves. Oh, he can turn it, move it forwards, or backwards slightly."[6] In view of the fact that computer technology was still in its early days in the 1960s, such a technology was far from being realistically conceivable. But recent research in Germany reveals that scientists are today working on just such wheelchairs with respective interfaces (Graimann et al. 2010, 1–27; dpa 2015).

The application of such a brain-computer interface in a healthy individual would already constitute a form of human enhancement, as there is no therapeutic foundation. It takes little imagination to repurpose such constructions for military applications, for example, if the interface is not used for steering one's own locomotion, but for controlling the launch of projectiles, selecting targets and defining the trajectories of drones.

Similar thoughts apply to pharmacological enhancement. A substance capable of affecting the body is not *per se* a so-called enhancer or therapeutic agent. It is its application and effect that permit a differentiation to be made, and it is often based on moral aspects. The terms curative and addictive drugs confront each other as a diametrical pair, although they may describe the very same substance. Attention is drawn, for example, to the long history of opium and the morphine derived from it, or to the use of steroids. They may be therapeutic for individuals who are ill, but can already be an enhancement for individuals who are healthy. A biologically active substance does not have an intrinsic quality that allows an *a priori* classification; the same applies to any technical enhancement of the body. Only its application and targeted intention permit a differentiation (Boldt and Maio 2009; Maio 2012, 321–334; Merkel et al. 2007, 289–382; Lin 2009).

Therefore, some authors suggest a wider definition of human enhancement that takes into account the various definition issues and consideration of the intention and social context: HE comprises medical or biotechnological innovations that do not have primarily therapeutic or preventive objectives; instead, they are aimed at changing people with respect to their capabilities or forms in a way that they are seen as improvements in the respective socio-cultural contexts (Biller-Andorno and Salathé 2012).

When applied to military conditions, this distinction is secondary in some respects because an interface, be it a prosthesis or an enhancement, may pose a security gap. The catchwords cyberwar and IT upgrading not only concern the military and civilian infrastructures, respectively, but also people. *Star Trek* shows in

[6] Star Trek [The Original Series] [Television]: "The Menagerie" (1966). USA (1966–1969), Gene Roddenberry (Creator).

multiple instances how a therapeutic device can become a security-relevant point of attack. Adversaries repeatedly use the eye prosthesis of the blind protagonist Geordi La Forge, the VISOR (Visual Instrument and Sensory Organ Replacement), for manipulating the wearer or infiltration purposes.[7] Retina prostheses, comparable to the VISOR presented in *Star Trek* in 1987, are meanwhile commercially available (Vongehr 2016a, b). La Forge's VISOR is also capable of receiving electromagnetic emissions outside the physiological spectrum. It is conceivable that the military sector will develop retina implants that can receive the infrared spectrum and turn emissions into nerve impulses and so make soldiers capable of seeing at night with their own eyes.

In addition to professional enhancement applications, there are now trends towards so-called body-hacking (Koch 2015, 44f.). Those involved in it are not necessarily scientific institutions or profit-oriented enterprises; they are also inventive individuals who are making use of the commercially available technology. The extensions range from portable computers that project information into the field of vision to sensor patches for the color-blind and equipment for monitoring vital parameters. They also include devices that transform ultrasound into acoustic impulses, facilitating echo location, and devices that can be directly connected to routine electronics (Dujmovic 2016; Meskó 2014, 164–170; Lin et al. 2013, 26f.). Enhancement and other technologies are leading to an extension of the human, the process summarized under the term transhumanism.[8]

Such civil developments are also relevant for MHE as they are gateways for adversary technologies: A society with electronic implants becomes vulnerable to certain types of weapons that operate with electromagnetic impulses or cyberattacks – like the Borg in *Star Trek*. This calls for renewed discussion of the extent to which new weapons technologies must be classified as weapons of mass destruction and consequently banned in international treaties.

14.4 Pharmacological Enhancement

"Rapid progress, to where humans learned to control their military with drugs."[9]
Q in the pilot movie of Star Trek: The Next Generation, 1987

Today, performance enhancement by means of pharmaceuticals usually concerns cognition and vigilance and in the military sector the will to fight and sustainability.

[7] Star Trek: The Next Generation [Television]: "The Mind's Eye" (1991). USA (1987–1994), Gene Roddenberry (Creator); Star Trek: Generations [Cinema]. USA (1994), Gene Roddenberry (Original Creator), Rick Berman, Ronald D. Moore, Brannon Braga (Creators).

[8] Due to space constraints, the subject of transhumanism is not addressed in more detail in this paper. See for example Kurzweil 2007.

[9] Star Trek: The Next Generation [Television]: "Encounter at Farpoint" (1987). USA (1987–1994), Gene Roddenberry (Creator).

There are various synonyms for the use of substances to enhance cognitive performance, for example neuro enhancers, neuro doping, psychostimulants or, a more popular one, "pep pills" and "psychopharmacological cognitive enhancement drugs" (PCEDs) (Soyka 2010; Merkel et al. 2007, 11–57). Today, pharmacological enhancement is mainly used to counter stress, pain and fatigue, i.e., to maintain fitness in critical situations (Koch 2015, 45f.). The USA's Defense Advanced Research Projects Agency, for example, is involved in the trial of wakefulness-promoting agents like Provigil (Modafinil) (Koch 2015, 44f; Auf dem Hövel 2007). This is of special importance for the maintenance of vigilance among military pilots (Röper and Disson 2012, 365–371).

Preparations for making use of pharmacological enhancement in the armed forces were already kept in store on quite a large scale during WWII and some were even administered.[10] The use of amphetamine preparations like Pervitin was not, however, an exclusive issue of the *Wehrmacht*, as this agent was available to the public before the war. In the 1930s, Pervitin with methamphetamine and Benzedrine and Aktedron (both containing amphetamine) were available from German pharmacies without a prescription. It was due to an increase in cases of psychosis and chronic intoxication that amphetamine preparations were only available on prescription in 1939, though pharmacies did not always adhere to this rule very strictly. For example, it was widely used by civilians, and it was not until 1941 that it was put under the Narcotics Law to contain the widespread misuse. Pervitin continued to be popular after the war and was only taken off the market in 1988. Since 2005, methamphetamine has regained popularity, though in the form of crystal speed or crystal meth (Schlick 2008, 315–323; Grönig 2008).

In addition to Pervitin with methamphetamine, combination preparations made of other highly effective alkaloids were used. The German *Kriegsmarine* is known to have conducted tests to find an ideal composition not only on soldiers, but in 1944 also on inmates of Sachsenhausen concentration camp. The substances involved were mainly preparations for suppressing sleep among the crews of very small submarines, who were meant to remain on operational deployment for several days without a break. Here is a note on the tests entered in the medical war diary (*Ärztliches Kriegstagebuch*) in 1944: "The military leadership is of the view that in this war, if it is necessary, damage due to highly effective medications must be accepted if it allows operations to be conducted."[11]

The use of pharmaceuticals as a means of enhancing performance among civilians and military personnel is a recurring topic in science fiction. In addition to the need to conduct a cost-benefit analysis when such substances are intended to be used, consideration must be given to the issue of soundness of mind. In the episode "Empok Nor", *Star Trek* presents a group of elite soldiers left behind by an alien

[10] Detailed references have been given on the use of Pervitin by the Germans and the tests conducted to find performance-enhancing substances. For details on this, see Hartmann 1994.

[11] Bundesarchiv/Militärarchiv Freiburg, BaMa RM 103/10, fol. 5r. Ärztliches Kriegstagebuch des Kommandos der K-Verbände für die Zeit vom 1. September 1944 bis 30. November 1944. For details on this, see Nöldeke and Hartmann 1996, pp. 207–212 and also Vongehr 2014, 496 f.

species to guard an abandoned space station. The soldiers had been conditioned for this task with a drug that intensified xenophobe tendencies and defense instincts. Such a drug that influences mental conditions and actions raises the question of whether the responsibility for any action taken lies completely with the soldier or whether the soldier can only be held responsible to a limited extent for any war crime committed and is protected from legal consequences (Koch 2015, 49). A hypothetical "berserker drug" could lead to a soldier being no longer able to distinguish between combatants and noncombatants (Michaud-Shields 2014, 31). Are the soldiers themselves responsible or indeed the medical team that changed them? (Shunk 2015, 96) Anyway, in the Star Trek episode, a crew member who murders a fellow soldier after involuntarily coming into contact with the drug goes unpunished – there is no more profound reflection on the matter.[12] The episode additionally presents the danger of civilians intentionally or unintentionally coming into contact with such a drug designed for military purposes.

The subject dealt with in *Star Trek* is not only performance enhancement, but also rather the aspect of control that can be exercised on the soldier with pharmacologically active drugs. This assigns pharmaceuticals another military function: the assurance of loyalty.

This subject is dealt with in the 1987 pilot episode of the extremely successful spin-off *Star Trek: The Next Generation.* In a key dialogue right at the beginning, the crew of the *Enterprise* attempts to convey to the superior being Q that mankind has made progress and drawn conclusions from the mistakes of earlier wars and no longer has any martial tendencies. Q doubts this progress, pointing out that it led to the following development in the twenty-first century: "Humans learned to control their military with drugs."[13] This highlights the warning and cautioning impetus of Roddenberry, the creator, who still had an influence on the series at the time.

The aspect of pharmacological control is also a subject in the later adaptation *Star Trek: Deep Space Nine*. It is even part of its central story arc: The Jem'Hadar, the combat troops of the Dominion – a superior adversary of the Federation –, are modified in their physiology in a way that their obedience towards their commanders is ensured by a rationed pharmacologically effective substance called Ketracel-White. It is vital for the soldiers' organisms and is issued to each one of them in a ritualized procedure containing the swearing of a pledge to the military leaders.[14] Besides the ethical problems surrounding such a massive invasion of the body, there is the question as to whether it is acceptable to destroy the facilities at which the

[12] Star Trek: Deep Space Nine [Television]: "Empok Nor" (1997). USA (1993–1999), Gene Roddenberry (Original Creator), Rick Berman, Michael Piller (Creators).

[13] Star Trek: The Next Generation [Television]: "Encounter at Farpoint" (1987). USA (1987–1994), Gene Roddenberry (Creator).

[14] The Jem'Hadar's addiction to the agent Ketracel-White is the subject of several episodes, with the fact being clearly made that it is not only an essential substance for a genetically modified metabolism, but also causes classical withdrawal symptoms in users who abstain from it. See Star Trek: Deep Space Nine [Television]: "Rocks and Shoals" (1997). USA (1993–1999), Gene Roddenberry (Original Creator), Rick Berman, Michael Piller (Creators).

vital Ketracel-White is stored or produced without the adversary soldiers dying? This matter is raised in at least one episode. The crew under Captain Sisko destroys such a storage facility, but no further questions are asked. Even the ship's surgeon Doctor Bashir is involved in the mission.[15] Our present understanding of international law is such that at least medical support facilities are not allowed to be destroyed, but no heed is taken of the use of MHE drugs. Does this as a consequence mean that in order to conform to international law standards, drugs or medical products that are suitable for both enhancement and therapy need to be stored differently to conventional medical supplies? How can a distinction be at all made between the categories of curative drugs and/or enhancement if the quality of medical supplies is ultimately determined by the way they are used? Is a field hospital no longer worthy of special protection because upgradings and repairs of enhancements can be conducted there? Is separate infrastructure or even additional medical personnel required as the special protection awarded to such personnel is not reconcilable with the maintenance or repair of enhancements?

14.5 Genetic Engineering

"Superior ability breeds superior ambition."[16]
Spock in "Space Seed", 1967

While there has already been extensive research on animals, genetic engineering and gene therapy in humans are still in their early days. As early as in 1953, James Watson and Francis Crick postulated the double helix structure of DNA, which marks the beginning of molecular genetics (Eckart 2011, 300). This allowed a better understanding of genetics to be acquired and opened up new possibilities in medicine and biology, at a time when the national socialist thought on so-called racial hygiene had not yet been forgotten and racial unrest was an everyday occurrence in the USA. *Star Trek* in 1967 presents a group of genetically enhanced individuals with superior mental and physical abilities led by the hypertrophic tyrant Khan. During the "Eugenics Wars", they are used on Earth as soldiers and eventually try to subjugate mankind. *Star Trek* repeatedly broaches the issues of the "Eugenics Wars" and in this way emphasizes the intention to be critical of society.[17] With regard to this

[15] Star Trek: Deep Space Nine [Television]: "A Time to Stand" (1997). USA (1993–1999), Gene Roddenberry (Original Creator), Rick Berman, Michael Piller (Creators).

[16] Star Trek [The Original Series] [Television]: "Space Seed" (1967). USA (1966–1969), Gene Roddenberry (Creator).

[17] Star Trek [The Original Series] [Television]: "Space Seed" (1967). USA (1966–1969), Gene Roddenberry (Creator); Star Trek II: The Wrath of Khan [Cinema]. USA (1982), Gene Roddenberry (Original Creator), Harve Bennett, Jack B. Sowards (Creators); Star Trek Into Darkness [Cinema]. USA (2013), Gene Roddenberry (Original Creator), J. J. Abrams et al. (Creators). The latest TV series also picks up on the events around Khan. Star Trek: Enterprise [Television]: "Borderland" (2004); "Cold Station 12" (2004); "The Augments" (2004). All USA (2001–2005), Gene Roddenberry (Original Creator), Rick Berman, Brannon Braga (Creators).

group's pronounced feeling of superiority, the Spock character states fittingly: "Superior ability breeds superior ambition."[18] After it becomes obvious that Khan must be classified as extremely dangerous, Kirk finally decides to exile the complete group on a planet. The final dialog nevertheless shows that the crew is optimistic about the way in which these now isolated superhumans will develop. Spock: "It would be interesting, Captain, to return to that world in a hundred years and to learn what crop has sprung from the seed you planted today." Kirk: "Yes, Mister Spock, it would indeed."[19] The spin-off *Enterprise* produced between 2001 and 2005, which is set in time before the original series, introduces the brilliant geneticist Dr. Arik Soong in this context. He vehemently argues in favor of the use of genetic engineering and regards the dilemma as rather being due to its improper use by humans. Though great harm has already been caused, Captain Archer finally states hopefully: "Maybe someday, we'll figure out how to use it to benefit humanity."[20]

In *Star Trek: Deep Space Nine*, the authors describe a long-lasting war with the Dominion, whose whole military structure is based on a rigorous MHE approach. Two groups of soldiers are literally made in the laboratory: the Vorta as military leaders and the Jem'Hadar as combat troops. The latter were exclusively designed for employment as fighting soldiers, with interfering needs like food intake and rest being minimized. They are clones with a significantly shortened ontogenesis that are bred at dedicated breeding centers. Their loyalty towards the rulers of the Dominion is assured through their genetically determined belief in their creators as godlike beings.[21]

The groups around Khan and the Jem'Hadar show that an elitist concept of soldiers can also be applied to their civilian lives, as the enhanced capabilities are retained after their retirement from military service, making them regard civilians as inferior (Koch 2015, 47). Soldiers discharged into civilian life could furthermore cause a distortion of the labor market in the real world, as their superior capabilities would then give them an edge in their competition with everyday individuals (Lin 2010, 2012; Klimas 2012).

---

[18] Star Trek [The Original Series] [Television]: "Space Seed" (1967). USA (1966–1969), Gene Roddenberry (Creator).

[19] Ibid.

[20] Star Trek: Enterprise [Television]: "The Augments" (2004). USA (2001–2005), Gene Roddenberry (Original Creator), Rick Berman, Brannon Braga (Creators).

[21] However, this assurance seems to be insufficient: "The Founders' ability to control the Jem'Hadar has been somewhat overstated. Otherwise we never would have had to addict them to the white." Star Trek: Deep Space Nine [Television]: "To The Death" (1996). USA (1993–1999), Gene Roddenberry (Original Creator), Rick Berman, Michael Piller (Creators). The Vorta are clones as well, although only one clone exists at any one time. After the predecessor's death, his memories are transferred by an unexplained procedure to the new clone, who can see to business matters seamlessly. Although this appears far from anything that can be done today, it is possible to derive from it the question of the extent to which one living being can be equipped with memories of another and so have its personal development prescribed.

This danger exists not only between soldiers and civilians, but also between soldiers who receive different enhancements for different tasks, which causes heterogeneity to arise within a group of soldiers. The *Star Trek* authors go into this: Among the Jem'Hadar, there are various versions of fighters, each adapted to the theater in which they are deployed. This not only creates a significant conflict potential among the soldiers themselves, as they see themselves as being differently suited for different tasks and they have corresponding ideas about their superiority.[22]

The Jem'Hadar's accelerated ontogenesis also raises the following ethical question: Is it legitimate to select breeding centers with embryos as a military target? Do they still have a right to live although they are only created for purposes of war and will definitely grow to become aggressive and dangerous beings? In *Deep Space Nine* this task of answering this question is elegantly left to the martial Klingons, so a potential debate about the perpetration of the human protagonists is evaded.[23]

In connection with genetic engineering, genetic improvement for military purposes is also depicted as being able to be of therapeutic benefit later. For example, Julian Bashir, the surgeon of space station *Deep Space Nine*, was born mentally disabled. With the aid of genetic engineering, he was given mental and physical abilities that were not equal to those of his peers, but that exceeded them by far. However, genetic manipulation is socially outlawed and regarded as punishable: "DNA resequencing for any reason other than repairing serious birth defects is illegal. Any genetically enhanced human being is barred from serving in Starfleet or practicing medicine."[24] Although Doctor Bashir could not by law have served in *Starfleet* or practiced medicine, an exception is made as it was after all his parents who had the illegal procedure performed and his father accepts the punishment. Unfortunately, this episode does not sufficiently reflect on the marginalization of genetically modified humans in society, even if they are not responsible for the enhancement they have undergone, and the issue is resolved by applying the *deus ex machina* principle. This subject is only taken up later.[25]

---

[22] Star Trek: Deep Space Nine [Television]: "One Little Ship" (1998). USA (1993–1999), Gene Roddenberry (Original Creator), Rick Berman, Michael Piller (Creators).

[23] Star Trek: Deep Space Nine [Television]: "The Abandoned" (1994) and "Once More Unto The Breach" (1998). Both USA (1993–1999), Gene Roddenberry (Original Creator), Rick Berman, Michael Piller (Creators).

[24] Star Trek: Deep Space Nine [Television]: "Doctor Bashir, I Presume" (1997). USA (1993–1999), Gene Roddenberry (Original Creator), Rick Berman, Michael Piller (Creators).

[25] Star Trek: Deep Space Nine [Television]: "Statistical Probabilities" (1997) and "Chrysalis" (1998). Both USA (1993–1999), Gene Roddenberry (Original Creator), Rick Berman, Michael Piller (Creators).

14.6 Technically Invasive Enhancement

"A difficult weapon to confiscate".[26]
Lt. Yar in "Symbiosis", 1988

The use of simple wooden and iron prostheses as limb replacements, for example, is a reminder that the integration of artificial structures into or onto a body is not a late industrial development. But as soon as the replacement parts are not only intended for use purely in the mechanical compensation of a physical disability, but also have an artificial intelligence or at least control circuits and increasingly penetrate the body or even interact with the nervous system, the question arises as to when a human is still a human or already something else (Le Blanc 2014, 24). Does it take one, two or a million circuits for something to be defined as a technical component? Or are particular organs the decisive factor? It is hardly possible to draw a satisfactory line.

Before the question of whether the enhancement is the property of an external organization or already belongs to the body of the patient or the "enhanced" is considered, a look should first be taken at the view of the individual concerned. Is the supplementation perceived as a foreign body or is it a welcome improvement (ibid., 24)?[27]

In *Star Trek*, the synthesis of technology and body, created for example by exchanging faulty organs by means of artificial replacements, is not an extraordinary procedure. Captain Jean-Luc Picard, for example, has an artificial heart, the transplantation of which is the subject of two episodes.[28]

The public associates the term Cyborg primarily with science fiction, although is it not originally from this genre, but was created by two scientists who wrote a paper – later picked up on by NASA – about how it would be better to adapt the human body to the hostile conditions of space rather than to provide it there with an artificial earth-like environment (Clynes and Kline 1960, 26f, 74–76; Gray 1995; Rid 2016).[29] Nevertheless, the term has spread widely within the

[26] Star Trek: The Next Generation [Television]: "Symbiosis" (1988). USA (1987–1994), Gene Roddenberry (Creator).

[27] In the pilot movie of the TV adaptation of Martin Caidin's novel *Cyborg*, injured astronaut Steve Austin is initially not very happy about the electromechanical prostheses that are to be transplanted as a consequence of his accident. The Six Million Dollar Man. [Television]: "The Moon and the Desert" (1973). USA (1973–1978), Martin Caidin, Henri Simoun, Richard Irving (Creators).

[28] Star Trek: The Next Generation [Television]: "Samaritan Snare" (1989) and "Tapestry" (1993). Both USA (1987–1994), Gene Roddenberry (Creator). Also cf. Luokkala 2014, 187; Stoppe 2014, 104. Limbs can also be replaced in the *Star Trek* universe. In contrast, the disabled veteran Nog has great difficulty in accepting his new artificial leg. Star Trek: Deep Space Nine [Television]: "The Siege of AR-558" (1998). USA (1993–1999), Gene Roddenberry (Original Creator), Rick Berman, Michael Piller (Creators).

[29] The idea and the term were picked up on by Martin Caidin in the 1972 novel *Cyborg* and a later TV series entitled *The Six Million Dollar Man*. Cf. also European Space Agency 2001, p. 34. For details on cybernetic organisms in science fiction and today's medical possibilities, see in particular Klugman 2001.

genre. Today, it is used as an expression that refers to the synergy of a machine and an organism (Spreen 1997, 1998, 1999, 2010, 2013).

The subjects of war and the medical care of soldiers raise several issues that are addressed by science fiction. MHE obviously includes the incorporation of weapons or devices used to control them. It seems that a Cyborg remains an organic being as long as it is comprising biological components.[30] In the future, both doctors and technicians will maybe care for a patient – and perhaps weapons engineers for a soldier?

The dilemma of whether a soldier "enhanced" this way shall be regarded primarily or secondarily as a human or a weapon is dealt with in a special scenario in *The Next Generation*. In one case, *Star Trek* introduces the Bynars, humanoid beings that always appear as twin-like pairs. Their brain functions are linked pairwise by a planetary mainframe computer. But the enhanced capability to process information also results in a dependence which makes social existence without the connecting computer impossible. The two individuals of such a pair are also connected by a form of synthetic telepathy generated by electronic means.[31] For this to happen, a surgeon removes the Bynar child's parietal lobe and replaces it with a synaptic processor. Consequently, this form of society rates participation in society higher than an individual's right to self-determination with respect to its own body and right to physical integrity. Humans' skeptical attitude towards such radical interference, however, only becomes a subject in a dialogue in the later *Star Trek* adaptation *Enterprise* between the alien ship's surgeon and a human crew member.[32] In *The Next Generation*, no critical comment is made on the enhancement of the Bynars.

The best-known example for technically invasive enhancement from the *Star Trek* canon are the "Borg", however. They extend their personal body by applying a procedure known as assimilation by means of which biological beings are provided numerous cybernetic implants with manifold functions. Some serve as weapons; others ensure communication with other Borg or synthesize all the substances required by the organic tissue. The body consequently no longer requires food and only needs electric energy – like a machine. The weapons arsenal and human organism merge into one.

Due to their synthetic telepathy, the Borg no longer have an individual consciousness and a will of their own and merely function as units of a large organism – the

---

[30] This consideration is extrapolated to the extreme in *Star Trek: The Next Generation*, when the android Lore wants to reform several Borg into fully artificial life forms in his similitude and has rigorous human experiments conducted for this purpose. The Borg are meant to dispense with all their organic components, which are ultimately the special feature of their kind. Star Trek: The Next Generation [Television]: "Descent I" (1993) and "Descent II" (1993). Both USA (1987–1994), Gene Roddenberry (Creator).

[31] Star Trek: The Next Generation [Television]: "11,001,001" (1988). USA (1987–1994), Gene Roddenberry (Creator).

[32] Lt. Reed: "What sort of people would replace perfectly good body parts with cybernetic implants?" Doctor Phlox: "You, of all people, should be open-minded about technology." Lt. Reed: "I don't have a problem with it, as long as it stays outside of my skin." Doctor Phlox: "If your heart was damaged, would you want me to replace it with a synthetic organ, or would you rather die?" Lt. Reed: "That's different." Star Trek: Enterprise. [Television]: "Regeneration" (2003). USA (2001–2005), Gene Roddenberry (Original Creator), Rick Berman, Brannon Braga (Creators).

Borg collective. This collective uses its soldiers to pursue the totalitarian objective of constant expansion alongside the subjugation of other cultures.[33]

Captain Picard, who is represented not only as a highly moral authority in the *Enterprise* crew, but also in *Starfleet* and the *Federation*,[34] once had the traumatizing experience of being kidnapped by the Borg and transformed into human-machine being called Locutus. He was only saved by deft action on the part of his crew.[35] The forced metamorphosis included not only the incorporation of several implants, but also his integration into a "hive" consciousness that interconnects all Borg. After this change, he was forced to disclose all the secrets he knew that protected his social environment and its values. He also had to participate in the mass murder of his companions.[36]

This plot reveals a security risk of MHE: The enhancement could force soldiers to lay open vital strategic data by eliminating their free will. So how should technical equipment enhancement be handled? Critical items of military equipment are usually kept under special supervision and subject to access restrictions. Does that mean enhancements have to be rendered unusable or forcefully removed when a soldier is off duty hours or retires from military service as they are military property? A soldier may no longer consent to further surgical interventions (Michaud-Shields 2014, 30). What is the situation regarding systems that are both weapons and control vital functions?

A question of much more relevance is that of what to do with prisoners of war whose organism is also a weapon system. In one *Star Trek* episode, the crew of the *Enterprise* encounters a Borg with life-threatening injuries. He is only taken aboard and administered medical care on the vehement insistence of the ship's surgeon.[37] Captain Picard, still traumatized by his earlier kidnapping, above all sees this captured Borg as a chance to exterminate the adversary once and for all. The young Borg is intended to receive a viral programming and be returned to the collective. A chain reaction in the cybernetic implants is intended to trigger a fatal cascade failure throughout the entire Borg collective. For Picard, the involuntarily enhanced young man is not an individual, but only an object – an It, destined to become his tool of destruction. The rescued or captured adversary soldier is consequently intended to be used directly as a weapon of mass destruction against his own people. When crew members express doubts about this plan, Picard reminds his subordinates of the fact that scientists must distance themselves from their lab animals in order to avoid emotional conflicts when they kill them for a test. He advises them not to forge personal ties with the prisoner (Barad 2001, 250).

---

[33] However, their primary motive is neither suppression or annihilation of alien life, but the achievement of perfection. Star Trek: Voyager [Television]: "The Omega Directive" (1998). USA (1995–2001), Gene Roddenberry (Original Creator), Rick Berman, Michael Piller, Jeri Taylor (Creators).

[34] Star Trek: The Next Generation [Television]: "The Measure Of A Man" (1989) and "The Drumhead" (1991). Both USA (1987–1994), Gene Roddenberry (Creator).

[35] Star Trek: The Next Generation [Television]: "Best of Both Worlds" (1990). USA (1987–1994), Gene Roddenberry (Creator).

[36] Star Trek: The Next Generation [Television]: "Family" (1990). USA (1987–1994), Gene Roddenberry (Creator).

[37] Star Trek: The Next Generation [Television]: "I, Borg" (1992). USA (1987–1994), Gene Roddenberry (Creator).

By planning this genocide, Captain Picard, who has actually has established himself as a high moral authority, is irritating, as he violates the previous *Star Trek* continuum of values at a sensitive spot.[38] It is only after Picard has hesitantly established contact with the Borg and opened up for talks with his crew that he show a readiness to recognize the original young man who was later transformed into a Borg. In the end, Picard leaves the decision of whether to return to the collective or remain on the *Enterprise* to the Borg. Picard lets him go free without any destructive programming – hoping that his newly gained feeling for individuality and the positive experience with the crew of the *Enterprise* will lead to a fundamental change in the Borg's nature (Barad 2001, 207–229).

Technologies almost allow beings like the Borg to be created today: The US military is conducting a research project on how sensory information can be exploited to obtain early warnings of potential dangers. The "Cognitive Technology Threat Warning System" is meant to directly communicate with the brain (Matthews 2015). In *Star Trek*, however, the artificial telepathy can be used not only as a communication tool, as it originally was by the Borg in the collective, but also as an instrument for controlling individuals. It is meant to prevent them from thinking for themselves and possibly having doubts about the sense of a mission. An enhancement thus becomes an instrument for carrying out a totalitarian indoctrination of subordinates.[39] An enhancement can also alter a soldier's understanding their role, as the new capabilities lead to the assignment of special tasks and duties that a normal soldier does not have. This could be, say, participation in particularly dangerous operations (Koch 2015, 47). It raises the question of how enhancement influences a soldier's personality (Shunk 2015).

14.7 Bioprinting

"What kind of species is born with a suicide gland?"
"Not this one. He was surgically enhanced, if you can call it an enhancement."[40]
Cdr. Tucker and Doctor Phlox in "Rajiin", 2003

[38] In a later movie, too, Picard only sees assimilated subordinates as adversaries and directs his crew to back off from the original fellow crew members who have been turned into adversary human-machines. "You may encounter Enterprise crewmembers who've already been assimilated. Don't hesitate to fire. Believe me you'll be doing them a favor." Although he is otherwise quite willing to see reason, Picard does not hesitate to truly massacre two assimilated crew members in order to get information that is stored in cybernetic implants. Star Trek: First Contact [Cinema]. USA (1996), Gene Roddenberry (Original Creator), Brannon Braga, Ronald D. Moore, Rick Berman (Creators).

[39] Star Trek: The Next Generation [Television]: "Descent I" (1993) and "Descent II" (1993). Both USA (1987–1994), Gene Roddenberry (Creator).

[40] Star Trek: Enterprise [Television]: "Rajiin" (2003). USA (2001–2005), Gene Roddenberry (Original Creator), Rick Berman, Brannon Braga (Creators).

Advancements in the area of 3D printing technology will render the use of 3D-printed or 3D-bred organs for civil and military purposes likely (Planungsamt der Bundeswehr 2013c).

In addition to the use of 3D printing in the area of conventional material technology, intensive research is being conducted on bioprinting: Research in this young field is looking into ways of producing tissue or whole organs artificially by firstly printing frames into which the living cells grow. This could offer future disfigured or transplantation patients new prospects. Applications in the enhancement sector are also conceivable for people who, for example, wish to have enhanced organs implanted or to strengthen their muscle – they will be put to military use for sure (Campobasso 2015; Peck 2015).

Star Trek already introduced potential applications: On the one hand, natural-like tissues can be produced for therapeutic purposes, while even complex parts of the central nervous system like the retina or spinal cord come into consideration for creation to replace damaged original ones.[41] Artificial organs that have functions beyond their actual physiology can also be implanted into the body. To evade captivity, the soldiers of the Xindi people, for example, have an additional gland that discharges a lethal neurotoxin into the body as required.[42]

14.8 Requirements for Enhancements

The potential dangers of MHE also demand enhancements to meet special requirements. Adversaries could attempt to remove forcefully an enhancement from the body of a prisoner of war in order to study its function or use it to their advantage. One requirement applicable to its design could therefore be to render the item unusable as soon as it is removed from a body (Michaud-Shields 2014, 31).

This on the one hand requires the design to be protected from manipulation by an adversary if the individual in which it is implanted is captured and, on the other, manipulations by the soldiers themselves to be prevented, say, in case they desert and try to remove their enhancements, sell them for profit or even try to hand them over to the adversary. It is therefore legitimate to equip implants with protective mechanisms? In the *Star Trek* episode "Legacy", soldiers carry an implant in their bodies that signals that an adversary is approaching. For security reasons, these implants are designed to function as an explosive charge and so kill both the soldiers in whom they are implanted and the operating surgeons if an unauthorized attempt is made to remove them from the soldiers' bodies.[43] The question here from a medi-

[41] Star Trek: The Next Generation [Television]: "Loud As A Whisper" (1989) and "Ethics" (1992). Both USA (1987–1994), Gene Roddenberry (Creator).

[42] Star Trek: Enterprise [Television]: "Rajiin" (2003). USA (2001–2005), Gene Roddenberry (Original Creator), Rick Berman, Brannon Braga (Creators).

[43] Star Trek: The Next Generation [Television]: "Legacy" (1990). USA (1987–1994), Gene Roddenberry (Creator).

cal ethics point of view is how is it possible to advocate the implantation of a device that is designed to inflict damage on the individual in whom it is implanted under certain circumstances. The implants presented in "Legacy" originally had a benevolent function, namely, to separate opposing parties and to prevent a situation from escalating into a civil war. They at the same time restrict personal freedom. Can a soldier who refuses to consent to having an implant be regarded as a destabilizing element or even as a warmonger? And must he therefore be sanctioned? Devices designed to monitor vital functions in the field can at the same time be used to monitor soldiers by transmitting their positions (Allhoff et al. 2009, 31f.).

The undesired transfer of technology to the adversary poses a military risk. Should there be a terminal emergency switch for this case? And also for the cases of a soldier deserting or defecting? Two examples from *Star Trek* are highlighted in the following: The Vorta, as diplomats and military commanders of the Dominion, have implants in their brain stems that allow them to commit suicide in the event of their being captured. This prevents the adversary from acquiring information of strategic value. The so-called termination implant is triggered by a particular touch of the lower jaw and ear.[44] "Section 31", the fictional secret service, uses a comparable device. The neuro-depolarizing device introduced here can even be triggered without physical contact and only by thought if the agent concerned is unable to move.[45]

There is no ruling out the possibility of a military power in the real world using such technology one day, for example, implanting them in particularly exposed soldiers and possibly also activating them by remote control to eliminate particularly endangered or captured soldiers for security reasons. Network centric warfare already allows decision-making levels higher up the chain of command to increasingly influence operational-level forces. Global real-time tracking of military operations is the status quo. This particularly became known due to one photograph: It showed President Barack Obama and his staff following Operation Neptune's Spear, in which Osama bin Laden was killed, on a screen. Real-time tracking of this operation was possible because the soldiers involved wore helmet cameras – a technically non-invasive MHE – that transmitted the events via satellite straight to the White House (Spreen 2015, 61–76; Crawford 2012).

14.9 Law of Armed Conflict: Help for the Adversary?

The scenario of a soldier with MHE – no matter whether friend or foe – in the care of medical service personnel causes several ethical dilemmas (Gross 2015; Vollmuth 2015, 2016). Generally, there is the question of whether a sick or wounded soldier

---

[44] Star Trek: Deep Space Nine [Television]: "Treachery, Faith and the Great River" (1998). USA (1993–1999), Gene Roddenberry (Original Creator), Rick Berman, Michael Piller (Creators).

[45] Star Trek: Deep Space Nine [Television]: "Extreme Measures" (1999). USA (1993–1999), Gene Roddenberry (Original Creator), Rick Berman, Michael Piller (Creators).

should be given privileged treatment if, for example, he has an expensive enhancement or if the enhancement constitutes a strategic advantage (Shunk 2015, 96).

This question becomes even more difficult with adversary soldiers: What are medical personnel supposed to do with captured soldiers who have enhancements (Koch 2015, 47f.)? Should wounded prisoners of war with enhancements be helped in exactly the same way as normal prisoners? Do they, for example, have a defective implant that also discharges vital functions of the organism? Must it or may it be repaired if it at the same time serves as a weapon control device? Or, on the contrary, should such soldiers and their implants be rendered safe (Shunk 2015, 97)? Are military personnel obliged to help a wounded prisoner of war even when their help possibly "repairs" the enhancement and the adversary thus gain a strategic advantage?

There is currently no international law basis for specifying how to deal with an individual as a prisoner of war who is part of a weapon – or is a weapon himself (Koch 2015, 47; Vollmuth et al. 2014). The Geneva Convention III of 1949 regulates that prisoners of war receive medical care. But what about soldiers where the enhancement does not allow medical care due to technical or medical reasons? Is it necessary to extend the competences of the medical service correspondingly even though they only help opposing soldiers because friendly forces do not have any or different enhancements (Michaud-Shields 2014, 31)?

The *Deep Space Nine* episode entitled "Hippocratic Oath" broaches a completely new issue: The station's surgeon Doctor Bashir is captured by a group of Jem'Hadar and tries to free them from their addiction to Ketracel-White. However, this would put the Federation at a decisive strategic disadvantage. The fact that he does not attempt to do so because he is forced to on account of being in captivity is revealed as Bashir even prevents a fellow soldier from trying to escape and orders him to cooperate. Despite the knowledge that they are adversaries, he wants to help the group.[46] When Doctor Bashir is later accused of collaboration, he counters: "They're not machines, they're sentient beings, and I couldn't just stand there and watch them die."[47]

Thought should also probably be given to international law in other respects: For example, the prohibition of torture is based on certain physiological conditions under which the human body works. These conditions may be defined differently for soldiers with enhancements, as hunger, the need for sleep and the ability to feel pain follow changed parameters (Lin 2013).[48] Furthermore, discussion is required on whether soldiers with enhancements are weapons themselves, as this is in violation of international law and must be punished as a war crime (Lin 2013; Dvorsky 2013).

[46] Star Trek: Deep Space Nine [Television]: "Hippocratic Oath" (1995). USA (1993–1999), Gene Roddenberry (Original Creator), Rick Berman, Michael Piller (Creators).

[47] Star Trek: Deep Space Nine [Television]: "Inquisition" (1998). USA (1993–1999), Gene Roddenberry (Original Creator), Rick Berman, Michael Piller (Creators).

[48] In one episode, genetic enhancements spare the protagonist the consequences of a forced interrogation. Star Trek: Deep Space Nine [Television]: "Inter Arma Enim Silent Leges" (1999). USA (1993–1999), Gene Roddenberry (Original Creator), Rick Berman, Michael Piller (Creators).

14.10 Follow-on Cost and Disarmament of Enhancements

In connection with the Borg, for whose viability artificial implants are absolutely indispensable, thought must be given to the cost factors, as eternal costs arise, for example, for the maintenance and functionality upkeep of implants or the administration of substances that have become essential for the organism. Who is to bear the costs for the repair of implants after those bearing them have retired from military service? It is the soldiers themselves or the state as the producer of the hybrid beings (Michaud-Shields 2014, 30)? Is there an entitlement to enhancements and the assumption of their costs by the community (Hick and Ziegler 2007)? The follow-on cost of MHE and consequences of its allocation cannot be estimated at present. The civil health care system may have to reserve capacities for providing former soldiers proper follow-on treatment after retirement (Lin 2010).

The question concerning the use of MHE in civilian life should correspondingly be extended to include consideration of whether an enhanced soldier is permitted to keep and use the enhancement during off-duty time or after retirement. Are the enhancements parts of a soldier's body or items of equipment like fatigues or uniforms that have to be returned? Can soldiers be forced to undergo risky surgery to demilitarize their bodies? Military status entails restrictions such as the obligation to accept examinations. However, medical personnel have the duty to provide soldiers special care because they belong to a group of particularly vulnerable patients (Vollmuth 2013; Vollmuth et al. 2013). Must soldiers be locked up or otherwise neutralized if they refuse such surgery? May an enhancement be removed if this means that the individual bearing it would die? This question is the subject of a *Star Trek* episode in which a crew member connects his brain with the ship's computer and is unable to sever the connection.[49]

In the discussion on MHE, the focus is usually on upgrading, while downgrading is rather neglected. A scenario for a future conflict is the increased or massed use of MHE in an arms build-up. But what happens after a conflict that has triggered a massive arms build-up? Do soldiers have to have their bodies returned to their original state when disarmament starts? What is the fate of soldiers whose enhancements are not easy to dismantle? Will they be an annoying remnant of a war that is wished to be quickly forgotten?

This question is also addressed in the 1990 episode entitled "The Hunted"[50]: The crew of the *Enterprise* examines the application of an advanced and seemingly friendly species for membership of the Federation. The process is disrupted by the escape of a prisoner from a penal colony. The crew is denied any information about the crimes the prisoners in the colony have allegedly committed. It then finds the

[49] Star Trek: The Next Generation [Television]: "The Nth Degree" (1991). USA (1987–1994), Gene Roddenberry (Creator); Robert Sekuler, Randolph Blake: Star Trek on the Brain. Alien Minds, Human Minds. New York 1998, p. 135 f.

[50] Star Trek: The Next Generation [Television]: "The Hunted" (1990). USA (1987–1994), Gene Roddenberry (Creator).

prisoners are not offenders, but soldiers who served in a past war. It learns that at the time of the war, the soldiers had been subjected to intensive physical and psychological conditioning and pharmacological manipulations.[51] After the war, the soldiers endeavored to reintegrate into society, but the manipulations conducted on them made them incompatible for peaceful conditions. Instead of receiving treatment for the conditioning, the soldiers were isolated in a remote penal colony. The episode entitled "The Hunted" also broaches the issue of soldiers with post-traumatic stress disorders (PTSD), which are particularly serious and difficult to treat due to the improvements in the soldiers' memories. A former soldier explains: "My improved reflexes have allowed me to kill 84 times. And my improved memory lets me remember each of those 84 faces. Can you understand how that feels?"

14.11 Conclusion

As the analysis has shown, *Star Trek* addresses and discusses a variety of aspects of MHE. *Star Trek* therefore provides a highly suitable basis for thought on medical ethics. Altogether, science fiction as a genre reflects on the effect of key emerging technologies on human life and society. The fiction not only functions as a comment on society but engages in a dialogue with the real-life world, since it must also be regarded as a mirror of its time. It can be said that ethical implications are pointed out in the aesthetic genre even before the technology concerned has been developed and produced and so even before there seems an obvious necessity for them to be debated by society. The intensive international activities in the area of MHE constitute challenges for military medical action and first indications of them have already been reflected in literature and film. They therefore anticipate necessary discourses on new dilemmas that are hardly predictable when thought is given solely to the *present state of affairs*.

MHE has been rigorously used in the past, for example, during WWII in the form of substances like amphetamine derivatives that kept individuals awake. New developments in electronics and genetics raise serious ethical issues in the MHE area as well, in both the civilian and the military sectors. Manifold rudimentary work is being done in MHE and each single line of it must be examined, with reversibility being an important criterion. Since advances in science and technology could overtake the efforts being made to establish some form of regulation, there is a need for an early discussion and analysis of MHE. This discussion must not be confined to malign and critical aspects of HE, but also take consideration of the potential the corresponding developments and interventions have for medicine.

[51] Counselor Troi in "The Hunted" (1990): "Roga Danar was an idealistic young man who answered his people's call to service. He joined the military to fight for the Angosian way of life. What he didn't realise was that by doing so he would have to give up that way of life forever. He's not the same man who left home to go to war. He's been through intense psychological manipulation and biochemical modifications."

Even if friendly forces are possibly not pioneering in the radical use of MHE, the medical service officer can still come into contact with them when taking care of allied or adversary soldiers. The civil health care system can also be affected when former soldiers need medical care.

Analyzing HE is not only a concern of the military, but is a matter that poses new challenges to society as a whole, as enhancements can change anthropological foundations. Science fiction proves to be a worthwhile object of medical ethics study because it anticipates possible developments in technology – or creates independent hypotheses – and outlines ethical implications arising from them. Due to its prevalence, the genre helps to promote reflection on the issue in society.

References

Allhoff, Fritz, Patrick Lin, James Moor, and John Weckert. 2009. *Ethics of human enhancement. 25 questions & answers*. San Luis: Obispo.

Auf dem Hövel, Jörg: Gehirn-Doping. 2007. Augen geradeaus. *Telepolis*, October 23. http://www.heise.de/tp/artikel/26/26412/1.html. Retrieved on 22 Mar 2016.

Barad, Judith A. 2001. *The ethics of Star Trek*. New York: HarperCollins World.

Bartz, Oliver, and Aimé Jäschner. 2015. Evaluation von Human Performance Enhancement (HPE) unter sportwissenschaftlichen Aspekten. *Wehrmedizinische Monatsschrift* 59: 311–315.

Biller-Andorno, Nikola, and Michelle Salathé. 2012. Human Enhancement. Einführung und Definition. In *Medizin für Gesunde? Analysen und Empfehlungen zum Umgang mit Human Enhancement*, ed. Akademien der Wissenschaften Schweiz, 10–18. Bern: Akademien der Wissenschaften Schweiz.

Boldt, Joachim, and Giovanni Maio. 2009. Neuroenhancement. Vom technizistischen Missverständnis geistiger Leistungsfähigkeit. In *Das technisierte Gehirn. Neurotechnologien als Herausforderung für Ethik und Anthropologie*, ed. Oliver Müller, Jens Clausen, and Giovanni Maio, 383–397. Paderborn: Mentis.

Bundesarchiv/Militärarchiv Freiburg, BaMa RM 103 / 10, fol. 5^{r}. Ärztliches Kriegstagebuch des Kommandos der K-Verbände für die Zeit vom 1. September 1944 bis 30. November 1944.

Campobasso, Theresa. 2015. Super soldiers: 3D bioprinting and the future fighter. *Small Wars Journal,* December 8. http://smallwarsjournal.com/jrnl/art/super-soldiers-3d-bioprinting-and-the-future-fighter. Retrieved on 15 May 2016.

Clynes, Manfred E., Nathan S [chellenberg] Kline. 1960. Cyborgs and space. *Astronautics* (September 1960), pp. 26f. and 74–76.

Crawford, Jamie. 2012. The bin Laden situation room revisited – One year later. *Security clearance*, May 1. http://security.blogs.cnn.com/2012/05/01/the-bin-laden-situation-room-revisited-one-year-later/. Retrieved on 24 Sept 2016.

Daum, Oliver, and Andreas Grove. 2015. Konzeptionelle Überlegungen zur wissenschaftlichen Evaluierung von Human Performance Enhancement in der Luftwaffe. *Wehrmedizinische Monatsschrift* 59: 301–305.

dpa. 2015. Gelähmte sollen Geräte über Gedankenkraft steuern können [dpa-Meldung]. *Rollingplanet, Portal für Behinderte und Senioren*, April 11. http://rollingplanet.net/2015/04/11/gelaehmtesollengeraeteuebergedankenkraftsteuernkoennen/. Retrieved on 18 May 2016.

Dujmovic, Jurica. 2016. *Biohackers implant computers, earbuds and antennas in their bodies*, February 10. http://www.marketwatch.com/story/biohackers-implant-computers-earbuds-and-antennas-in-their-bodies-2016-02-10. Retrieved on 14 Aug 2016.

Dvorsky, George. 2013. It could be a war crime to use biologically enhanced soldiers. *Io9*, January 22. http://io9.gizmodo.com/5977986/would-it-be-a-war-crime-to-use-biologically-enhanced-soldiers. Retrieved on 12 Oct 2016.

———. 2014. *DARPA's new biotech division wants to create a transhuman future,* April 2. http://io9.gizmodo.com/darpas-new-biotech-division-wants-to-create-a-transhum-1556857603. Retrieved on 15 Oct 2016.

Eckart, Wolfgang Uwe. 2011. *Illustrierte Geschichte der Medizin. Von der französischen Revolution bis zur Gegenwart*. Berlin/Heidelberg/New York: Springer.

European Space Agency. 2001. *Innovative technologies from science fiction for space applications*. Noordwijk: ESTEC.

Friedl, Karl E. 2009. Overview of the HFM-181 symposium programme. Medical technology repurposed to enhance human performance. In *Human performance enhancement for NATO military operations (Science, technology and ethics)*, ed. Research and Technology Organisation (NATO). Conference proceedings, RTO Human Factors and Medicine Panel (HFM) Symposium held in Sofia, Bulgaria, on 5–7 October 2009. http://www.dtic.mil/cgi-bin/GetTRDoc?Location=U2&doc=GetTRDoc.pdf&AD=ADA562561. Retrieved on 25 Aug 2016.

Graimann, Bernhard, Brendan Allison, and Gert Pfurtscheller. 2010. Brain-computer interfaces: A gentle introduction. In *Brain-computer interfaces. Revolutionizing human-computer interaction*, ed. Bernhard Graimann, Brendan Allison, and Gert Pfurtscheller, 1–27. Berlin/Heidelberg: Springer.

Gray, Chris Hables, ed. 1995. *The cyborg handbook*. New York: Routledge.

Grönig, Nils-Steffen. 2008. *Zur Geschichte des Arzneistoffs Methamphetamin (Pervitin)*. University thesis. Marburg.

Gross, Michael L. 2015. Kameraden zuerst? Militärische vor medizinischer Notwendigkeit. *Ethik und Militär* 2015–01: 27–36. http://www.ethikundmilitaer.de/fileadmin/inhalt-medizinethik/Den_Gegner_retten_Militaeraerzte_und_Sanitaeter_unter_Beschuss_Ethik_und_Militaer_2015_1.pdf. Retrieved on 15 May 2016.

Hartmann, Volker. 1994. Pervitin – Vom Gebrauch und Mißbrauch einer Droge in der Kriegsmarine. *Wehrmedizinische Monatsschrift* 38: 137–142.

Haupt, Oliver, and Christoph Friedrich. 2016. Zur Geschichte der Dopingmittel. *Pharmakon* 4: 8–16.

Hick, Christian, and Andreas Ziegler. 2007. Mittelverteilung im Gesundheitswesen. In *Klinische Ethik*, ed. Christian Hick, 227–249. Heidelberg: Springer Medizin.

Hoffmann, Klaus-Peter. 2011. Einführung in die Neuroprothetik. In *Medizintechnik*, ed. Rüdiger Kramme, 645–651. Berlin/Heidelberg: Springer.

Hughes, James J., and John D. Lantos. 2001. Medical ethics through the Star Trek lens. *Literature and Medicine* 20: 26–38.

Juengst, Eric T. 2009. Was bedeutet Enhancement? In *Enhancement. Die ethische Debatte*, ed. Bettina Schöne-Seifert and Davinia Talbot, 24–45. Paderborn: Mentis.

Klimas, Liz. 2012. *Will drugs, technology lead to pressure for the creation of a superhuman workplace?* http://www.theblaze.com/stories/2012/11/07/willdrugstechnologyleadtopressureforthecreationofasuperhumanworkplace/. Retrieved on 24 Aug 2016.

Klugman, Craig M. 2001. From cyborg fiction to medical reality. *Literature and Medicine* 20: 39–54.

Koch, Bernhard. 2015. Es geht noch besser! Medizin und die Debatte um Human Enhancement bei Soldaten. *Ethik und Militär* 2015–01: 44–50. http://www.ethikundmilitaer.de/fileadmin/inhalt-medizinethik/Den_Gegner_retten_Militaeraerzte_und_Sanitaeter_unter_Beschuss_Ethik_und_Militaer_2015_1.pdf. Retrieved on 15 May 2016.

Kurzweil, Ray. 2007. Der Mensch, Version 2.0. Werden wir in zwanzig Jahren mit künstlich verbesserten Körpern leben? *Spektrum der Wissenschaft Dossier* 2007: 77–82.

Le Blanc, Thomas, ed. 2014. *Die Zukunftsideen der Science Fiction Literatur ... und welche bereits Wirklichkeit wurden*. Wetzlar: Phantastische Bibliothek.

Ley, Stefan. 2010. Infanterist der Zukunft – Erweitertes System. Die Kampfausstattung auf dem Weg zur Realisierung. *Strategie & Technik* 53: 18–23.

Lin, Patrick. 2009. Therapy and enhancement: Is there a moral difference? Drawing a principled line between the two is complicated, if it even exists. *GEN. genetic engineering & biotechnology news* 29. http://www.genengnews.com/gen-articles/therapy-and-enhancement-is-there-a-moral-difference/2959. Retrieved on 24 Aug 2016.

———. 2010. Robots, ethics & war. *Stanford, The Center for Internet and Society,* December 15. http://cyberlaw.stanford.edu/blog/2010/12/robots-ethics-war. Retrieved on 24 Aug 2016.

———. 2012. More than human? The ethics of biologically enhancing soldiers. *The Atlantic,* February 16. http://www.theatlantic.com/technology/archive/2012/02/more-than-human-the-ethics-of-biologically-enhancing-soldiers/253217/. Retrieved on 24 Aug 2016.

———. 2013. Could human enhancement turn soldiers into weapons that violate international law? Yes. *The Atlantic,* January, 4. http://www.theatlantic.com/technology/archive/2013/01/could-human-enhancement-turn-soldiers-into-weapons-that-violate-international-law-yes/266732/. Retrieved on 24 Aug 2016.

Lin, Patrick, Maxwell J. Mehlman, and Keith Abney. 2013. *Enhanced warfighters. Risk, ethics, and policy*. San Luis: Obispo.

Luokkala, Barry B. 2014. *Exploring science through science fiction*. New York: Springer.

Maio, Giovanni. 2012. *Mittelpunkt Mensch. Ethik in der Medizin*. Stuttgart: Schattauer.

Matthews, William. 2015. Supersoldiers: Can science and technology deliver better performance? *ARMY Magazine* 65. http://www.armymagazine.org/2015/04/20/supersoldiers-can-science-and-technology-deliver-better-performance/. Retrieved on 24 Aug 2016.

Merkel, Reinhard, Gerhard Boer, Jörg Fegert, Thorsten Galert, Dirk Hartmann, Bart Nuttin, and Steffen Rosahl. 2007. *Intervening in the brain. Changing psyche and society*. Berlin/Heidelberg/New York: Springer.

Meskó, Bertalan. 2014. *The guide to the future of medicine: Technology and the human touch*. Budapest: Webicina Kft.

Michaud-Shields, Max. 2014. Personal augmentation – The ethics and operational considerations of personal augmentation in military operations. *Canadian Military Journal* 15: 24–33.

Nöldeke, Hartmut, and Volker Hartmann. 1996. *Der Sanitätsdienst in der deutschen U-Boot-Waffe und bei den Kleinkampfverbänden. Geschichte der deutschen U-Boot-Medizin*. Hamburg: Mittler.

Peck, Michael. 2015. Can 3-D bioprinters create Captain America? *The National Interest,* December 19. http://nationalinterest.org/feature/can-3-d-bioprinters-create-captain-america-14687. Retrieved on 08 May 2016.

Perkowitz, Sidney. 2016. Science fiction: Boldly going for 50 years. *Nature* 537: 165–166.

Planungsamt der Bundeswehr. 2013a. *Future Topic. Weiterentwicklungen in der Robotik durch Künstliche Intelligenz und Nanotechnologie. Welche Herausforderungen und Chancen erwarten uns?* Berlin: Planungsamt der Bundeswehr.

———. 2013b. *Future Topic. Human Enhancement. Eine neue Herausforderung für Streitkräfte?* Berlin: Planungsamt der Bundeswehr.

———. 2013c. *Future Topic. Potenziale additiver Fertigungsverfahren. Was können 3D-Drucker?* Berlin: Planungsamt der Bundeswehr.

Pomidor, Bill, and Alice K. Pomidor. 2006. "With great power..." the relevance of science fiction to the practice and progress of medicine. *The Lancet* 368: 13–14.

Reschke, Stefan, Jan B. van Erp, Anne-Marie Brouwer, and Marc Grootjen. 2009. Neural and biological soldier enhancement. From SciFi to deployment. In *Human performance enhancement for NATO military operations (Science, technology and ethics)*, ed. Research and Technology Organisation (NATO). Conference proceedings, RTO Human Factors and Medicine Panel (HFM) Symposium held in Sofia, Bulgaria, on 5–7 October 2009. http://www.dtic.mil/cgi-bin/GetTRDoc?Location=U2&doc=GetTRDoc.pdf&AD=ADA562561. Retrieved on 25 Aug 2016.

Rid, Thomas. 2016. *Maschinendämmerung. Eine kurze Geschichte der Kybernetik*. Berlin: Propyläen Verlag.

Rogotzki, Nina, Thomas Richter, and Helga Brandt ed. 2009. *Faszinierend! STAR TREK und die Wissenschaften*. Kiel: Ludwig.
Röper, Gerhard, and Andreas Disson. 2012. Flugmedizin. In *Taktische Medizin. Notfallmedizin und Einsatzmedizin*, ed. Christian Neitzel and Karsten Ladehof. Berlin/Heidelberg: Springer.
Schlick, Caroline. 2008. *Apotheken im totalitären Staat – Apothekenalltag in Deutschland von 1937 bis 1945*. Stuttgart: Wissenschaftliche Verlagsgesellschaft.
Schöne-Seifert, Bettina. 2007. *Grundlagen der Medizinethik*. Stuttgart: Kröner.
Shunk, Dave. 2015. Ethics and the enhanced soldier of the near future. *Military Review* 95: 91–98.
Soyka, Michael. 2010. Neuro-Enhancement – eine kritische Annäherung. *Arbeitsmedizin. Sozialmedizin. Umweltmedizin. Zeitschrift für medizinische Prävention* 45: 242–246.
Spreen, Dierk. 1997. Was ver-spricht der Cyborg? *Ästhetik & Kommunikation* 26: 86–94.
———. 1998. *Cyborgs und andere Techno-Körper. Ein Essay im Grenzbereich von Bios und Techne*. Passau: Edfc.
———. 1999. Sinne und Sensoren. Zur Rekonstruktion der Sinne. In *Referenzgemetzel. Geschlechterpolitik und Biomacht. Festschrift für Gerburg Treusch-Dieter*, ed. Barbara Ossege, Dierk Spreen, and Stefanie Wenner, 39–46. Tübingen: Konkursbuch.
———. 2010. Der Cyborg. Diskurse zwischen Körper und Technik. In *Die Figur des Dritten. Ein kulturwissenschaftliches Paradigma*, ed. Eva Eßlinger, Tobias Schlechtriemen, Doris Schweitzer, and Alexander Zons, 166–179. Berlin: Suhrkamp.
———. 2013. Prothesen, Aufschreibesysteme, Cyborgs. *Ästhetik & Kommunikation* 43: 89–96.
———. 2015. Upgradekultur. In *Der Körper in der Enhancement-Gesellschaft*. Bielefeld: Transcript Verlag.
Sternbach, Rick, and Michael Okuda. 1991. *"Star Trek": The next generation – Technical manual*. London: Boxtree Ltd.
Stoppe, Sebastian. 2014. „Tee, Earl Grey, heiß." Star Trek und die technisierte Gesellschaft. In *Technik und Gesellschaft in der Science Fiction*, ed. Jan A. Fuhse, 94–111. Berlin: LIT.
Tucker, Patrick. 2014. The cyborg medicine of tomorrow is inside the veteran of today. *Defense One*, November 10. http://www.defenseone.com/technology/2014/11/cyborg-medicine-tomorrow-inside-veteran-today/98651/. Retrieved on 24 Aug 2016.
Vollmuth, Ralf. 2013. Das berufliche Selbstverständnis im Sanitätsdienst im historischen und ethischen Kontext. *Wehrmedizin und Wehrpharmazie* 37: 31–33.
———. 2015. Die Gefahr der „schiefen Ebene" – Sanitätspersonal zwischen Medizinethik und militärischem Auftrag. *Ethik und Militär* 2015–01: 74–78. http://www.ethikundmilitaer.de/fileadmin/inhalt-medizinethik/Den_Gegner_retten_Militaeraerzte_und_Sanitaeter_unter_Beschuss_Ethik_und_Militaer_2015_1.pdf. Retrieved on 15 May 2016.
———. 2016. Sanitätsdienst zwischen Medizinethik und militärischem Auftrag. *Wehrmedizinische Monatsschrift* 60: 113–117.
Vollmuth, Ralf, André Müllerschön, and Friederike Müller-Csötönyi. 2013. Therapiefreiheit, Gehorsamspflicht und Patientenwille – ein unauflösbares Problem? Eine klinisch-ethische Falldiskussion. *Wehrmedizinische Monatsschrift* 57: 45–49.
Vollmuth, Ralf, Erhard Grunwald, Rufin Mellentin, and André Müllerschön (eds.). 2014. *150 Jahre Schlacht bei Solferino*. Vorträge des 1. Wehrmedizinhistorischen Symposiums vom 22. Juni 2009. Munich.
von Engelhardt, Dietrich. 2002. Star Trek im Urteil der Medizinethik. *Focus MUL* 19: 47–53.
Vongehr, Frederik F. 2014. *Geschichte der deutschen Marinepharmazie. 1871–1945. Die pharmazeutische Versorgung der Kaiserlichen Marine, der Reichsmarine und der Kriegsmarine*. Stuttgart: Wiss. Verl.-Ges.
———. 2016a. 50 Jahre Cordrazin und Inaprovalin. Bemerkungen zur Geschichte der Sternenflotten-Medizin anlässlich des 50. Geburtstages des Raumschiffs Enterprise. *Geschichte der Pharmazie* 68: 15–22.
———. 2016b. Pharmacy beyond the final frontier: Star Trek between science fiction, forecast and science fact. In *The exchange of pharmaceutical knowledge between East and West*, ed. Afife Mat, Halil Tekiner, and Burcu Şen, 175–179. Istanbul: Eczacılık Tarihi Araştırma Derneği.

Chapter 15
Supersoldiers and Superagers? Modernity Versus Tradition

Paul Gilbert

15.1 Superagers and Supersoldiers

Old people like myself are constantly bombarded with advice on how to stay young, with a recent newspaper article reporting that 'the elixir of youth has finally been discovered, but couch potato Britain is not going to like it. A lifetime of vigorous exercise will let you keep the body of a 20-year old well into your 70s, scientists have found' (*The Times* 9/3/18). So-called 'superagers', who have followed such a regime, live longer lives, which, it is presumed, is what everybody wants. But for those of us who have not been so energetic, surely some pill that has the same effects is simply awaiting discovery? Advances in medicine and pharmaceuticals have increasingly led us to think that our bodies can be improved to slow down ageing. However, 'there is', writes Michael Hauskeller, 'a traitor waiting within who will eventually open the gates to the enemy, to heart failure or cancer, to Alzheimer's or to other forms of dementia, and this traitor is our body' (Hauskeller 2015, 95). We worry about our bodies and we seek some medical intervention to delay their treachery.

It is indeed in old age that most people in the Western world face the greatest risk of disease and consequent death, and experience the greatest anxiety about this. They can attempt to reduce their anxiety by adopting a healthy diet, taking regular exercise and being prescribed statins and other appropriate drugs. Death cannot be avoided but it can be pushed as far into the future as possible and this is what we so desperately desire.

It is this Western fear of death that, in the quite different environment of war, was mocked by Al Qaeda in its notorious slogan, 'You love life, but we love death'. In the modern world, war is the other situation in which Western people may face

P. Gilbert (✉)
University of Hull, Hull, England
e-mail: P.H.Gilbert@hull.ac.uk

D. Messelken, D. Winkler (eds.), *Ethics of Medical Innovation, Experimentation, and Enhancement in Military and Humanitarian Contexts*, Military and Humanitarian Health Ethics,
https://doi.org/10.1007/978-3-030-36319-2_15

a much greater than usual risk of trauma and death, and by the steps they take to avoid them reveal their fear. In *The City of God* St Augustine wrote that what constitutes a collection of individuals into a people is 'a common agreement on the objects of their love' (Augustine 1972, xxix 24). Al Qaeda's dictum follows Augustine's methodology in distinguishing the unbelieving people of the West as loving al-dunya, immediate or earthly life, by contrast with al-akhira, the deferred life of those who live and die in the service of Allah. Thus, while radical Islamists fight recklessly and engage in suicide attacks, Western forces secure troop protection to the extent of employing Predator drones whose controllers are safely 8000 miles away from their targets.[1]

Now I am not concerned with the contrast between those who believe that death ends everything and those who believe in a life after death, though in the modern western world the former are probably in a majority. Rather it is the contrast between those who, whatever their beliefs, show their fear of death in their anxiety to take steps of a certain sort to avoid or to delay it – roughly speaking technological steps- and those without such a marked desire or capacity to take such steps. For I wish to place within the context of this kind of fear and its manifestations the demand for the old to become superagers or for combatants to become 'supersoldiers'.

Starting with the old, we can see that superagers are not normal in Western populations, though apparently quite common among Amazonians. They are not normal, that is, in the sense that their abilities are not the norm within this age range, but they in no way go beyond human norms, being comparable to 20-year olds. I shall stipulate that any medical interventions needed to produce such superagers will count as therapy, not enhancement. For these will be instances of curative or preventative medicine. It would be quite otherwise if the aim was to extend life beyond any human norms. This would, by contrast, count as enhancement, though I shall not be considering such so-called extensionism.

It is evident, therefore, that I shall also count as enhancement programmes like that of the US Defense Advanced Research Projects Agency to produce 'stronger, faster and more deadly' soldiers by means other than conventional training and equipment. And I am concerned only with what may be termed medical enhancements, by contrast with the provision of exoskeletons, bionic eyes and so on. The relevant enhancements are those that have effects on human physiology, changing human beings, if only temporarily, into something resembling an alien species. Such enhancements will typically augment a soldier's biological energy, and thus his fighting capacity, combat fear and improve cognitive functioning (Orend 2015, 134–5). We can thus distinguish within these measures those that seek primarily to protect the soldier, to armour him, we may say, and those aimed at increasing his 'lethality', to arm him, so to speak, internally. Suffering less from fatigue will fulfil both functions and both will reduce a soldier's risk of death. It is to the ethics of augmenting invulnerability and military capability through medical enhancements that I shall now turn, albeit rather indirectly.

[1] See Gilbert 2015.

15.2 The Modern World View

In order to tackle this question, I shall contrast two approaches to moral judgment, roughly following in the path of Alasdair MacIntyre in his groundbreaking book *After Virtue* (MacIntyre 1985).[2] I shall link these approaches to a more general characterization of the condition of modernity, on the one hand, and of traditional society, on the other, as adumbrated by such sociologists as Anthony Giddens.[3] We may think of the two approaches as reflecting different world views, though there is no suggestion that particular subjects may not combine elements of each, even though they are, considered as a whole, sharply divergent.

Anachronistically I shall start with what I shall call the modern world view, in the sense of the view that has developed in the West since the Enlightenment and with the advance of science and technology on the one hand and of political and economic liberalization on the other. I have already sought to locate contemporary attitudes to death as a feature of this world view and to suggest that conceiving the possibility and desirability of supersoldiers is only intelligible against this background. But why should there be this kind of fear of death in the modern world – a fear also manifested, I claimed, in the usually forlorn hopes of the old to be superagers?[4]

It is, I suggest, because, as Daniel Zupan captures the modern view, of all our values autonomy is 'the most basic' (Zupan 2004, 29). Autonomy is the power of each individual to choose for him or herself how to live their lives. It is the power to choose this, not in the light of standards that are given from outside ourselves, but to choose in the light of standards that we ourselves choose for ourselves. We make, as the etymology of 'autonomy' implies, our own laws. What counts as a good life is, for a modernist as I shall call someone who holds this kind of view, just what someone who lives it counts as good. It will in fact be good to the extent to which it actually matches what he or she so chooses, so that what naturally consorts with a high valuation of autonomy is a desire for control, and anything that threatens this control is conceived of as a risk. Thus in what Ulrich Beck terms the modern 'risk society' (Beck 1992) there may actually be fewer dangers than in the past, but because of the demands we make for control of our lives it is as risks to our control that they are to be conceptualized and, other things being equal, to be minimized.

One of the factors over which we in the modern world try to exert control is our body, as if it were the sort of machine over which we do have such control. One aspect of the Western fear of death is, I suggested, that at death our control of our body is irretrievably lost. The treacherous body opens its gates to the old enemy. But

[2] What follows is a very rough adaptation of MacIntyre's account, for he distinguishes within what I later term the traditional world view between 'heroic societies' and later ones influenced by Aristotle's account of the virtues.

[3] See Giddens 1990, 1991.

[4] MacIntyre himself describes 'the attempt by the medical profession in the United States to use its technologies to postpone death for as long as possible' as 'the counterpart to the general loss of any conception of what it would be to have completed one's life successfully and so to have reached a point at which it would be right for someone to die' (MacIntyre 1999, 255).

it is, I think, this final loss of all control that is the fundamental fear here. For if what is ultimately of value is my being able to choose autonomously how to live my life, then at death this choosing ends. So what I fear is not just that certain projects I have chosen will not go through. It is that all effective choice is extinguished. And here, parenthetically, we may think of the fear of dementia as a fear closely allied to the fear of death. In both one fears extinction as an autonomous agent.

I have tried to sketch out some features of what I claimed as the modern world view to suggest how it might generate an irremediable fear of death and a consequent desire to treat or to enhance the human body and mind so as to avoid or to delay their demise or to allay disabling anxiety about this. On the face of it it looks as if whatever can be done to achieve this will be good, for each of us must surely welcome anything which increases our control and secures our continued existence as autonomous agents, leaving aside, to repeat, the issue of extensionism. One proviso will therefore be that any psychotropic effects, to relieve our anxiety, say, should not compromise our autonomy, as would be the case if these were analogous to the effects of brainwashing. For autonomy would be compromised if the situation the old person or soldier was really in was obscured or misrepresented as a result of treatment, so that they were not choosing to act on the basis of a reasonable assessment of it. There is in any event a considerable tension here. For to the extent to which I wish to remain fully autonomous my impending death will confront me with horror while to the degree to which I want to avoid that horror I will seek to avoid contemplating my death.

But are superagers and supersoldiers really so desirable? There is an aspect of the modern world view that may lead one to qualify this conclusion. For although under it each of us chooses their own conception of the good we have to acknowledge the rights of others to do likewise, to enjoy their own autonomy so long as this does not infringe the rights of others. The modern world is a regime of rights and its social morality is a morality of upholding rights and condemning their violation. So, for example, when it comes to treating the elderly so that they can live longer healthier lives, resources are consumed which might be better spent on, say, the education of younger people to which they have a right so that they can take control of their own lives. This is the sort of ethical question which may arise about both superagers and supersoldiers under modernism, but now I shall turn to modernism's antithesis, the traditional world view.

15.3 The Traditional World View

To call it traditional is not only to gesture at this world view's origins and prevalence in a pre-modern environment – in a world before the Enlightenment, before the widespread possibility of manipulating things, including human bodies, through scientific knowledge and before society was conceived of as consisting basically of individuals each pursuing his or her personal goals. To call it traditional is, furthermore, to indicate how this world view is transmitted and received. Just like modernists, traditionalists, as we shall call the holders of this sort of view, may not be thoroughgoing in their adherence to it, admitting elements of modernism in some of their attitudes. Like modernism too, the traditional world view may be largely tacit, apparent only in its followers' actions and reactions. But while the fact that the modern world view is modern provides no reason for adopting it, with the traditional world view things are otherwise. That the attitudes it fosters are those that have traditionally been held is taken as a reason for sticking with them, and this is itself an aspect of the view.

The traditional world view starts from a picture of people as already enmeshed in a web of relationships and occupying roles, initially, for most, in families, and later in occupations. These relationships and roles are defined in part by rules to which their participants adhere, and conscientious adherence to which marks them out as good sons, good soldiers and so forth. A good life is one where these relationships and roles have in this and other ways gone well. What makes for a good life is not, then, decided by the person who lives it as under modernism, but by standards external to them, standards transmitted through each role's traditions, where, that is, the role is being carried on in its customary, its traditional way.

While each role of this sort carries its own ethical standards we might, if we wished to find some overarching value that plays a similar role in the traditional view as autonomy does in the modern one, suggest that it is service – service to the family and to the community of which each individual is inextricably a part, and to the various institutions in which he or she pays a role, such as the army, the medical profession or whatever. It is, however, the good of the family, the community and so on which is the most basic value and which service to them promotes. This is, therefore, under traditionalism, a value not to be thought of as accruing because it aggregates together the individual goods of members, but because, on the contrary, these consist in participation in a wider good.

Since the traditional world view regards people's lives as good to the extent to which they carry out the requirements of their roles dutifully and with reasonable success it is evident that the body may be adequate or inadequate to the task, a locus of power or of vulnerability. But traditionalism counsels a measure of acceptance, of rejoicing in the body's capabilities and regretting its failings, but not bemoaning the ultimate lack of control we have of it as human beings fated to flourish and to decline. Death, on this view, is to be feared only if my life has not been good and death deprives me of opportunities to come to terms with this; or it is to be feared for what it prevents me doing in the performance of a role, since the traditional world view assigns a place for death either as the natural end of life or as a hazard of a particular role.

For the ageing, then, a healthy old age ending in a peaceful death is a blessing, and that is the best we can expect. Whatever therapies conduce to this are to be welcomed, but the desire to live longer or with greater powers than is normal risks falling into attitudes of self-assertion or of narcissism which are inconsistent with accepting the place that traditionalism has assigned to the old – a place that is honourable not least because it is from within it that their traditions can be handed down.

The role of the soldier is, of course, a hazardous one, and one purpose of enhancement would be to secure greater protection, better armour, as I have put it, for him. What, on the traditional world view, as I have sketched it, should be the reaction to this possibility? An answer can come, I suggest, only from the point of view of soldiers themselves – soldiers who will have absorbed the standards of their role through their induction into the army and their relationships with comrades, For even today this role is, I believe, to a large extent a traditional one, with standards passed down from generation to generation. There is a complicated history here, but we can see that from the principles that have come down in codified form, namely the rules of jus in bello, that these standards antedate the modern world view. It is true that Michael Walzer attempts to underpin them from a theory of rights (Walzer 1977, 72) and in the context of modern international law this is now how they may be seen, but from the point of view of the soldier it is not this that provides the reason for adopting them, but rather that it is such rules that are handed down to him in his formation as a soldier and which shape his self-conception as one.

The in bello rules, whether codified or not, are, I suggest, determined by a tacit agreement, stretching back into the past, between the soldiers of 'civilised nations', as the international law phrase has it, on what they are prepared to do and to undergo in battle.[5] And this, in turn, reflects their conception of themselves as sent out by their community to fight in order to defend it. They undertake not to employ their arms for purposes beyond the pursuit of victory nor to expose themselves to the risk of death except as obstacles to the other side's chance of winning. That is why, for example, there is agreement on the taking of prisoners in preference to the killing of opponents, as well as on the principles of discrimination and proportionality with respect to the civilians who are being defended on both sides. The question we need to ask about supersoldiers, then, is what sort of agreement there might be on whether their use would be acceptable to soldiers performing their traditional role.

15.4 Supersoldier or Simple Soldier?

It would, if I am right, be impertinent on the traditional view to try to answer this question from outside the ranks of the military, for it is they alone, I have suggested, who can assess the ethical requirements and limitations of their role. We can, however, consider the factors of which they might take account. Performance enhancements should, according to US sources, produce both 'iron bodied and iron willed personnel' (Robbins 2013, 127). In respect of their bodies supersoldiers will be less vulnerable and more lethal than simple soldiers, as we may call them.

So I imagine soldiers being in a position of ignorance about whether and how it is they or the enemy who would be enhanced. Then they will need to ask two questions: would they be prepared to encounter such forces in battle if they themselves were not so enhanced, and would they be willing to undergo such enhancement, possibly to fight against unenhanced troops? Here questions about the gross inequality in casual-

[5] For development of this contractualist account see Gilbert 2005.

ties likely to arise from such encounters will be raised, and soldiers may neither wish to be on the receiving end of them nor, perhaps, to inflict them upon other soldiers.

Similar questions need to be posed about psychological enhancement. It is such 'iron willed' supersoldiers that President Putin no doubt had in mind when calling for an international ban on the biomedical creation of soldiers who 'fight without fear, compassion, regret or pain' (*Daily Mail* 7/2/17). While we do not know what the mental state in battle of such supersoldiers would be, we can, I think, be fairly confident that simple soldiers of a traditional sort would be reluctant to confront them in battle or have such mental states themselves. The reason is that, as I have claimed, soldiers on each side performing their traditional role, think of themselves as fighting in a limited way in defence of their community. To fight fearlessly and perhaps therefore ruthlessly would be incompatible with this attitude, since it does not show the respect for opponents which this attitude entails. Nor, if it implies as it would seem to do, the absence of any fear of dying in battle, does it betray the concern for the community being fought for which a preparedness to die for them reveals. The mindset of psychotropically enhanced supersoldiers seems quite out of kilter with that which is presupposed in the traditional military role.

I do not have time here to adequately consider the military virtues which would seemingly be unavailable to supersoldiers, though one way of characterizing the attitude that I have suggested they would lack is chivalry. Courage is possible only because there are restraints on what one can do to an enemy and its people to protect oneself and one's own. For courage implies deliberately running risks that might have been avoided. But if these risks were minimized by the use of excessive violence no such courage would be displayed, just as it would not if invulnerability were assured. Indeed such violence would show a lack of humanity. It is because certain virtues of these sorts are available in the military that it can provide a life of satisfaction and fulfillment, that there can be, as Alasdair MacIntyre puts it, 'a happiness…which belongs peculiarly to the military life' (MacIntyre 1985, 64). It is quite unclear, by contrast, why anyone would wish to take up the life of a supersoldier. Or rather, it is quite unclear why anyone should want to from within the perspective of the traditional world view. I return, therefore, to considering the ethics of supersoldiers from the contrasting viewpoint of modernism.

Now the account of the proper conduct of war that I have given is that of what has been called orthodox just war theory, which is, I claim, that implied by the traditional world view. This orthodox view has recently been challenged by so-called revisionists such as Jeff McMahan.[6] A key feature of revisionism is the denial of what Michael Walzer terms the moral equality of soldiers (Walzer 1977, 34–47), whether or not the war they fight is just. For revisionists deny the view that soldiers do not incur an adverse moral judgment and its consequences just because they are fighting on the unjust side. The moral equality thereby denied, or something like it, is implied by the traditional account that I have given. Under revisionism, however, it is because rights to life are violated by unjust attackers that these attackers are morally in the wrong and are liable to defensive killing, for by their actions they

[6] See especially McMahan 2009.

have forfeited their own rights to life. Indeed, because they are autonomous agents they should have refused to fight if they thought their cause was unjust.

On this revisionist account war is seen as a conflict between individuals, albeit individuals formed into collective organizations, rather than between collectivities whose combatants play a traditional role in acting for collective goals. Thus in what McMahan describes as the 'deep morality' of war individual rights of general applicability underpin the judgments we should make about what is right and wrong in war, rather than some ethical code specific to it. Revisionism is, therefore, a reflection of many of the features I have ascribed to the modern world view. How then should it regard the creation of supersoldiers?

Under revisionism someone who is convinced that his cause is just should surely welcome the greater protection for his soldiers that enhancement brings, since it prevents yet further violations of their right to life. And their greater lethality would be justified by the lack of a right to life for their unjust opponents. Prima facie, therefore, revisionism favours supersoldiers, though a caveat might be that in the hands of an unjust opponent they would wreak even more injustice. Revisionists might, then, take refuge in a further feature of their position, namely the claim that since both sides usually believe that their cause is just, for these pragmatic reasons the laws of war should apply in the same way to both rather than reflecting 'deep morality'. In that case they might be able to along with Putin's suggestion after all. My strong suspicion, however, is that modernism will lead governments to push for supersoldiers, while soldiers themselves-influenced by the traditions which provide them with their ethical standpoint- will resist this.

References

Augustine, St. 1972. *The city of god.* Trans. H. Bettenson. Harmondsworth: Penguin.

Beck, U. 1992. *Risk society: Towards a new modernity.* London: Sage.

Giddens, A. 1990. *The consequences of modernity.* Cambridge: Polity.

———. 1991. *Modernity and self-identity.* Cambridge: Polity.

Gilbert, P. 2005. Proportionality in the conduct of war. *Journal of Medical Ethics* 4: 100–107.

———. 2015. You love life but we love death: Suicide bombings and drone attacks. In *Making humans*, ed. A.D. Ornella, 129–139. Oxford: Inter-Disciplinary Press.

Hauskeller, M. 2015. Messy bodies or why we love machines. In *Ornella* (supra), 93–106.

MacIntyre, A. 1985. *After virtue.* London: Duckworth.

———. 1999. Some enlightenment projects reconsidered. In *Questioning ethics*, ed. R. Kearney and M. Dooley. London: Routledge.

McMahan, J. 2009. *Killing in war.* New York/Oxford: University Press.

Orend, B. 2015. Framing the issues in moral terms II: The Kantian perspective on Jus in Bello. In *The Ashgate research companion to military ethics*, ed. J.T. Johnson and E.D. Patterson. Farnham: Ashgate.

Robbins, L.R. 2013. Refusing to be all that you can be. In *Military medical ethics for the 21st century*, ed. M. Gross and D. Carrick, 127–138. Farnham: Ashgate.

Walzer, M. 1977. *Just and unjust wars.* New York: Basic Books.

Zupan, D.S. 2004. *War, morality and autonomy.* Aldershot: Ashgate.

Printed by Printforce, the Netherlands